BOOK I

FUNDAMENTALS OF
RADIO-VALVE TECHNIQUE

This book is reprinted by the kind permission of
N.V. Philips Gloeilampenfabrieken
of Eindhoven, The Netherlands.
It was first published in 1949 in
four languages: Dutch, German, French and English.

Philips no longer manufacture the valves
referred to in this publication and regrets it
cannot engage itself in answering any enquiries from
those who build the circuits.

First Reprint Edition 1999

Copyright © 1949
N.V. Philips Gloeilampenfabrieken, Eidhoven (Holland)

Reprint edition published by
Audio Amateur Incorporated
Peterborough, New Hampshire 03458-0876
United States of America

Distribution Agents
Old Colony Sound Laboratory
Post Office Box 876
Peterborough, New Hampshire 03458-0876

Printed in the United States of America

ISBN 1-882580-23-0
Library of Congress Catalog Number 99-095263

FUNDAMENTALS OF
RADIO-VALVE TECHNIQUE

621.385:621.396.694

by

J. DEKETH

Graduate Federal Technical University
Zurich (Switzerland)

An introduction to the Physical Fundamentals,
Properties, Designs and Applications of Radio-
receiver and Power-amplifier Valves covering
the technical development reached
up to December 1947,

WITH 384 ILLUSTRATIONS
AND APPENDIX

1949

PUBLISHED BY
N.V. PHILIPS' GLOEILAMPENFABRIEKEN
EINDHOVEN (NETHERLANDS)

TECHNICAL AND SCIENTIFIC LITERATURE DEPARTMENT

0—4000

03

Translated by

F. G. GARRATT

View of the Philips' Laboratories for Scientific Research with, behind,
part of the Radio Works at Eindhoven

PUBLISHER'S NOTE

This work has been published in Dutch, English, French, German and Swedish.
The Italian, Spanish and Finnish editions are in preparation.

Preface

The aim of this work is to give engineers and technicians not specialised in radio and allied techniques an impression of the construction and functioning of radio valves and their applications in receiving sets and other electronic apparatus. It is hoped, however, that those who have already considerable experience in electronic techniques may also find this book useful as a reference work. It constitutes, moreover, an introduction to subsequent volumes dealing with characteristics and circuits of numerous valve types.

Electronic valves have found a wide-spread application, not only for communication purposes, but also in industries of all kinds. A profound knowledge of these valves and their circuits will open new horizons for perfecting industrial processes and measurements.

First the physical fundamentals of electronic valves are given, followed by a brief description of their construction and manufacture; valves of very recent design, the all-glass Rimlock valves, are included. These valves have particularly small dimensions and are manufactured by entirely new processes and they will not fail to arouse considerable interest.

After these chapters the author deals with the valve properties, the action of the various grids and the types of valves designed for various functions as well as the conditions to which they must be adapted. The author explains such notions as valve noise, short-wave properties, low-frequency inverse feed-back, but he always emphasizes the more important aspects.

The sequence of the chapters has been so chosen that a logical development of ideas is built up. Where the clarity of the text made it necessary, the author has not refrained from repeating himself.

It was judged necessary to deal in some detail with the physical fundamentals, because a profound knowledge of certain facts, as, for instance, potential distribution within the valve, influence of space charge, thermal emission, secondary emission, contact potential, etc., is indispensable for acquiring a complete picture of the mode of operation of electronic valves, their properties and possibilities of application. The author, however, had to restrain himself considerably because much more could be told of these interesting physical topics.

A chapter has been devoted to valve capacities and several others to interfering effects which may occur. This latter topic has been treated in greater detail in this English edition, at the request of several design engineers, and the same applies to the subject of Chapter XVI, namely the mathematical analysis of transfer characteristics of valves.

Not only the conventional electrostatic and electromagnetic units have been used, but also the units of the excellent rationalized system of Giorgi, a short explanation of which with formulae and a table for converting units of the electrostatic, electromagnetic and practical systems into units of the rationalized Giorgi system are provided.

At the end the reader will find an appendix which gives a rather important collection of definitions, formulae, tables and graphs which may be helpful in designing electronic apparatus. Attention is also drawn to the list of Philips publications on radio and allied subjects. A great number of them have been printed in "Philips Technical Review".

This work is dedicated to all those who are designing and manufacturing equipment based on electronic valves, those who are concerned with repairs and maintenance and those who are studying and experimenting. The author used in part publications and articles of the Physical Laboratories of the N.V. Philips' Gloeilampenfabrieken and of the Design Departments of the Philips Valve Works. He wants to mention particularly the publications of C. F. Veenemans, M. J. O. Strutt, J. L. H. Jonker, H. van Suchtelen, B. D. H. Tellegen, A. J. Heins van der Ven, B. G. Dammers, J. M. van Hofweegen, P. H. J. A. Kleynen and others.

The author wishes to acknowledge gratefully the encouragement and assistance, the ever-ready advice and counsel of Dr. E. Oosterhuis, Dr. W. de Groot, Dr. C. F. Veenemans, Prof. Dr. M. J. O. Strutt, Prof. Ir. B. D. H. Tellegen, Dr. J. L. H. Jonker, Ir. H. van Suchtelen, Ir. P. H. J. A. Kleynen, Ir. G. van Beusekom, Ir. J. M. van Hofweegen, Ir J. D. Veegens and many others. He is furthermore indebted to Mr. Harley Carter of the Mullard Electronic Products Ltd. in London and Dr. E. J. B. Willey, Consulting Physicist, London, for checking the English translation and their numerous linguistical and technical observations.

December 1948.   J. Deketh

Contents

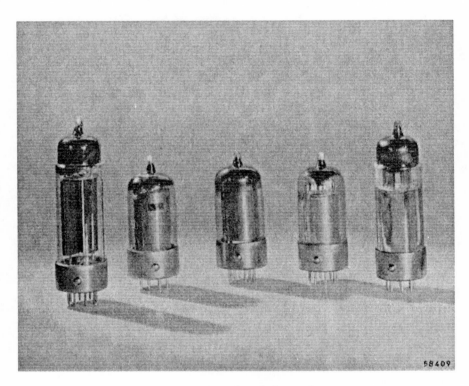

58409

A set of modern A-technique valves.

Introduction

Radio valves are at once the most vital and the most interesting components of a radio receiver, not only on account of the functions they perform, but also because their successful design, manufacture and application depend upon intimate knowledge of a large number of physical phenomena, combined with technological processes adjusted to the highest standards of precision.

The importance of radio valves in a receiving set will be realized when it is stated that, in addition to other no less essential functions, valves perform the task of amplifying the very small voltages induced in the receiving aerial until they are powerful enough to reproduce the programme with sufficient volume of sound.

The voltages induced in the receiving aerial by the programme from a transmitter of average power are of the order of a few thousandths of a volt only; and in order to reproduce the programme at normal volume for domestic listening the voltage at the loudspeaker transformer must be some 20 to 40 volts, representing a magnification of some ten thousand-fold. In terms of power, the comparison is still more striking, for the minute amount of energy, of the order of 4×10^{-12} watts, intercepted by the aerial, must be amplified to a total of some 4 watts to provide full volume in the domestic loudspeaker—an amplification of thousand billion-fold!

With these facts in mind, it is easy to realize that, but for the development of the electronic valve, radio communication would never have been possible in all its present-day perfection, but would have remained a most primitive process, possibly of interest only as a scientific curiosity.

But great as has been the influence of the electronic valve already, progress and development are not at a standstill. New achievements in radio broadcasting, in communications and, more recently, in television owe their success to further development and perfection of valve technique.

In other fields, too, including scientific measuring apparatus and the automatic control of industrial machinery, the electronic valve is playing a vital and ever-increasing part. Let us not forget the important part the electronic valve is playing now in sea and air navigation, accomplished by the various radar systems.

A Modern Radio Valve, showing External Appearance and
Internal Construction.

CHAPTER I

Basic Principles of the Action of a Radio Valve

1. Electrodes

A radio valve consists essentially of a glass or metal vessel, known as the envelope or bulb, from which, by pumping and other processes, almost all the air and other gases have been withdrawn, and containing two or more insulated metal parts termed electrodes. The electrodes are connected, by metallic conductors called the lead-in wires passing through the wall of the valve, to external contacts which may be the pins or side contacts of the base, or a contact cap at the top of the valve. By means of these contacts the electrodes are connected to the electric circuits outside the valve.

The electrode system of a valve always comprises at least two electrodes, a **cathode** and an **anode,** which are normally connected respectively to the negative and positive poles of a source of direct voltage.

If two metal plates, mounted in an evacuated vessel, are connected to a source of direct voltage, i.e. act as cathode and anode, no current will normally flow through the vacuous space between them [1].

Current will flow, however, if the cathode is composed of one or other of a number of special materials and is heated to a temperature at which thermal electron emission occurs [2]. In these circumstances, and given a sufficiently high voltage between cathode and anode, an electron current will flow from the cathode to the anode within the valve [3] and is termed the **anode current** (see Fig. 1). The amount of power (anode input power) required to produce this current is equal to the product of the applied voltage and the cathode-to-anode current.

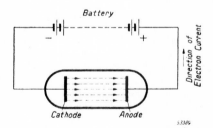

Fig. 1

Evacuated glass vessel containing an anode and an emitting cathode. A positive voltage between anode and cathode causes an electron current to flow from the cathode to the anode across the vacuous space.

---

[1] This statement does not hold when the voltage between the electrodes is very high, or when one of the electrodes is photo-sensitive.

[2] The conceptions of electrons, electron currents and electron emission are explained more fully in later chapters.

[3] In this chapter the direction of the current is considered to be the same as the direction of the electron flow.

2. Rectifying Action

Because the anode is not heated, and therefore does not emit electrons, no current will flow through the valve if the applied voltage is reversed, i.e. if the cathode is connected to the positive pole of the external source and the anode to the negative pole.

If, therefore, an alternating voltage is applied between the cathode and anode, current will only flow at those times when the anode is positive with respect to the cathode, and no current will flow at those times when the anode is negative.

This is known as the **rectifying action** of the valve, and is employed in the process known as detection, in which the incoming signal is rectified to render it suitable for reproduction in the loudspeaker. The rectifying action is also utilized for converting alternating current from the mains into direct current for operating the valves in a radio receiver.

3. Amplification by Anode-current Control

The amplifying action of a valve is obtained by varying the current (anode current) flowing from the cathode to the anode, and is effected by a third electrode, the **control electrode**, located between the cathode and anode. If a voltage is applied between the control electrode and the cathode the intensity of the electron current from cathode to anode is varied (see Fig. 2). The control electrode, on account of its form, is called the **grid**, or the **control grid** to distinguish it from other similarly formed electrodes. It is also sometimes referred to as the **input grid**.

If an alternating voltage is applied to the control grid (see Fig. 3), the anode current will alternately increase and decrease, and the current in the **anode circuit**, i.e. between the anode and cathode outside the valve, will also increase and decrease periodically. Because of the uni-directional conductivity (rectifying action) of the valve, the anode current can never be reversed,

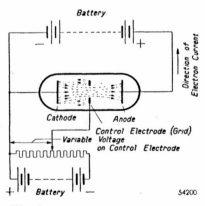

Fig. 2
Evacuated glass vessel containing an anode, a control electrode and an emitting cathode. By varying the negative voltage of the control electrode with respect to the cathode, the intensity of the electron current flowing from the cathode to the anode is similarly varied. If the control electrode is positive with respect to the cathode it will draw an electrode current in the same way as the anode.

4

i.e. flow in a direction opposite to that of the arrow, so that the anode current is a pulsating one and can be considered as an alternating current superimposed upon a direct current. By inserting suitable elements, such as a **resistance** or **impedance** (A.C. resistance), in the anode circuit as indicated in Fig. 4, the alternating current component of the anode current may be manifested as an alternating voltage of much greater value than the alternating voltage applied to the control grid (the **grid alternating voltage**). The grid alternating voltage is thus reflected as an amplified voltage in the anode circuit of the valve. In other words, **voltage amplification** has been obtained, the voltage gain being equal to the anode alternating voltage divided by the grid alternating voltage.

Generally some resistance or impedance is present in the grid circuit. If this resistance is represented by R_g and the effective value of the grid alternating voltage by V_{in}, the energy absorbed by the grid circuit, apart from the additional factors described later, will, according to Ohm's Law, amount to:

$$\frac{V_{in}^2}{R_g}.$$

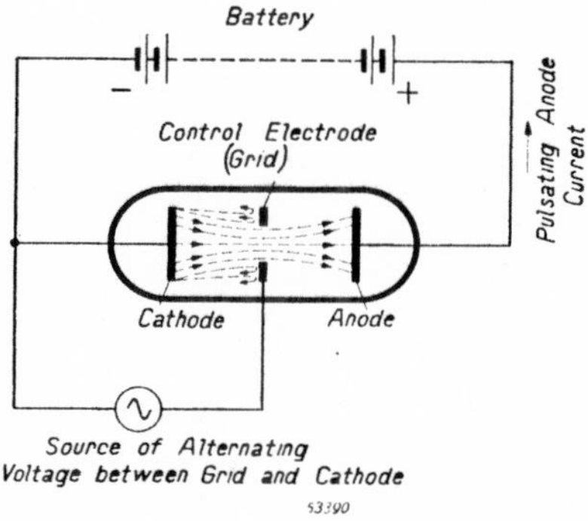

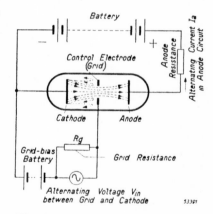

Fig. 3
Evacuated glass vessel containing an anode, a control electrode and an emitting cathode. An alternating voltage is applied between the control electrode and the cathode and produces in the anode circuit a pulsating direct current which can be resolved into a steady direct current with an alternating current superimposed.

Fig. 4
Evacuated glass vessel containing an anode, a control electrode and an emitting cathode. A resistance is interposed in the anode circuit. Between the control electrode and the cathode there is a resistance R_g, a source of negative grid bias voltage and a source of alternating voltage V_{in}.

Similarly, if the alternating current component of the anode current has an effective value of I_a r.m.s., and the resistance in the anode circuit be represented by R_L, and if the peak value of the alternating current component is less than the direct-current component, the anode current will never fall to zero, but will vary similarly to the grid alternating current. As a rule the power ($I_{a\ r.m.s.}^2 \times R_L$) in the anode circuit will be greater than the power $\dfrac{V_{in}^2}{R_g}$ supplied to the grid circuit. In other words, **power amplification** has been obtained, and the power gain is equal to the anode power output divided by the grid power input, both terms being derived from the respective alternating currents and voltages present.

4. Negative Grid Bias for Amplifying Valves

It was assumed in Section 3 above that power equal to $\dfrac{V_{in}^2}{R_g}$ was required to control the anode current by means of an alternating voltage

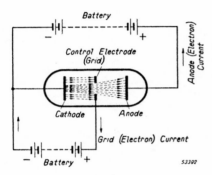

Fig. 5
If the control electrode is positive it will also draw current. Grid current then flows between grid and cathode.

applied to the grid. This is true only when the voltage at the grid is negative with respect to the cathode. If the control grid were to become positive during part of the grid-alternating-voltage cycle, the grid would function as an anode during this part of the cycle and current (**grid current**) would flow from the cathode to the grid, as indicated in Fig. 5.

In order to supply this current considerably more power is needed as a rule than that represented by the grid power $\dfrac{V_{in}^2}{R_g}$ mentioned above, and as a result the power amplification is considerably reduced. To avoid grid current, a negative direct voltage (**negative grid bias**) is applied to the grid as shown in Fig. 4. The negative grid bias is of such a value that when the alternating grid voltage is superimposed on it the resulting voltage can never become positive with respect to the cathode.

CHAPTER II

Physical Conceptions of Electrons and Electric Currents

5. Electrons

Investigation of the phenomena of electric conduction in liquids (electrolysis) and in rarefied gases indicates that an electric charge may be composed of a number of minute elemental charges of a definite magnitude. The existence of these elemental charges is revealed in the **cathode rays** as negatively charged particles called **electrons**, each carrying a charge of e = 4.8 × 10⁻¹⁰ electrostatic units, equivalent to 1.60 × 10⁻¹⁹ coulombs (practical and mks units) or 1.60 × 10⁻²⁰ electromagnetic units.

The cathode ray is, in fact, simply a stream of electrons travelling at great velocity from the cathode to the anode in a valve. They can be rendered visible by causing them to impinge upon one of a series of substances, such as zinc silicate, which possess the property of emitting light when bombarded with electrons travelling at high velocity. The cathode-ray tube, used for television reproduction and also in many modern measuring devices (cathode-ray oscillographs), operates on this principle.

By means of such a tube it has been possible to examine the deflection of an electron beam in a magnetic field, and from the extent of the deflection to calculate not only the velocity of the electrons but also the relation $\dfrac{e}{m_e}$ between the charge and the mass of an electron. This relation amounts to 5.27 × 10¹⁷ electrostatic units per gramme, or 1.76 × 10⁸ coulombs per gramme and in mks units (rationalised Giorgi system) [1]) 1.76 × 10¹¹ coulombs per kilogramme, from which, by substituting the value (e = 1.60 × 10⁻¹⁹ coulombs) given above, the mass of an electron is shown to be 9.1 × 10⁻²⁸ grammes (9.1 × 10⁻³¹ kilogrammes), or approximately 1840 times lighter than the lightest chemical atom—the atom of hydrogen.

6. Structure of the Atom

All matter is composed of atoms. Modern theory visualises an atom as an aggregation of smaller components, namely an **atom nucleus** in which is concentrated the major part of the mass of the atom, and a number of electrons. The atom nucleus carries a positive charge which

---

[1]) See Appendix I A 2.

is a multiple of the elemental or electron charge e, and the number of electrons attached to each atom is such that their combined charges just neutralise the positive charge of the nucleus, so that the charge of a normal atom is zero.

An atom is pictured with the positive nucleus at the centre with the electrons revolving around it at high velocity, much as the planets in the solar system revolve around the sun. The orbits on paths of the electrons are arranged in groupings known as shells; in the heaviest atoms there are seven such shells.

If one or more electrons are removed from the outermost shell of an atom, the atom is left with a deficit of negative charge—in other words with a surplus of positive charge. It is then termed a **positive ion**. Generally speaking, energy has to be expended in order to remove an electron from an atom because, among other things, it is necessary to overcome the electrostatic attraction between the negative electron and the positively charged remainder of the atom. This energy is termed the **ionisation energy.**

There are also negative ions, which are formed when an extra electron is added to a neutral atom.

Under certain conditions two or more atoms or ions can combine to form a **molecule.** Thus, the gases hydrogen, oxygen and nitrogen consist of molecules H_2, O_2 and N_2, each composed of two identical atoms. An example of an **ionic molecule** is the vapour of household salt (sodium chloride, NaCl), which is formed by the union of a positive sodium ion (natrion), Na^+, with a negative chlorion, Cl^-.

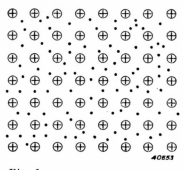

Fig. 6
Diagrammatic representation of the ion grid of tungsten metal. The circles with the positive sign represent metal ions. The dots represent the negative electrons of the electron gas (two electrons per ion).

7. Structure of Solid Matter

Some solids can be pictured as aggregations of uniformly arranged molecules. In many cases, however, particularly with metals and certain simple compounds including metallic salts and oxides, the atoms themselves and not the molecules constitute the structural elements, and these, too, are not in the neutral state but are ionized. In metals such as tungsten, for instance, the atoms are positive ions arranged in a regular pattern (the **ion grid**), with "free" negative electrons existing in the spaces between the ions. Fig. 6 is a pictorial representation of the metal tungsten, the dots representing the "free" electrons, which are able to move about in the metal—they are often referred to as an **electron gas,** contained between the ions of the metal.

8. How an Electric Current Originates; Electric Resistance

The free electrons in a metal can be set in motion under the influence of an electric field. This is the phenomenon which is known as an electric current.

The electrons, being negatively charged, move towards the positive pole of the voltage source producing the electric field. This direction is the reverse of that which is conventionally termed the positive direction of the electric current. In order to avoid confusion, therefore, the conventional positive direction will be termed positive throughout the remainder of this book, and it must be remembered that this is the reverse of the direction of the electron flow as described in Chapter I.

Electric resistance is the result of the obstruction of the electron movement by the positive ions, whereby kinetic energy, imparted to the electrons by the electric field, is transferred to the ions, which, in consequence, increases the thermal movement of the ion grid. This explains the heating of an electric conductor during the passage of an electric current.

Just as the "free" electrons in a metal move through the spaces between the ions under the influence of an electric field, so, in the presence of an electric field, any electrons existing in a vacuous space will move in a direction opposed to that of the field force lines, i.e. from the negative pole or cathode to the positive pole or anode; and this is the principle underlying the flow of electric current in a radio valve.

CHAPTER III

Behaviour of Electrons in Electrostatic and Magnetic Fields; the Space Charge

A radio valve contains, among other elements, a cathode, which when heated emits electrons. If the temperature of the cathode is sufficiently high a number of electrons leave the cathode material and arrive within the vacuous space in the valve. The nature of electron emission is discussed in the following chapter; it is first of all desirable to consider the behaviour of electrons under the influence of electrostatic or magnetic fields.

9. Electrons in an Electrostatic Field

Fig. 7 represents two plates mounted parallel to and at a distance d from each other in a vacuum. If a voltage V is applied between the plates an electric field is set up, the lines of force being parallel to each other except at the boundary regions of the plates. The field strength F at any point in that part of the field where the lines are parallel is equal to $\dfrac{V}{d}$.

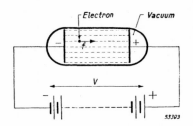

Fig. 7
A free electron between two parallel plates mounted in a vacuum. A voltage V existing between the plates produces an electrostatic field, the lines of force of which are parallel.

If an electron carrying charge —e is present at some point in the field, it will be subjected to an electrostatic attraction by the positive electrode. The attraction is determined by the field strength F at the point at which the electron is located. The attractive force f is represented in Fig. 7 by an arrow, which represents the direction of the attraction, and is equal to:

$$f = e \times F \ldots . \text{ dynes,} \qquad (1)$$

where e and F are expressed in electrostatic units, or:

$$f = e \times F \times 10^7 \ldots . \text{ dynes,} \qquad (2)$$

where $e = 1.60 \times 10^{-19}$ coulombs and F is expressed in volts/cm, or:

$$f = e \times F \ldots . \text{ newtons,} \qquad (2a)$$

where $e = 1.60 \times 10^{-19}$ coulombs, F is expressed in volts/metre and one newton = 10^5 dynes (Giorgi or mks units).

The effect of the attractive force upon the electron is to accelerate the electron in the direction of the force. Because force is equal to the product of mass and acceleration, the electron will be subjected to an acceleration of:

$$g = \frac{e \cdot F}{m_e} \ldots \text{m/sec}^2, \tag{3}$$

where m_e = mass of the electron, and m_e, e and F are expressed in mks units.

If the electron was originally at rest at the negative plate and travels across a potential difference V, it will have attained a kinetic energy equal to $\frac{1}{2} m_e v^2$, where v is the velocity of the electron in m/sec. This kinetic energy is equal to the work (eV) done by the field on the electron. The two may thus be equated:

$$\tfrac{1}{2} m_e \cdot v^2 = e \cdot V, \tag{4}$$

whence:

$$v = \sqrt{2 \frac{e}{m_e} V} \ldots \text{m/sec} \tag{5}$$

when e, V and m_e are again expressed in mks units.

As $\dfrac{e}{m_e}$ = 1.76 × 10^{11} coulombs/kg, the final velocity of the electron after travelling across a potential difference V is:

$$v = 5.93 \times 10^5 \times \sqrt{V} \ldots \text{m/sec.} \tag{6}$$

Thus, if the potential difference between the cathode and anode is 200 volts, the final velocity of an electron will be 8.4 × 10^6 m/sec, or roughly 5,200 miles per second. If the electron had traversed a voltage difference of 1 volt only, its final velocity would be 593 km/sec, which is equivalent to a kinetic energy of:

$$\tfrac{1}{2} m_e v^2 = 1.60 \times 10^{-19} \ldots \text{joule.}$$

This quantity of energy is termed an **electron volt** (eV), and the term is often employed as a unit even in cases where no movement of electrically charged particles in an electrostatic field are involved.

If there is a small opening in the positive electrode, permitting the electron to pass through to a region where no electric field exists, the electron, no longer influenced by a field, will continue to travel at uniform velocity in a straight line, its speed being determined by the difference of potential

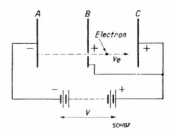

Fig. 8
An electron projected through an opening in the positive electrode B, and entering a field-free space, moves at uniform velocity.

between the negative and positive plates. This arrangement is illustrated in Fig. 8, where A is the negative plate, B the positive plate with aperture, and C a third plate located behind B and maintained at the same positive potential, so that no field of electrostatic force exists between B and C.

As such an arrangement of plates "shoots" the electron through the aperture it is sometimes termed an **electron gun.**

If an electron which has been projected by an electron gun arrives in a space where, instead of no electric field, there exists an electrostatic field of which the lines of force are at right angles to the direction of motion of the electron, as shown in Fig. 9, the electron will again be accelerated, but this time in the direction of the positive plate of the

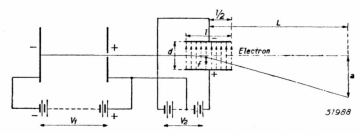

Fig. 9
When an electron, originally travelling in a straight line, enters a transverse field (i.e. a field at right angles to the original direction of travel) the electron is deflected and it leaves the field in a new direction.

new field. As a result, the electron path becomes parabolic, in the same way that a projectile which has been fired horizontally describes a parabolic trajectory due to the attraction of the earth. The electron thus leaves the deflecting field with a new direction of travel. The electrostatic deflection of the beam in cathode-ray tubes is based upon this principle.

The amount of the deflection a (see Fig. 9) at a distance L from the deflecting plates can be easily calculated and is approximately equal to:

$$a = \tfrac{1}{2} \times \frac{L}{V_1} \times \frac{l}{d} \times V_2. \tag{7}$$

It will thus be seen that the amount of the deflection depends upon the voltage V_1, on the distances L, l and d, and upon the deflecting voltage V_2.

It is once more emphasized that the deflection of electrons in an electrostatic field is in a direction opposed to that of the lines of force.

In amplifying valves conditions are not usually so simple, and the field distribution is so complex that exact calculation of the electron paths is difficult. Special apparatus has, however, been devised by means of which it is possible to determine the paths followed by the electrons in the valve [1]).

10. Electrons in a Magnetic Field

An electron in motion can be considered as an electric current. A current of 1 A flows in a wire if 6.25×10^{18} electrons per second pass each cross-section of the conductor in the direction of the positive pole.

An electron moving through the vacuous space in a valve also constitutes an electric current,

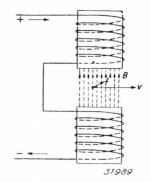

Fig. 10
The force f acting upon an electron moving in a magnetic field is at right angles both to v and to B, and is in a direction away from the reader.

albeit a very small one. If the electron moves through a magnetic field a force is exerted on it in the same way that a force acts on a current-carrying conductor in a magnetic field. This force is at right angles to the direction of travel of the electron and to the direction of the field.

The force f exerted on an electron travelling with a velocity v in a magnetic field of flux density B, the direction of which is at right angles to the direction of travel of the electron, is given by the formula:

$$f = B \cdot e \cdot v \ldots \text{newtons}, \tag{8}$$

where v, e and B are expressed in mks units.

This continuously acting deflecting force causes the electron to change its direction of travel continuously and to follow a circular path. This path cannot be shown in Fig. 10 as it is directed away from the reader. Its radius can be calculated, since the centripetal force must be equal to the (magnetic) force f:

$$\frac{m_e \times v^2}{r} = B \cdot e \cdot v, \tag{9}$$

whence:

$$r = \frac{m_e}{e} \times \frac{v}{B} \cdots \text{m}, \tag{10}$$

where v, e, m_e and B are expressed in mks units.

---

[1]) See Philips Technical Review: **2** (1937) p. 338; **4** (1939) p. 223; and **5** (1940) p. 131.

As an example, if the kinetic energy of an electron is equivalent to 100 e V, or 1.6×10^{-17} watt-seconds (joules), when it enters a magnetic field the lines of force of which are at right angles to the direction of travel of the electron, and the field strength is 100 gauss or $^1/_{100}$ Vsec/m², the radius of the path followed by the electron will be:

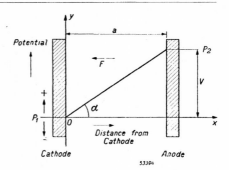

Fig. 11
Diagram of potential distribution in the space between the cathode and the anode of a valve having two parallel flat electrodes (two-electrode valve or diode).

$$v = \frac{9.1 \times 10^{-31} \times 5.93 \times 10^5 \times \sqrt{100}}{1.60 \times 10^{-19} \times 10^{-2}}$$

= 0.0034 metre or 0.34 cm approximately.

If, on the other hand, the initial direction of travel of the electron were parallel to the magnetic field instead of across it, it would not be deflected at all. If, however, the electron enters the magnetic field at an angle to the direction of the lines of force, it will follow a helical path, the component of the velocity normal to the lines of force providing the circular component of the helix and the component parallel to the lines of force providing the axial component of the helix. Even very weak magnetic fields are sufficient to produce considerable deflection of the electron path, so that the sensitivity to magnetic influence is very great.

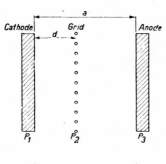

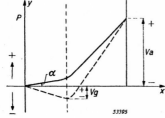

Fig. 12
Diagram of potential distribution over a cross-section between two grid wires in a valve having a negatively charged grid.

11. Potential Diagrams

Electrostatic fields and their distribution play a predominant part in radio valve technique, in addition to which interference is often caused by external magnetic and electric fields. It is therefore very necessary to form a correct picture of the electrostatic distribution of potential.

Fig. 11 represents a flat cathode and an anode mounted parallel to it at a dis-

14

tance a. The cathode is at a low potential P_1 and the anode at a high potential P_2; the voltage difference between them is therefore $V = P_2 - P_1$.

In Fig. 11 distances from the cathode are plotted horizontally on the x-axis and potential differences are plotted vertically on the y-axis. In the diagram the potential of the cathode is assumed to be zero in conformity with the usual convention in which all voltages in a valve are referred to the cathode as zero. The potential of the anode is thus equal to V.

Because the potential at points between the cathode and anode increases linearly with the distance from the cathode, the sloping line from the cathode to the anode represents graphically the increasing potential as a function of the distance from the cathode. Therefore the field strength F in the space between the cathode, which, by definition is $\dfrac{V}{a}$, is equal to the tangent of the angle a between the potential line and the horizontal axis. If an electron is present in the space between the anode and the cathode, a force f will be exerted upon its negative charge in a direction opposed to the field strength, i.e. towards the anode.

Control of the electron current from the cathode to the anode can be exercised by a grid consisting, for example, of parallel inter-connected wires of small diameter, represented in cross-section in Fig. 12.

The potential difference between a grid wire and the cathode is equal to $P_2 - P_1 = V_g$, the grid bias, and the potential difference between the anode and a grid wire is $P_3 - P_2 = V_a - V_g = V_a + |V_g|$.

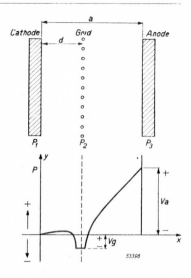

Fig. 13
Diagram of potential distribution over a cross-section intersecting a grid wire in a valve having a negatively-charged grid.

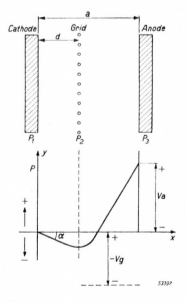

Fig. 14
Diagram of potential distribution in a three-electrode valve when a very large negative bias is applied to the grid.

15

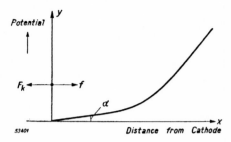

Fig. 15
The tangent of the angle a between the potential curve and the x-axis at the surface of the cathode is a measure of the field strength F_k at the cathode and of the force f acting upon the electron.

If V_g is negative with respect to the cathode, it would be expected that the potential curve would be of the shape shown by the dotted line in Fig. 12. In practice this is not the case because the potential in the space between the grid wires is greater than P_2, due to the penetration of the anode-cathode field. The potential distribution is therefore of the form indicated by the full line in Fig. 12. This curve, however, only holds for the potential distribution in a plane midway between two grid wires. In a plane which intersects a negative grid wire the potential distribution is approximately of the form indicated in Fig. 13.

The field strength at any point within a valve is indicated by the slope of the potential curve at that point. Fig. 14 is the potential curve in a valve where the negative voltage at the grid with respect to the cathode is high, and for a plane midway between two grid wires, and it is seen that in the region between the cathode and the grid the potential line runs below the x-axis, and the angle a is therefore negative. This means that in this region the field strength is in the direction of the anode, and an electron present in this region will travel towards the cathode instead of towards the anode—in other words, the negative grid bias blocks the path of the electrons through the grid. In these circumstances no electron current can flow to the anode. This statement is not strictly true for small negative values of the angle a, as explained in Chapter IV, Sections 16 and 17.

The tangent of the angle a between the potential curve and the horizontal axis at the surface of the cathode is a measure of the field strength F_k acting at the cathode surface (see Fig. 15), and this field strength determines the force acting in the direction of the anode and exerted on an electron leaving the cathode.

The average field strength F_k at the cathode surface of a valve having three electrodes, one of which is a control grid, is given by the formula:

$$F_k = \frac{p\left(V_g + \dfrac{V_a}{\mu}\right)}{d},\tag{11}$$

where μ is a factor called the **amplification factor** and depends upon the dimensions of the valve (distances between anode and cathode and between grid and cathode; diameter of grid wires; and number of grid wires per centimetre of grid length) and d is the distance between grid and cathode.

The magnitude of p is usually only a little less than unity. The voltage

$$p\left(V_g + \frac{V_a}{\mu}\right)$$ may be considered as the

voltage between the anode and the cathode of an imaginary valve having two electrodes spaced at a distance d and producing the same field strength F_k at the cathode surface as the three-electrode valve with control grid mentioned above.

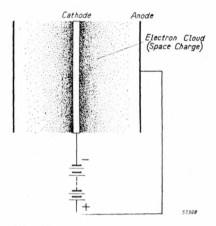

Fig. 16
The cathode is surrounded by a cloud of electrons (elemental negative charges), called the space charge.

For the sake of simplicity this section has dealt only with valves having flat electrodes, and it must be mentioned that most modern valves do not possess flat electrodes but concentrically mounted electrodes.

12. The Space Charge

Section 11 above deals with the conditions in the region between the cathode and the anode when no electrons or only a small number of electrons are present in this space.

If, however, the cathode emits a large number of electrons the conditions have to be reviewed.

Considering first the case of a two-electrode valve, the cathode of which emits a large number of electrons into the vacuous space surrounding the cathode, it has already been shown that, under the influence of the field existing at the cathode due to a positive anode voltage, the electrons will move in the direction of the anode.

At first sight it would be expected that in these circumstances all the electrons leaving the cathode would reach the anode and, further, it is reasonable to expect that if a negative voltage is applied to the anode the electron current will be entirely blocked.

In practice however, the potential distribution is quite different from that based only on the geometry of the electrodes and the voltages at the electrodes. The reason is that the electrons leaving the cathode constitute a negatively charged cloud in the space between the cathode

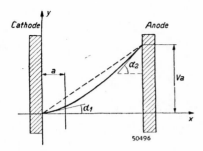

Fig. 17
Potential (vertical axis) as a function
of the distance from the cathode
(horizontal axis) for a valve with two
parallel flat electrodes.
Full line — with space charge present.
Broken line — no space charge present.

and the anode. As indicated in Fig. 16, this cloud of electrons is very dense in the neighbourhood of the cathode. It has a very considerable influence on the electron current.

The charge of the electron cloud, the **space charge**, is negative because the individual electrons are negatively charged, and this results in some of the electrons being forced back to the cathode, since the negative space charge repels the negatively charged electrons leaving the cathode. This action limits the electron current to the anode.

The anode current thus has a definite value for every value of anode voltage, and is equal to that current which maintains a negative space charge exactly neutralising the field produced at the cathode by the positive anode voltage. This statement is not strictly true because, as explained in Chapter IV, the majority of the electrons do not leave the cathode with zero velocity. If the anode voltage is increased the velocity of the electrons in the electron cloud increases, and a greater number of electrons are able to leave the cathode. As a result the anode current also increases, and at the same time the density of the electrons (i.e. the space charge) also becomes greater.

In the neighbourhood of the cathode the velocity of the electrons is small, but it increases very rapidly as the electrons approach the anode. Electric-current density is simply the amount of charge which is displaced in unit time through a given cross-section. If it is assumed that the current in the valve for a constant cross-section is the same at all points, it is clear that the density of the charge at any point will be greater if the charge moves slowly than it would be if the charge moves quickly. For this reason the density of the space charge is high in the neighbourhood of the cathode and decreases with the increasing velocity of the electrons as the anode is approached.

As the result of the space charge, the potential curve between the anode and the cathode is not a straight line but curved. This is represented in Fig. 17, where the negative space charge causes the potential at a distance a from the cathode to be lower (on the full line) than it would be if there were no space charge (broken line).

Because both the full and the broken lines in Fig. 17 must terminate at the same point at the anode, the potential distribution under **negative**-space-charge conditions will be represented by a line bent downwards. The shape of the curve is determined by the space-charge distribution between the cathode and the anode, and the curvature of the potential line at any point on the graph is a measure of the space-charge density at that point, this density being expressible in coulombs per cubic metre.

Now, let it be assumed that the cathode is capable of emitting an unlimited number of electrons. A given anode voltage will result in the formation of a negative space charge between the cathode and anode and the potential distribution will be bent downwards. If the angle a_1 (see Fig. 17), the tangent of which is a measure of the field strength at the surface of the cathode, is positive, the number of electrons leaving the cathode will increase — always provided that the cathode is able to emit an unlimited number of electrons. As a result the density of the space charge will increase until the angle a_1 becomes zero, and it is only when this occurs that a state of equilibrium is attained, resulting in a definite density of the space charge and a definite value of the anode current depending upon it. This means that for a given area and spacing of the electrodes there is for every value of anode voltage a definite value of the electron current, determined by the space charge. This current is said to be **space-charge limited**. At the state of equilibrium the potential curve has a characteristic shape; once the dimensions of the electrodes and their spacing have been fixed, the characteristic shape of the curve is the same for any anode voltage; if the voltage is altered only the voltage scale of the diagram is affected. This means that the density of the space charge is proportional to the anode voltage. As, however, the velocity of the electrons also changes with change of anode voltage, and this change is proportional to the square root of the voltage [see Equation (5)], the current will be proportional both to the voltage and to the square root of the voltage, and hence proportional to the $^3/_2$ power of the voltage. A current which is limited by the space charge is expressed by:

$$I = k \cdot V^{3/2}, \tag{12}$$

where k is a factor depending upon the dimensions of the valve. For two flat parallel electrodes

$$k = \frac{\sqrt{2}}{81 \cdot 10^9 \cdot \pi} \sqrt{\frac{e}{m_e} \cdot \frac{S}{d^2}} \cdot \cdot \cdot \cdot (A/V^{3/2}), \tag{13}$$

where d is the distance in metres between the electrodes,
   S is the area of the electrodes in square metres.

Substituting the numerical value of $\sqrt{\dfrac{e}{m_e}} = \sqrt{1.76 \cdot 10^{11}}$ and introducing the value of k thus obtained in Equation (12) we obtain:

$$I = 2.33 \cdot 10^{-6} \frac{S}{d^2} \cdot V^{3/2} \ldots \text{(A)}. \qquad (14)$$

For a valve with a control grid as shown in Fig. 12 the anode current is:

$$I_a = k \left[p \left(V_g + \frac{V_a}{\mu} \right) \right]^{3/2} = k' \left[V_g + \frac{V_a}{\mu} \right]^{3/2}, \qquad (15)$$

where V_g = control-grid voltage,

V_a = anode voltage,

k′ is a factor depending upon the dimensions of the valve

and μ is the "amplification factor" of the valve.

Thus, in a three-electrode valve the space-charge-limited anode current is proportional to the $3/2$ power of the potential $\left[V_g + \dfrac{V_a}{\mu} \right]$. The potential $p \left(V_g + \dfrac{V_a}{\mu} \right)$ is the mean potential in the plane of the control grid, and will be referred to as the **effective potential** of the valve. Equation (15) indicates that the anode current limited by space charge depends upon the value of the grid bias V_g. In normal valves even a small variation of the voltage on the grid causes a relatively large change in the anode current. It is upon this fact that the control action of the grid is based.

CHAPTER IV

Electron Emission

In order that an electron current can flow through an evacuated valve it is necessary that the valve contains a surface of metal or of some other material which will give off "free" electrons. Such a surface has already been referred to as a cathode. The release of free electrons from a surface is called **electron emission.**

Electrons may be made to leave the surface of a metal or other suitable material in any of the following ways:

1. By heating a metal or a metal coated with a suitable material to a high temperature corresponding to red to yellow-white heat. This is known as **thermal emission**.
2. By the action of light rays upon the surface, known as the **photo-electric effect.**
3. By bombarding the surface with electrons—**secondary emission.**
4. By bombarding the surface with positive ions.
5. By applying an extremely high field strength, of the order of ten million V/cm, to the surface—**cold emission.**

Of these only (1), thermal emission, and (3), secondary emission, fall within the scope of this book.

13. Electrons "Bound" within the Metal

The electrons of the "electron gas" move about within the metal at widely varying but usually very high velocities. Notwithstanding their high velocities, however, the electrons are not able to leave the metal at ordinary room temperature. Even when they are not bound to specific atom nucleï the electrons are attracted by the positive ions in whose vicinity they happen to be at the moment.

Inside the metal an electron is in the electrostatic field produced by the positive ions and by all the other electrons. Generally speaking, the forces created in this manner and acting upon an individual electron neutralise each other within the metal, since the electron is surrounded on all sides by approximately equal numbers of ions and electrons. This, however, is no longer the case when an electron is at, or has penetrated, the surface of the metal, for the attractive forces exerted by the ions upon the electron are then not entirely neutralized by the remaining electrons and therefore tend to draw the electron back into the metal.

Those electrons having the highest velocities will occasionally penetrate the surface layer of ions and remain outside the metal for a short time, that is, until their velocity has been checked by the attractive force of the ions forming the surface layer, and the electrons return to the metal. During the time that they are outside the metal the electrons form part of the negatively charged cloud consisting of all the electrons which have similarly escaped from the metal. The negative charge existing at any instant outside the metal has thus been drawn from the outer layers of the metal surface, leaving a positive charge on these layers. There is thus formed the so-called "double layer" consisting of the surface of the metal and the negative electron cloud at the surface, and in the double layer there exists a potential gradient.

In order that a stationary electron can be released from the metal, a definite quantity of energy, E_a, must be imparted to it to enable it to pass through the double layer.

14. Velocity or Energy Distribution of the Electrons, and the Work Function

The free electrons move through the ion grid within the metal. In the course of its movement through the space between the ions the velocity of a particular electron will not remain constant because the electron will be accelerated and retarded and will undergo changes of direction. At a given instant, therefore, all the free electrons will be moving in different directions at different velocities, and they will have, therefore, different kinetic energies.

At a temperature of 0 °K the most rapid electrons will possess a kinetic energy E_0, and to enable them to leave the metal additional energy equal to at least $E_a - E_0$ must be supplied. This additional energy is termed the **work function** and is somewhat analogous to the ionization energy of an atom: the work function is usually expressed in electron volts (eV, see Chapter III, Section 9), and is represented by the symbol φ. The work function for metals with densely arranged ions is greater than for metals with wider spacing between the ions. For tungsten, for example, it is 4.54 eV, for caesium 1.96 eV, and for barium 2.70 eV. The magnitude φ is sometimes called the **work potential.**

15. Electron Emission and Saturation Current

At a temperature T a certain number of electrons in the metal will possess a kinetic energy greater than E_0, and some of these may have sufficient kinetic energy to perform the quantity of work E_a, and thus be able to leave the metal.

The energy distribution of the electrons can be determined statistically, and from this can be calculated the number of electrons which will leave the metal per second. The formula is:

$$I_s = \frac{4\,\pi\,e\,m_e\,k^2}{h^3}\,T^2\,\varepsilon^{-\frac{e\,\varphi}{kT}} \ldots (A/m^2), \qquad (16)$$

where:

m_e = mass of an electron = 9.1×10^{-31} kg,
k = Boltzmann's constant = 1.38×10^{-23} joules per °K,
h = Planck's constant = 6.6×10^{-34} joules sec,
T = absolute temperature in °K,
ε = base of natural logarithms = 2.72 . . . .,
e = charge of one electron in coulombs and
φ = the work function in volts.

Substituting A_0 for $\dfrac{4\,\pi\,e\,m_e\,k^2}{h^3}$ amperes per m² per °K² in (16):

$$I_s = A_0\,T^2\,\varepsilon^{-\frac{e}{kT}}\ A/m^2. \qquad (17)$$

The theoretical value of A_0 for all metals is 120.4 $\dfrac{A}{m^2\,°K^2}$.

The current I_s is termed the **saturation current.** It means that, at a given temperature, only that number of electrons can leave the surface of the metal per second which represents the value of the saturation current as calculated from Formula (17), which gives the relation between temperature and saturation current.

In order to measure the saturation current the valve must contain a positive anode to collect the electrons emitted by the cathode, and the positive voltage at the anode must be so great that no space charge can exist. In other words, the potential difference between the anode and cathode must be such that the saturation current as measured does not increase if the potential difference is increased.

Curves can be plotted showing the current passing through a two-electrode valve as a function of the voltage V_a at various values of the cathode tem-

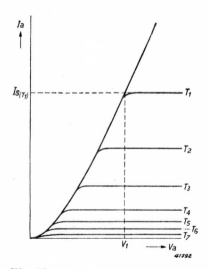

Fig. 18
Curves showing the relation between the current I_a through the valve and the voltage between anode and cathode at various cathode temperatures T.

perature T. Such a series of curves is reproduced in Fig. 18. At a cathode temperature T_1 and with increasing anode voltage, the current first increases as the $^3/_2$ power of the voltage in accordance with the law of space-charge-limited current. At a certain value of the anode voltage, however, the curve becomes horizontal, that is to say the current remains practically constant for a further increase in anode voltage, and this constant value is the value of the saturation current I_s at temperature T_1. At lower temperatures, T_2, T_3 etc., the value of the saturation current is smaller.

16. Influence of Electron Emission upon the Potential Diagram

When discussing potential diagrams in Chapter III it was assumed that the cathode was capable of emitting a large number of electrons and that these electrons had no initial velocity.

The known velocity distribution of electrons in metals shows, however, that if the cathode temperature is sufficiently high the electrons after leaving the cathode will have a residual velocity. When the saturation current of the cathode at a given temperature is greater than the space-charge-limited current for the prevailing voltage difference between anode and cathode it is because, in consequence of their initial velocity, more electrons are able to enter the space between cathode and anode than those corresponding to the normal space-charge-limited current. The potential as a function of the distance from the cathode must therefore be lower than the values corresponding to the potential distribution for space-charge-limited current (when the electrons possess no initial velocity).

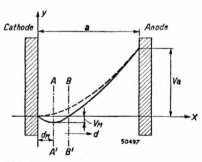

Fig. 19
Broken line: Potential-distribution diagram when electrons have no initial velocity.
Full line: Potential-distribution diagram for a given cathode temperature when electrons have initial velocity. At a distance d_M from the cathode a potential minimum V_M occurs.

The broken line in Fig. 19 shows the potential distribution when the electrons have no initial velocity, and it will be observed that the curve starts in a horizontal direction from the cathode. It is clear, therefore, that when the space charge becomes denser as the result of the initial velocity of the electrons, the potential in the vicinity of the cathode must become negative, and the potential-distribution curve will be of the form shown in full line in Fig. 19. It will be seen from this curve that at a distance d_M from the cathode the potential reaches a minimum value V_M. This value is known as **Epstein's minimum.**

All the electrons emitted are thus subjected to a negative potential gradient V_M, and the field between the potential minimum and the cathode tends to force the electrons back to the cathode. Only those electrons will reach the anode which (a) have overcome the "binding" forces of the metal and (b) possess sufficient kinetic energy, corresponding to their velocity components perpendicular to the surface of the cathode, to overcome the potential minimum.

A state of equilibrium will be reached when the potential minimum, which is dependent upon the geometry of the valve and upon the anode voltage, reaches a value at which all surplus electrons are forced back to the cathode.

17. The Potential-barrier Current

It will now be obvious that, due to the initial velocity of the electrons, some electron current must flow even when the potential of the anode is zero or slightly negative with respect to the cathode. If the anode potential were zero all the electrons emitted would reach the anode were it not for the space charge, and the value of the electron current would be the same as the saturation current at the prevailing cathode temperature. However, because the cloud of electrons emitted sets up a space charge, a potential minimum is formed.

If, in Fig. 19, the anode voltage is reduced, first to zero and then to a negative value, as shown in Fig. 20, it is found that with increasing negative anode voltage the value of the potential minimum V_M also increases, and its distance from the cathode is increased. This is seen by comparing the dot-and-dash curve with the full curve in Fig. 20. With further reduction of the anode voltage a condition is

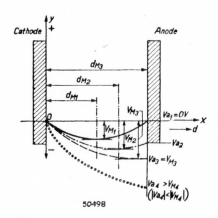

50498

Fig. 20
Full line: Potential distribution in the space between cathode and anode when the anode voltage is zero.
Broken line: Potential distribution when the anode voltage has a negative value such that the potential minimum occurs at the anode.
Dot-dash line: Potential distribution when the anode voltage has a less negative value than that corresponding to the broken line.
Small circles: Potential distribution when the anode voltage has a more negative value than that corresponding to the broken line.

reached at which the potential minimum occurs at the anode, that is to say $V_M = V_a$. This is shown in the broken-line curve in Fig. 20, when $V_{M3} = V_{a3}$. (In Fig. 19 the potential minimum would have occurred at the anode if the anode were at the position AA′.) Under this condition the curvature of the potential-distribution graph is extremely small, indicating that the space charge is of very low density.

At a still greater negative anode voltage the curvature and the density of the space charge have further diminished.

The current obtaining with negative values of V_a thus overcomes a rather considerable potential barrier. The current which flows when the absolute value of V_a is equal to or greater than the value of the potential minimum V_{M3} (located at the anode) will hereafter be referred to as the **potential-barrier current.** Where a potential-barrier current flows the space charge is practically non-existent.

The formula expressing the potential-barrier current I_a as a function of the anode voltage V_a is:

$$I_a = I_s \, \varepsilon^{\frac{eV_a}{kT}}, \qquad (18)$$

where I_s is the saturation current in amperes per m² at the prevailing absolute temperature T of the cathode, as in Equations (16) and (17), and e, V_a and k are expressed in mks units.

From the form of Equation (18) it is clear that the potential-barrier current can never become zero. It will, however, be very small when V_a reaches a sufficiently high negative value.

It can easily be established experimentally that an anode current flows in a two-electrode valve when the anode voltage is slightly negative. Similarly it is possible to measure the current to the control grid of a multi-electrode valve when the control grid is at a small negative potential. It is thus clear that in a valve containing a control grid it is possible that a relatively large number of electrons may reach the grid during that portion of the grid-alternating-voltage cycle when the grid is only slightly negative, unless the direct nega-tive grid voltage (grid bias) is

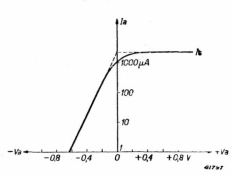

Fig. 21

Potential-barrier current and saturation current, both plotted to a logarithmic scale, as function of the anode voltage plotted to a linear scale. If the straight line portions of the curves are extended through the transitional region, as in the broken lines, they must intersect on the vertical axis.

sufficiently great. As a result grid current will flow and will have a deleterious effect upon amplification. In order to avoid this the grid should never be allowed to become less than 1.3 V negative. In most valves the grid current at this voltage is less than 0.3 μA and this value is usually innocuous.

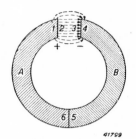

Equation (18), $I_a = I_s \, \varepsilon^{\frac{eV_a}{kT}}$, can be plotted as a graph to show I_a as a function of V_a. It is usual to plot the potential-barrier current on a logarithmic scale and the voltage on a linear scale, giving a curve of the form shown in Fig. 21. The curve first rises in a straight line, but at small negative values of the anode voltage it turns over and becomes a horizontal line. Once the horizontal part has been reached the anode current remains constant and is equal to the saturation current. If the two straight portions of the graph are extended, the extensions intersect on the I_a-axis, that is to say at $V_a = 0$. The fact that the potential-barrier current when plotted

Fig. 22
When two metals, A and B, having different values of work function, are brought into contact with each other (points 5 and 6), a potential difference is set up in the space outside the metals (for instance between 2 and 3). This potential difference is equal to the difference between the work functions of the two metals, the metal with the lower work function being positive with respect to the metal with the higher work function.

on a logarithmic scale as a function of V_a must be a straight line follows from Equation (18), for:

$$\text{Log } I_a = \log I_s + \left(\frac{e}{kT} \log \varepsilon \right) V_a = K_1 + K_2 V_a \cdots \quad (19)$$

$$K_1 = \log I_s = \text{a constant,}$$

$$K_2 = \frac{e}{kT} \log \varepsilon = \text{a constant.}$$

Therefore log I_a depends linearly upon V_a.

The slope of the potential-barrier-current line is $\dfrac{e}{kT} \log \varepsilon$, and thus is inversely proportional to the cathode temperature. The extension of this line must cut the I_a axis (where $V_a = 0$) at $I_a = I_s$.

18. The Contact Potential between Metals (the Volta Effect)

A phenomenon which has an important bearing upon the behaviour of radio valves is the electric field existing between two electrodes as a result of the **contact potential** or the so-called Volta effect. This potential difference is superimposed upon the externally applied potentials (anode or grid voltages), and its magnitude depends upon the metals used and upon the nature of their surfaces.

When two metals, A and B (see Fig. 22), are mutually contacted (theoretically before they actually come into contact, thus already when their surfaces are brought into close proximity) free electrons begin to pass from the metal with the smaller work function φ_1 (e.g. the metal A in Fig. 22) to that with the larger work function φ_2 (metal B). In this way the metal B receives a negative charge, whilst A loses negative charge and thus, as it were, receives a positive charge. These charges are distributed over the surfaces of the metals. If A and B are shaped as in Fig. 22, for instance forming a ring with an air gap, then mainly the surface 1 of A that is facing B has a positive charge and the surface 4 of B facing A has a negative charge. This gives rise to an electric field between point 2 just outside A and a point 3 just outside B. The potential at 2 is positive compared with the potential at 3, while the potential difference between 2 and 3 equals the difference between the work functions φ_1 and φ_2. This phenomenon is called the Volta effect and the potential difference the contact potential (V_{cp}). Hence:

$$V_{cp} = \varphi_1 - \varphi_2. \tag{20}$$

We have purposely confined our statement to saying that a charge passes from A to B, and, in so far as mention was made of a potential difference, we only considered the potentials outside the metal. Nothing is said about the electrons inside the metal (see also Sections 15 and 16), for it would lead us too far to go into that more closely. Suffice it to say that the condition inside the metal is such that an electron can pass through the plane between 5 and 6 without any opposition. In principle electrons can also pass from 1 to 4 without any gain or loss in energy, since the difference between the work functions is set off by the energy resulting from the Volta potential difference. At room temperature, however, this will not take place of its own accord.

Upon the ring being closed, bringing 1 and 4 together, the external field disappears and the condition becomes like that at 5 and 6. An electron can then travel through the whole of the ring without any gain or loss of energy. Thus no current is generated in the ring, at least not so long as the temperature is uniform throughout. If there is a difference in temperature between the points of contact then, as is known, thermal currents arise. The potential differences arising in the vacuum of an electronic valve due to the presence of different metals — like the potential difference between 2 and 3 — and the

electric fields resulting from those differences all influence the movement of electrons in the system of electrodes.

The practical result of the Volta effect is that, in a two-electrode valve
for example, a Volta potential is set up between the anode and the
cathode, its value being equal to the difference between the work
functions of the metals used for the anode and cathode. The metal
of the connecting leads between the electrodes, and also to the source
of the externally applied voltage (e.g. the battery), play no other role
than that of charge-carrier.

If the work function of the cathode metal is lower than that of the
metal used for the anode or for the grid, the contact potential of the
cathode will be positive with respect to the anode or to the grid of the
valve. As the value of the contact potential is constant, this means
that the anode-current-anode-voltage curve of a two-electrode valve, as
represented in Fig. 21, will be displaced to the right.

In a three-electrode valve both the grid voltage and the anode voltage
will be affected by the contact potentials used for the grid and for
the anode; the grid, if negative, becomes more negative and the anode,
if positive, becomes less positive.

Because the grid-cathode and anode-cathode Volta potentials for a
given three-electrode valve are constant, they may be combined in
a single term, V_{cp}:

$$V_{cp} = V_g' + \frac{V_a'}{\mu}, \tag{21}$$

where V_g' is the contact potential of the grid with respect to the cathode,
V_a' is the contact potential of the anode with respect to the cathode,
and μ is the amplification factor of the valve.

Because μ is always much greater than unity, the influence of the
contact potential of the anode with respect to the cathode will be
extremely small and, in fact, negligible, so that only the contact
potential of the grid with respect to the cathode need be taken into
account.

19. Electron-emission Efficiency

The formula already given for the saturation current indicates that
the temperature of the cathode is an important factor in the electron
emission, the emission increasing with increasing temperature. It is
clearly desirable, therefore, to operate the cathode at the highest
possible temperature. A high working temperature confers a further
advantage, namely, that because a certain amount of power (in watts)
is required to heat the cathode (the actual method of heating need

not be discussed for the moment), the higher the temperature the greater the number of electrons emitted per watt of heating power. The quotient $\dfrac{\text{electron emission in milliamperes}}{\text{heating power in watts}}$ is termed the **electron-emission efficiency** of the cathode.

As the temperature rises this efficiency increases rapidly, due to the fact that the increase in electron emission at higher temperatures is much greater than that of the heat lost by the cathode by radiation and conduction. (The heating power must cater for the heat losses and can be put equal to the total energy given off in the form of heat.)

The various metals and special materials which might be considered suitable for use as cathodes have widely differing maximum permissible working temperatures, so that it may happen that certain metals with a low work function, and, therefore, on the face of it desirable, have, by reason of low melting point and high rate of evaporation, a much lower electron-emission efficiency than other materials with a higher work function, and for this reason may be of no practical use at all. Moreover, the choice of cathode material must always be greatly influenced by the extent to which the material is amenable to processing, by its durability, and by other properties.

20. Electron Emission from the Surface of a Pure Metal

The rate of electron emission by a pure metal is shown, by the formula previously given, to be exponential. The alkali metals sodium, potassium, rubidium and caesium, although having the lowest work functions of all metals, are of no practical value as cathode material for thermal emission because their melting points are too low and their rates of evaporation too high. Caesium, for example, melts at 26 °C, and at that temperature gives a saturation current I_s of approximately 5×10^{-21} A/m² [1]).

This value is far too low, as the cathode current in normal amplifying valves is of the order of some milliamperes and in output valves of the order of 20 to 100 milliamperes, while the area of the cathode is only a few square centimetres. Again, metals such as zirconium and platinum have melting points which preclude their consideration as cathode material. Only metals having a very high melting point, such as tungsten, tantalum and molybdenum, can be considered suitable.

Among other factors of importance when selecting cathode material is the behaviour of the metal in the presence of traces of gas remaining

---

[1]) Value computed from $A_0 = 162$ A/cm² °K² and $\varphi = 1.87$ V (see Reimann: "Thermionic Emission" — John Wiley & Sons, Inc., New York. 1936, page 99).

in the valve. In spite of most careful pumping and other processes applied with the object of achieving a perfect vacuum, there is always a certain quantity of residual gas left in the valve, and other gases are subsequently released from the various component parts of the valve. These residual gases, which include oxygen, nitrogen, carbon dioxide, water vapour and so on, may react with the cathode material to form chemical compounds which so change the emitting surface as seriously to reduce its power to emit electrons. Oxygen, for instance, may form a thin layer of oxide on the cathode, in consequence of which electron emission is almost entirely prevented. These effects of residual gases are termed **"poisoning" of the cathode.**

The life of the cathode may be adversely affected by the presence of rare gases such as argon, helium and neon, and also by mercury vapour, even though these gases do not directly affect the electron emission. Given a sufficiently high voltage between one of the other electrodes and the cathode, these gases become ionized and the positive ions strike against the cathode. This ion bombardment releases atoms from the cathode and speedily destroys it. Choice of cathode metal must therefore be governed by its behaviour in the presence of residual gas.

Of the three metals previously mentioned as suitable for use as cathodes, tungsten is the most important, for the following reasons:

1. It has a high melting point (3655 °K), which ensures long life at a high working temperature (2500 °K).

2. Its electron emission is not affected by residual gases because its compounds with these gases—e.g. tungsten oxide—are extremely volatile at high temperatures, so that the surface is kept clean. Even a layer one atom thick of oxygen, which would seriously impair the electron emission, cannot exist at this high working temperature.

Molybdenum has, it is true, a higher specific electron emission than tungsten, but on account of its lower melting point (2895°K) it has a short life at the desired working temperature. Tantalum has a melting point of 3300°K, and in this respect is more suitable than molybdenum, but its electron emission is greatly affected by residual gases, as its compounds, such as tantalum pentoxide, are only slightly volatile at normal working temperature.

The table below gives the electron emission of tungsten, in A/cm^2, and the emission efficiency in milliamperes per watt of heat dissipation, for various cathode temperatures in degrees absolute.

TABLE I

Metal: Tungsten

Temperature (°K)	Electron emission (A/cm²)	Emission efficiency (mA/W)
1500	0.102×10^{-6}	1.85×10^{-5}
1600	0.102×10^{-5}	1.28×10^{-4}
1700	0.812×10^{-5}	7.58×10^{-4}
1800	0.490×10^{-4}	3.47×10^{-3}
1900	0.257×10^{-3}	1.38×10^{-2}
2000	0.112×10^{-2}	4.67×10^{-2}
2100	0.427×10^{-2}	0.140
2200	1.41×10^{-2}	0.369
2300	4.37×10^{-2}	0.920
2400	12.3×10^{-2}	2.13
2500	30.2×10^{-2}	4.27
2600	0.776	8.92
2700	1.74	19.6
2800	3.74	27.5
2900	7.75	43.7
3000	14.9	65.0
3100	28.1	89.2
3200	50.5	112
3300	87.7	135
3400	149	155

21. The Electronic Emission of Oxide Cathodes

By applying a coating of suitable oxides to the metal of a cathode it is possible to produce a cathode which is far more efficient than those composed of pure metals. Cathodes composed of a metal or metal alloy covered with a thin layer of alkaline earth oxide are termed **oxide cathodes**. They give a much greater emission than those of pure metal, at a relatively low temperature of the order of 700 °C to 800 °C.

(a) Activation of Oxide Cathodes

Oxide cathodes for radio valves consist of a metal core of tungsten or nickel wire or a strip or small tube of nickel, coated with a mixture of barium oxide and strontium oxide from 20 to 80 microns thick

(1 micron $= 10^{-4}$ cm). In the manufacture of these cathodes the wire or nickel tube is first coated with a mixture of barium carbonate ($BaCO_3$) and strontium carbonate ($SrCO_3$). The cathode prepared in this way is activated during the pumping and "burning-out" process after the assembly of the valve has been completed. By means of high-frequency eddy currents, induced by a coil of a few turns surrounding the valve, all the internal metal parts of the valve are raised to a red heat, and at the same time an electric current is passed through the tungsten or nickel wire (or through a heating element located in the nickel tube), thus raising the cathode temperature to approximately 1000 °C.

At this high temperature the carbonates with which the cathode has been coated are decomposed, with the evolution of carbon dioxide (CO_2), which is removed from the valve by pumping, leaving a coating of oxide on the metallic core. When the pumping, which is performed simultaneously with the heating, has been completed, the envelope is sealed off by fusion. But the cathode is not yet fully activated, for the electron emission has not reached its maximum and is not uniform over the whole area of the cathode. Activation is completed by the process known as "burning out". After the valve has been sealed off the cathode filament (or the heater element of a tubular cathode) is raised to a high temperature by applying from one-and-a-half times to twice the normal working filament or heater voltage. Simultaneously electron emission is drawn from the cathode by applying, to both anode and grid, voltages which are positive with respect to the cathode. During this process the electron emission from the cathode steadily increases. After a certain time the "burning out" is continued for a final period with increased filament or heater voltage, but the control-grid voltage is reduced to zero, so that the emission during the final stage is relatively low.

The activated oxide cathode contains a certain percentage of metallic barium and strontium. The number of free barium atoms has been found to be approximately 0.2% of the number of barium oxide molecules. Some of these barium atoms are adsorbed at the surface of the barium oxide crystals [1]), but it is presumed that the majority of them are contained within the oxide crystals. The free barium present in the activated cathode plays an important part in the emission, as is proved by the fact that if oxygen is admitted into the valve for a

---

[1]) By "adsorption" of metal atoms at a surface is meant the phenomenon of foreign metal atoms being "bound" to the surface of a metal (or oxide). Foreign metal atoms may often also be "bound" as ions.

short time, thus oxidising the free barium, the emission is considerably reduced. When the barium is thus oxidised the cathode is said to be "poisoned" by the oxygen. Other gases which do not combine with barium do not poison the cathode. Water vapour, however, is extremely injurious to the cathode because it not only forms a compound with the barium but also converts the oxide to a hydroxide.

Another proof of the importance of barium in the emission of electrons lies in the fact that an oxide cathode can be activated by distilling barium onto the cathode from other parts of the valve. In point of fact the activating process itself appears to consist exclusively in the formation, through various causes, of metallic barium during the period of pumping and burning out.

Various explanations have been put forward to account for the manner in which electron emission takes place from an oxide cathode. The most recent conception, which seems to approach nearest to the facts of the case, is that expounded by J. H. de Boer in his "Electron Emission and Adsorption Phenomena", pp 353 et seq., published by the University Press, Cambridge (1935), to which the reader is referred.

(b) Saturation Current of the Oxide Cathode

The saturation current per square metre can be expressed with sufficient accuracy by the formula:

$$I_s = A'T^2 \varepsilon^{\frac{-e\varphi'}{kT}}. \tag{22}$$

It appears that the work function φ' of a properly activated oxide cathode amounts to $1.0 - 1.3$ volts. The factor A' is then from 0.2×10^4 to 10^5 A/(°K m)^2.

Compared with that of a tungsten cathode, the work function of an oxide cathode is remarkably low. As a result, the emission of the oxide cathode per watt dissipated in heating is relatively high, that is to say the electron-emission efficiency is high, amounting at the normal working temperature of 1050 °K to as much as 1,000 mA per watt. As already mentioned, the oxide cathodes used in practice usually consist of a mixture of barium oxide and strontium oxide, because such a mixture gives a much greater emission than either of the oxides alone.

(c) Potential-barrier Current of Oxide Cathodes

Equation (18) gives the formula for the potential-barrier current, and this formula applies equally for oxide cathodes. The value of I_s is then given by Equation (22). In Equation (18) the term $\dfrac{kT}{e}$, the dimension of which corresponds to the dimension of a voltage, can be replaced by a magnitude V_T, which is called the **voltage equivalent of the temperature.**

Equation (18) then becomes:

$$I_a = I_s \, \varepsilon^{\frac{V_a}{V_T}}. \tag{23}$$

With most oxide cathodes the voltage equivalent of the temperature, V_T, has a value of 0.1 volt.

(d) Contact Potential with Oxide Cathodes

A contact potential exists between an oxide cathode and the control grid, just as between a pure-metal cathode and the control grid. Because, however, emission of electrons from an oxide cathode occurs in quite a different manner from that in the case of a pure-metal cathode, and the work function φ' has, in consequence, quite a different significance, it cannot be said that:

$$V_{cp} = \varphi_{grid} - \varphi_{cathode}.$$

Nevertheless, it may be conceived that here, too, the contact potential is due to the nature of the cathode and grid metals and to the presence of barium on their surfaces. It may also be that the contact potential is influenced by the barium ions which are always present in the oxide coating during emission. The contact potential between control grid and cathode results in a shifting of the grid bias and of the starting point of grid current.

Generally the value of the cathode-to-control-grid contact potential lies between 0.5 and 1.0 volt, and is negative with respect to the cathode. Before the oxide cathode is activated, that is to say, before the burning-out process, the value of V_{cp} is between 1.5 and 2.0 volts, because at that stage no barium has been deposited on the grid.

CHAPTER V

Secondary Emission

The impact of very swiftly moving electrons or ions upon the surface of a substance, either conducting or non-conducting, results in the release of electrons from the surface. This phenomenon is termed **secondary emission.** Thus, a substance which is bombarded with electrons emits electrons itself, and sometimes the number of secondary electrons emitted exceeds the number of primary or bombarding electrons.

Secondary emission is an important phenomenon for many reasons. Its effect in radio valves may have very undesirable consequences. On the other hand, in the manufacture of certain valves having special characteristics the principle of secondary emission is turned to good account. This chapter deals with only some of the main features of secondary emission [1]).

22. The Suitability of Surfaces for Secondary Emission

The term "secondary-emission factor" is used to express the average number of secondary electrons released from a surface by each primary electron with which the surface is bombarded. It is represented by the Greek letter δ.

The secondary-emission factor depends upon the velocity of the primary electrons at the moment of impact at the surface, and the impact velocity of the primary electrons, as shown by Formula (6), Chapter III, is related to the voltage between the primary cathode and the secondary emitting surface:

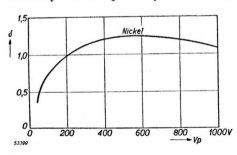

Fig. 23
Secondary-emission factor δ as a function of the impact velocity of the primary electron expressed in terms of the voltage V_p.

$$v = 5.93 \times 10^5 \times \sqrt{V_p} \ldots \text{m/sec,}$$

where V_p is the voltage between the cathode and the secondary emitting surface.

When the secondary-emission factor δ is measured for different metals and plotted as a func-

---

[1]) For further information see H. Bruining's thesis, Leiden 1930; H. Bruining's article in Philips Technical Review No. 3, 1938, pp 80 et seq; and H. Bruining's "Die Sekundäremission fester Körper", published by Julius Springer, Berlin 1941.

36

tion of the impact velocity of the primary electrons (expressed in volts of potential difference traversed), all the curves thus obtained show a maximum value for the factor δ. Fig. 23 shows such a curve for nickel, and Fig. 24 gives curves for barium and for barium oxide. From Fig. 23 it is seen that the maximum value of δ for nickel occurs at about 500 volts and that the secondary electron current is then approximately 1.25 times the primary electron current.

With barium the secondary electron current is smaller than the

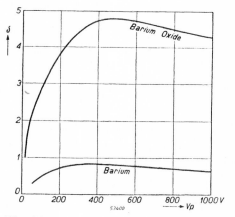

Fig. 24
Secondary-emission factor δ for barium and for barium oxide, as a function of the impact velocity of the primary electron expressed in terms of the voltage V_p.

primary electron current, but barium oxide gives a secondary-emission current five times as great as the primary electron current. It is clear, therefore, that various substances differ greatly as regards the secondary emission of electrons.

When determining δ as a function of the impact velocity of the primary electrons by employing various values of V_p, the measurements are made with small currents so that practically no space charge is formed.

23. Energy Distribution of Secondary Electrons; Reflection of Primary Electrons

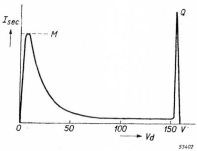

Fig. 25
The number of secondary electrons emitted by silver, determined by measuring the current, plotted as a function of the velocity of emergence, expressed in terms of the voltage V_d. The impact velocity of the primary electrons is that corresponding to $V_p = 160$ V.

It is sometimes of importance to know the velocities or the kinetic energies of the secondary electrons, and particularly the most frequently occurring velocities, i.e. the velocities possessed by the majority of the secondary electrons. Velocities and their distribution can be determined by several methods, one of which is the counter-field method by which is determined what proportion of the secondary electrons can overcome a given potential difference V_d, and what proportion, therefore, possesses a kinetic energy greater than eV_d. This method, therefore, gives an insight into the distribution of the kinetic energy of the secondary electrons although it does not give directly an energy-distribution curve. Other methods, such as the transverse-magnetic-field method [1]), give immediately a velocity-distribution curve which can be converted into an energy-distribution curve.

Fig. 25, for example, shows the number of secondary electrons within a small energy range emitted by silver and plotted as a function of the energy. In this example the impact energy of the primary electrons is 160 V [2]). This curve shows that most of the electrons have kinetic energies in the region 0 to 30 V, while the maximum M occurs between 4 and 10 V. At the same time it will be seen that there is a very sharp maximum at Q (about 160 V). These are electrons which are emitted with a velocity equal to the impact velocity, and are apparently primary electrons which have been reflected without loss of energy.

It has been ascertained that the velocity of the primary electrons does not appear to affect the situation of the maximum, that is to say, the most frequently occurring velocity of the secondary electrons is not dependent upon the velocity of the primary electrons. Further, it has been established that as the velocity of the primary electrons increases so does the number of elastically reflected electrons diminish.

24. Some Practical Values of the Secondary-emission Factor

In the two preceding sections a brief explanation has been given of some of the characteristics of secondary emission. The figures now to be quoted are derived from practical experiments and give a quantitative impresssion of the secondary emission from metallic and other surfaces. Table II, for instance, gives the value of δ for a number of metals and for carbon, at $V_p = 150$ V, and also the maximum values of δ as well as the corresponding impact velocities of the primary electrons expressed in terms of V_p. The same table also gives the impact velocity when δ equals unity, and finally the work function φ.

Metals having a low value of work function φ might be expected to have a high δ factor, but the contrary is the case. As a rule the secondary-emission power of metals with high values of work function φ is greater than that of metals with a low value of φ. This may be explained by the fact that metals with a low value of φ generally have a low ion density, so that the high-velocity primary electrons are able to

---

[1]) R. Kollath, "Annalen der Physik", Volume 27, 1936, pp. 731 et seq.
[2]) By this is meant the kinetic energy imparted to an electron while traversing a potential difference of 160 V.

TABLE II

Secondary Emission of Pure Metals and of Carbon

Material	Form of test surface	Impact velocity V_p, when $\delta = 1$ (V)	δ when $V_p = 150$ V	Max. secondary-emission factor δ_{max}	Impact velocity V_p, when $\delta = \delta_{max}$ (V)	Work function φ (V)
Silver	Sheet	165	0.95	1.47	800	4.61
Aluminium	Sheet exposed to air	50—30	1.65—2.1	2.1—2.6	400	—
Gold	Sheet	165	0.96	1.46	800	4.90
Barium	Precipitated in vacuo	—	0.63	0.83	400	2.70
Caesium	ditto	—	0.55	0.72	400	1.91
Copper	Sheet	205	0.85	1.27	600	4.26
Potassium	Precipated in vacuo	—	0.72—0.88	0.8—0.94	300	2.24
Magnesium	ditto	—	0.9	0.95	300	2.74
Molybdenum	Sheet	150	1.0	1.25	400	4.15
Nickel	ditto	180—160	0.94—0.98	1.22—1.34	500	5.01
Platinum	ditto	150	1.0	1.78	750	5.29
Tantalum	ditto	275	0.76	1.29	600	4.12
Titanium	ditto	—	0.75	0.90	275	3.95
Tungsten	ditto	250	0.75	1.33	650	4.5
Zirconium	ditto	180	0.96	1.09	350	4.1
Carbon	Paste	300	0.9	1.0	300	—

penetrate deep into the surface, while the slow-moving secondary electrons are absorbed in the relatively thick layer of metal they have to pass through before they can be emitted from the surface. It should also be mentioned that the secondary-emission factor δ of metal surfaces is greatly influenced by inadequate degassing and by oxide coatings on the surface.

25. Influence of the Nature of the Surface upon the Secondary-emission Factor

The suitability of a surface for emitting secondary electrons depends, to a large extent, upon the nature of that surface. If the surface of a conductor is rough, as, for example, carbon in the form of soot, or tungsten in the state known as "black tungsten", which consists of very finely divided tungsten, the secondary emission will be lower than normal, on account of the so-called "labyrinth effect" which is represented diagrammatically in Fig. 26. Once a secondary electron has overcome the work function at a smooth surface there are no further

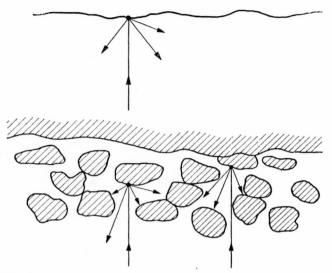

Fig. 26
Diagrammatic representation of the labyrinth effect occurring
with secondary emission from rough surfaces.

obstacles to its emergence from the surface (see top portion of Fig. 26).
Where, however, the surface is rough and labyrinthine it may happen
that, after having emerged at the surface, the electrons are retained by
the particles lying immediately in front of the surface (see lower
portion of Fig. 26) and in consequence the number of secondary
electrons actually emitted is considerably reduced. Moreover, the external electric field is unable to penetrate sufficiently into the interstices
to attract the secondary electrons.

CHAPTER VI

The Principal Components of a Radio Valve

26. The Cathode

The cathode is one of the most important components of a radio valve, because it has to supply the electrons necessary for the valve to function. As explained in Chapter IV, the cathode consists of a metallic wire or strip, or a wire, strip or tube coated with metallic oxide, and is heated to the temperature at which sufficient velocity is imparted to the electrons to permit them to be emitted from the cathode. Although, in principle, it is immaterial in what manner the cathode is heated, for practical reasons this is always done by means of an electric current termed the **heating current.**

A distinction must be made between **directly-heated cathodes** and **indirectly-heated cathodes.** A directly-heated cathode consists of a metallic filament, of suitable emitting properties, through which the heating current is passed. This cathode-filament is raised to the necessary temperature by virtue of its electrical resistance, according to Joule's law.

An indirectly-heated cathode, on the other hand, consists of a thin-walled oxide-coated metal tube, in the interior of which is an insulated heating filament. An electric current of sufficient strength to raise it to the required temperature is passed through the filament, and the heat so produced is transferred to the cathode proper through the insulation, which is capable of withstanding the high working temperature.

(a) Directly-heated Cathodes

As mentioned above, a directly-heated cathode consists of a filament through which the heating current is passed. In small receiving valves the filament is extremely fine; in the most modern types the diameter may be as small as 10 μ (0.01 mm). In large transmitting valves, however, the filament diameter is much greater and may be of the order of 1 mm. As it is desired to release as many electrons as possible for a given consumption of heater power (heater current $\times$ heater voltage), the choice of cathode material is very important (see Chapter IV), and for this reason the oxide cathode is now used almost exclusively for receiving valves.

A directly-heated oxide cathode, therefore, consists of a core composed

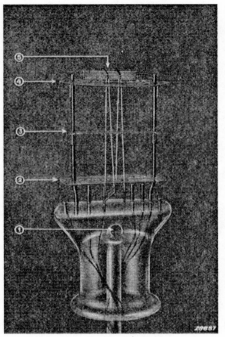

Fig. 27
Directly-heated cathode consisting of an oxide-coated tungsten filament held taut by springs.
(1) Opening of the exhaust tube
(2) Lower mica plate
(3) Mica plate pressed against the filament to avoid lateral vibration of the wire (microphony, see Chapter XXIX)
(4) Upper mica plate
(5) Tension springs

of a metallic filament, to which an oxide coating is applied. The core may be of tungsten or nickel, or a nickel alloy in the form of a wire; nickel is also often used in tape form. For valves with extremely low heater consumption, such as modern battery valves whose heater consumption is only 0.035 watt, tungsten wire is used for the core because this metal can be easily drawn down to a diameter of 10 μ. In less modern battery valves, and also in directly-heated rectifier valves, nickel wire or tape is often found as the core material. Whereas in the older types of receiving valves and in modern directly-heated rectifier valves, the oxide coating is from 20 to 80 μ thick, in the latest battery valves it is only 10 μ thick, so that the over-all diameter of the cathode (core diameter plus twice the thickness of the oxide coating) is only 30 μ.

This may be compared with earlier types of battery valves having similar characteristics, in which the overall diameter of the cathode is 85 μ. It is this reduction in cathode diameter which has made possible the reduction of heater consumption to a minimum.

The oxide cathode gives sufficient emission at a relatively low temperature. Usually these cathodes are operated at a temperature between 700 and 800 °C, corresponding to a cherry-red colour of the filament.

The activation of oxide cathodes has already been described in Chapter IV, Section 21 (a), but it should be added that the working temperature of the cathode is usually so chosen that variations of that temperature due to normal voltage fluctuations of the heater current source, which produce over-heating, or under-heating, do not appreciably affect the performance of the valve.

The filaments of directly-heated cathodes must be anchored very carefully, especially if the core is of very fine wire. Generally the filament is stretched in the form of an M or a V by means of tension springs (5 in Fig. 27), which prevent the limbs of the filament from sagging sideways as they would otherwise do as a result of expansion when heated. Held in this manner, the cathode filaments are maintained in their correct position relative to other parts of the electrode system in all circumstances.

Figs 27 and 28 show typical directly-heated cathodes. Fig. 27 illustrates the cathode of a battery valve with a relatively low emission current, while in Fig. 28 is shown the cathode of a directly-heated rectifier valve, with nickel tape core, and designed to give a large emission current. In many instances the tension of the

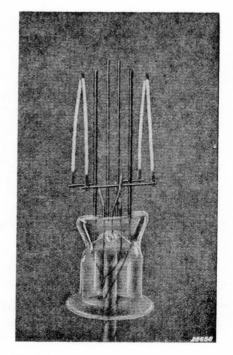

Fig. 28
Cathode of a directly-heated rectifying valve, consisting of nickel tape coated with oxides (the white portion is the oxide coating).

Fig. 29
Filament formed by a fine wire which is coiled and folded to an M. It is covered by a coating of heat-resisting material.

anchor springs is insufficient to prevent the filament from vibrating in response to mechanical vibration of the valve, and it is then necessary to fit a sheet of mica, shown at 3 in Fig. 27, so that it is pressed against the centre of the filament, thus damping the vibrations. This construction has, however, a disadvantage in that the filament is cooled at the centre of each limb.

Pure tungsten filaments, with no oxide coating, are no longer used as electron-emitters in radio-receiving valves.

43

In valves whose cathodes are heated from batteries direct heating is almost exclusively employed on account of the low filament consumption. In valves to be heated by alternating current, however, direct heating is not so suitable, because of the introduction of hum. The periodic voltage variations at the ends of the filament produce variations in the electron current from the cathode to the anode, with the result that an alternating current of frequency equal to the alternating-mains frequency is superimposed upon the anode current. Moreover, the filament temperature and, therefore, the electron emission will also fluctuate at the mains frequency.

One drawback with direct heating is that, on account of the voltage drop across the filament, the potential is not the same at all points in the cathode. To minimise the potential gradient along the cathode the heater voltage is kept as low as possible, and in modern battery valves it is 1.4 volts (in earlier valves 2.0 volts or 4.0 volts).

(b) Indirectly-heated Cathodes

Indirectly-heated cathodes are invariably of the oxide-coated type, as the heater consumption of an indirectly-heated pure metal cathode would be excessive. The cathode consists of a small metal tube, usually of nickel or a nickel alloy, and the oxide coating is applied to a portion of its exterior surface. Practically no electron emission takes place from those parts of the tube which are not so coated, and this fact is sometimes turned to account in order to limit the emission to a given area or in a certain direction. The filament of the indirectly-heated cathode is called the **heater** and is usually of tungsten wire, wound helically as illustrated in Fig. 29.

The filaments actually used for heating the metal cathode tubes are formed either by a straight wire of tungsten folded as a reversed **V** or an **M** or by a very thin wire wound as a coil of very small diameter. This coil is again folded at one or more places. Fig. 29 shows a coil folded as an **M**.

For valves having a low heater-voltage rating (6.3 volts for instance) and a high heater current (power valves) a folded straight wire is used. Valves having a low heater-current rating are provided with a coiled and folded filament.

Before sliding the heater in its cathode tube the filament is covered by a coating of heat-resisting material (aluminium oxide). This coating produces an electrical separation between the cathode and the filament. Fig. 30a shows a complete modern cathode of a high-frequency amplifying valve and also the cathode tube alone as well as the coiled and folded filament which is inserted in the cathode tube. Fig. 30b shows a complete cathode of less recent design and also the cathode tube alone and a coiled-coil filament. The extremely thin wire of this filament was wound to a coil and the coil thus obtained was again wound to a coil around a supporting rod of heat-resisting material. This type of heater is actually no more used.

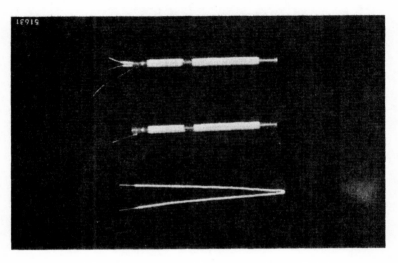

Fig. 30a
Modern cathode designed for indirect heating. *Above:* The complete cathode which is formed by a small nickel tube in which the heater filament is inserted. *Centre:* The cathode tube alone. The active emissive layer is applied at a certain length. *Below:* Filament which is formed by a very thin coiled wire. This coiled wire is folded in two (reversed V) and covered by a coating of heat-resisting material. It is slid into the cathode tube.

In earlier types of indirectly-heated cathodes the spiralised heater filament was supported on a small rod of magnesium oxide, and the whole was enclosed in an insulating tube of magnesium oxide inserted in the nickel cathode tube. When, later on, the magnesium-oxide tube and rod were dispensed with, the heat capacity of the cathode was considerably reduced, thus appreciably shortening the heating time taken to raise the cathode to the working temperature, i.e. the time between switching on the heater current and the commencement of adequate emission. The cathode could thus be made smaller and its heater consumption less. In valves for high heater voltages, such as the DC/AC mains valves, the greater length of heater filament required is accommodated by using a double helix consisting of extremely fine wire which is first coiled into a spiral and then re-coiled upon itself (coiled-coil heater).

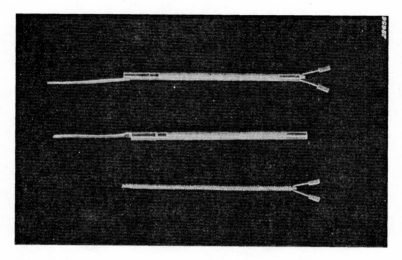

Fig. 50b
Indirectly-heated cathode of less recent design. *Above:* The complete cathode. *Centre:* The cathode tube alone. *Below:* The filament which is formed by a very thin coiled wire, the coil being again coiled around a thin rod of heat-resisting material.

Indirectly-heated cathodes are particularly suitable for use in valves heated by alternating current. One of its greatest advantages is the fact that the whole cathode surface is at the same potential. The cathode has also a considerable heat capacity, so that the emission does not vary in accordance with the mains frequency. The fact that the heater is electrically insulated from the cathode offers great advantages as re-

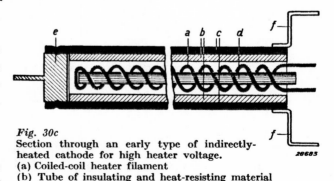

Fig. 30c
Section through an early type of indirectly-heated cathode for high heater voltage.
(a) Coiled-coil heater filament
(b) Tube of insulating and heat-resisting material
(c) Cathode tube
(d) Mandril of heat-resisting material
(e) Centering piece with spigot for locating the cathode
(f) Metal straps for fixing and making electrical connection to the cathode.

gards circuit design, and for this reason nearly all modern valves for alternating-current operation are made with indirectly-heated cathodes.

Another advantage of the indirectly-heated cathode is that, because of its compact dimensions and rigid form, the other electrodes can be mounted much closer to the cathode, thus reducing appreciably the dimensions of the electrode system and improving the electrical characteristics of the valve—e.g. the transconductance.

The heating power of an indirectly-heated receiving valve is, of course, considerably greater than that of a small, directly-heated battery valve, but in recent years a great improvement has been made in this direction. Whereas earlier indirectly-heated cathodes needed a heating power of 4 watts, in modern valves this has been reduced to 1.26 watts.

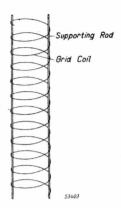

Fig. 31
Grid coil with supporting rods.

27. Grids

Surrounding the cathode there are one or more grids and the anode. These electrodes are cylindrical and are mounted concentrically around the cathode on rigid supports. The cross-section of these electrodes is usually circular or elliptical. Grids are generally made of fine molybdenum wire or some metal-alloy wire, wound helically over one or more supporting rods. The length of the grid cylinders is approximately the same as that of the cathode or filament assembly. The electrons emitted by the cathode pass through the spaces between the grid wires (the **grid meshes)**.

In a valve with only one grid this usually functions as the control electrode and is therefore called the **control grid**. In a **multi-grid** amplifying valve there are, in addition to the control grid, other grids such as the **screen grid** and the **suppressor grid**. Some receiving valves have two control grids, between which are other electrodes, including a screen grid.

The control grid is usually the first grid, mounted immediately surrounding the cathode, the distance between the two electrodes being generally very small, of the order of a few tenths of a millimeter, in order to obtain the maximum degree of control.

There is a possibility that, due to the heat radiated from the cathode, the grid may become hot enough to emit electrons itself. This is more likely to occur in valves with oxide cathodes, as part of the oxide may be evaporated and deposited on the grid, which may then emit an electron current. This matter is discussed at greater length in Chapter XXX, Section 184 (b). In order to minimise the heating of the grid, the grid supports are often made of copper or other good conductor of heat. Another solution is to fix blackened cooling plates, termed **cooling fins,** to the grid supports; in this way a large proportion of the heat reaching the grid is radiated by the fins and the temperature of the grid does not rise excessively.

28. The Anode

The whole, or at least the greater part, of the electron current leaving the cathode flows to the anode—a portion of the electron stream may flow to the screen grid. The electrons strike the anode at great velocity, thereby heating it, and the rise in temperature must be kept within a specified permissible limit. The heating of the anode depends upon the number of electrons striking the anode and upon their velocity, i.e. upon the anode current and the anode voltage. The product of these two quantities is the **anode dissipation** and, together with the heat radiated from the cathode, determines the total heating.

The anode is the outermost electrode, enclosing all the other electrodes, and like, them, it is usually cylindrical. As a rule the anode is made from sheet nickel, but sometimes of wire gauze. Wire gauze has the advantage that heat rays emitted by the cathode can pass through the meshes. Further, it gives the anode a much greater radiating surface. Against this advantage, however, there is the drawback that a large number of electrons are able to pass through the meshes of a wire-gauze anode and to strike against the glass wall of the valve, thus giving rise to secondary emission. Gauze anodes are chiefly used in high-power output valves. Usually the anode is blackened to increase its heat radiation and thus to reduce temperature rise.

29. Internal Carbonising of the Bulb

When electrons pass outside the electrode system they may strike against the glass walls of the bulb or against insulated parts of the valve, and release secondary electrons therefrom. This secondary emission produces various undesirable effects such as signal distortion and fluctuations in the degree of amplification, which can be ascribed to positive charges in the glass wall, as explained in Chapter XXXI. In order to minimize these effects, the interior surface of the bulb is usually coated with carbon, which, as stated in a previous chapter, has a low secondary-emission factor.

30. Screens

In order to prevent, as far as possible, electrons from passing outside the electrode system, screens are fitted at both ends of the assembly, often in the form of metal caps. When several electrode systems are contained in one bulb similar screens are fitted between the systems, in order to limit interaction between the systems due to stray electrons or alternating fields.

31. The "Getter" or Gas-binder

After a valve has been pumped to produce as high a degree of vacuum as possible, the residual gases are absorbed by evaporating, within the valve, a piece of suitable material—usually magnesium or barium. This material is termed the **getter,** and it is placed either in a small metallic receptacle or on a small plate, as shown in Fig. 33, mounted in a convenient position inside the valve. After the pumping

Fig. 32
Construction of the electrode system of a triode-heptode (type ECH 21) mounted on a base of pressed glass. The screening of the base and the screening cage are still present, only the bulb being removed.

operation the getter is heated by high-frequency eddy currents, as a result of which it evaporates and is deposited on the opposite glass wall of the valve as the familiar internal mirror of the bulb.

The getter receptacle is located in such a position that as little as possible of the evaporated material is deposited on the mica or ceramic plates used for securing the electrodes. This is to avoid short circuits, leakage currents and reduction of the high-frequency resistance between the electrodes.

32. The Metallic Coating

In order to prevent external electric fields from affecting the performance of the valve, the external surface of the valve is often provided with a **metallic coating** of zinc or copper. The metallic coating may be earthed direct, or connected to the cathode. It is protected by a layer of enamel and can be recognised by its gold or red colour.

A similar metallic coating may be applied to output valves, but only to the lower part of the bulb because a complete metal coating would result in too great a temperature rise. In output valves the metallisation serves to neutralise capacitively the undesirable effect of charges on the wall of the bulb as explained in Chapter XXXI.

33. Screening Cage

In modern receiving valves the metallic coating is replaced by a screening cage within the bulb. This usually consists of a cylinder of perforated sheet metal surrounding the electrode system as illustrated in Fig. 32. This cylinder has an electrostatic screening action similar to that of Faraday's cage, and at the same time it collects any stray electrons which may have escaped from the electrode system. In this way the effect of charges on the glass wall are to a great extent prevented. Perforated sheet is preferable to solid sheet as it affords better heat radiation.

CHAPTER VII

The Construction of Radio Valves

The various component parts described in Chapter VI must be fitted into the bulb to produce a complete valve. The electrodes are first assembled to form an **electrode assembly** before being inserted in the bulb, and provision is also made for the electrodes to be connected to the external circuits by means of **leading-in wires** which pass through the glass wall of the bulb.

With less recent valves a **base** made from the artificial-resin product "Philite" (a plastic material) is fitted to the bulb. This base is provided with metal contacts to which the leading-in wires are attached. The object of the base is to permit the valve to be conveniently connected to the electrical circuits in which the valve operates. These circuits terminate in a **valve-holder** into which the valve is inserted in the same way that a table lamp or household electrical appliance is plugged into a wall socket. The use of a base and valve-holder has the advantage that a valve can be readily removed for replacement, which would not be the case if the connecting leads of the valve were soldered directly to the wiring of the receiver.

The bulb is usually made of glass, but some valves have iron bulbs. The two forms are known as **glass valves** and **metal valves** respectively. Both glass and metal valves are made with several forms of construction; further some of the earlier constructions also exist side by side with the newer techniques. Description of all the forms of construction is precluded by considerations of space, but in the following sections details are given of the more important constructions used by Philips.

34. Glass Valves with "Pinch"

Fig. 33 shows the interior construction of an early type of valve employing a "pinch". The valve is an octode—a frequency-changer with eight electrodes. The electrode assembly is carried on two metal **support rods** marked (10) in the illustration, which form the backbone and are pressed into the **pinch** or **press** (4) made from glass tubing. The electrode assembly is mounted between two horizontal metal plates or supporting discs (6) and (7), which also act as screens. In these plates are a number of slots through which the grid supports (14), (15) etc. and the cathode pass, small mica or ceramic plates being provided to insulate these parts from one another and from the two metal discs. In some constructions, however, the discs are made entirely of insulating

Fig. 33
Octode, type AK 2, with bulb sectioned to show internal construction.

(1) Connecting cap for grid No. 4

(2) Wall of bulb

(3) Base

(4) Pinch

(5) Getter plate

(6) Lower supporting plate

(7) Upper supporting plate. The electrode assembly is mounted between these two discs

(8) Mica plate for centering the electrode assembly in the bulb

(9) Connecting lead from grid No. 4 to top cap.

(10) Support rod for electrode assembly

(11) Connecting lead passing through the pinch

(12) Anode (sectioned)

(13) Cathode

(14) Support rod for grid No. 6

(15) Support rod for grid No. 5

(16) Grid No. 6.

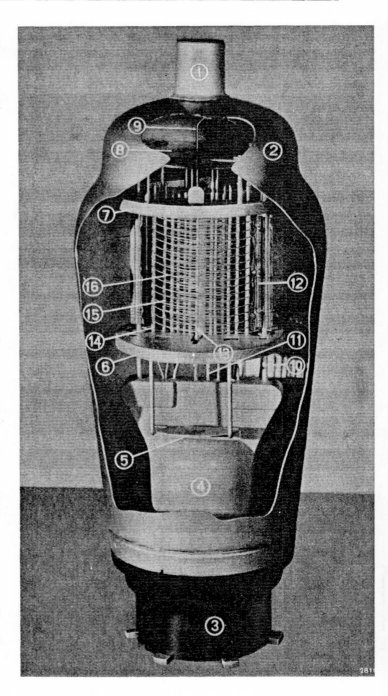

material, mica or ceramic. The anode (12) is a metal cylinder secured to the two **support rods.** In the illustration the front half of the anode has been cut away to expose the other electrodes.

The pinch is also provided with the necessary number of **connecting leads** (11) which are welded to the various electrodes or to their support rods. In this way each electrode is provided with a conductor which passes to the exterior of the valve via the pinch. The support rods of the electrode assembly and those of the individual electrodes are pressed into the glass of the pinch while it is white hot. An airtight joint is thus formed. As these supports, however, must be of fairly large diameter in order to give them sufficient rigidity, and because this makes it difficult to obtain an airtight closure in the

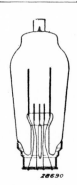

Fig. 34
Section
through a
bulb, showing
pinch and base.

glass, a thin wire is welded to the end of each support, and this forms a perfect airtight connection with the glass and ensures that the vacuum within the valve is maintained. (Fig. 34).

The lower part of the pinch is flared out to form a flange or rim which is fused into the neck of the bulb. Through the pinch also passes a glass **pumping or exhaust tube,** giving communication between the interior of the bulb and the exterior. When the pinch has been sealed to the bulb, the air within the bulb and the gases released from the electrode metals can be pumped out through the pumping or exhaust tube, which, when evacuation is completed, is sealed off to make an airtight closure.

For connecting the fourth grid to the external circuit a small funnel-shaped hole is left in the top of the bulb. Through this is passed

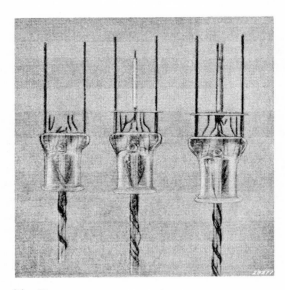

Fig. 35
Three stages in building up the octode type EK 3. *Left.* Pinch with pumping tube, support rods and leading-in wires. *Centre.* The same pinch with lower supporting plate and cathode in position. *Right.* The same pinch with grid No. 1 in position.

Fig. 36
Section of R.F. pentode, type EF 9 (actual size)
  (1) Getter plate
  (2) Bulb
  (3) Cathode leading-in wire
  (4) Support rod for electrode assembly
  (5) Pinch
  (6) Cement securing the base to the bulb
  (7) "Philite" base
  (8) Base contact
  (9) Top cap for grid connection
  (10) Sealed aperture in bulb for grid-connecting lead
  (11) Pumping tube
  (12) Fused joint between bulb and pinch
  (13) Grid-connecting lead
  (14) Metallic coating of bulb

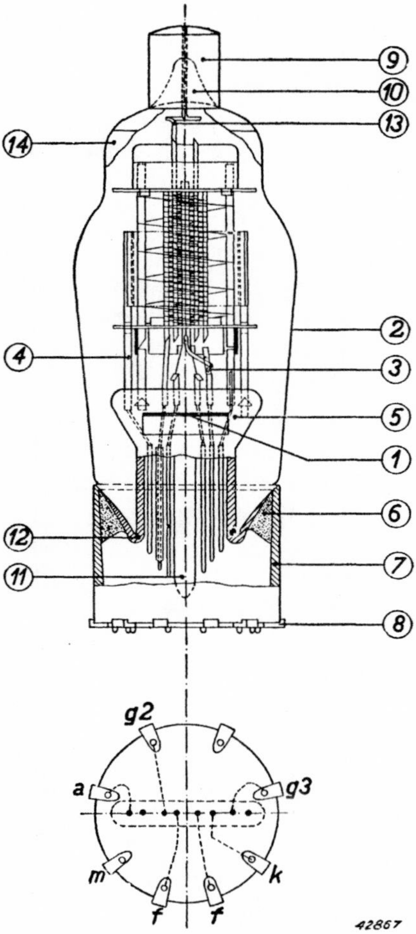

the top-connecting lead (9) and after the electrode assembly has been mounted in the bulb the glass is fused around this lead to make an airtight joint. The manner in which the other parts of the valve are mounted in the valve is clearly shown in Fig. 33. The lower supporting disc (6) carries below it the concave plate (5) for the getter which has already been described, and the upper supporting disc (7) and the support rods carry the mica disc (8) for centering the electrode assembly in the bulb.
The connecting leads (11) from the pinch are soldered to the contacts in the base (3), and the base itself is cemented to the bulb as shown in Fig. 34. The top-connecting lead passing through the bulb is soldered to the metal contact cap (1), which is also cemented to the glass.

Such, in principle, is the method by which a valve having its electrode assembly mounted on a pinch is constructed. It is the construction which, up to several years ago, was adopted for nearly all Philips valves, being modified to meet the special requirements of each type.

A later construction is shown in the sectional drawing reproduced in Fig. 36. This is the pentode type EF 9, a valve of very compact dimensions. Here, again, can be seen how the electrode unit is mounted on a glass pinch. There are the same two support rods, but instead of

Fig. 37
Internal construction of a directly-heated output valve
of early design.

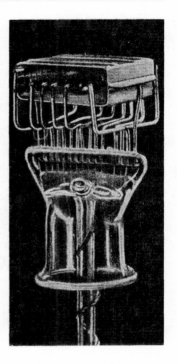

the metal supporting discs used for the
octode, only mica discs are employed. This
valve, also, has a top connection—in this
case for the first grid. By connecting the
control grid at the top of the bulb the capacitance between that electrode and the anode
can be kept small. If the control-grid connection were led out through the pinch its
lead would run parallel and very close to
the anode connecting lead, and the mutual
capacitance would be considerable. A coupling between the control grid and the anode
in a high-frequency valve is most undesirable;
the control-grid connection in high-frequency, intermediate-frequency and frequency-
changing valves [1]) is at the top, and the anode
connection in
the base. In
the case of earlier valves, such as type
E 446 and AF 2, the grid connection was
in the base and the anode connection at
the top of the bulb.

Fig. 36 shows another detail, namely that
each lead-in wire below the pinch is enclosed in an insulating sleeve to avoid
the risk of short-circuit between wires.

In early battery valves the method of
mounting the electrode assembly on the
pinch is somewhat different, the filament
being fixed horizontally and the other
electrodes mounted around it in that
position as illustrated in **Fig. 37**.

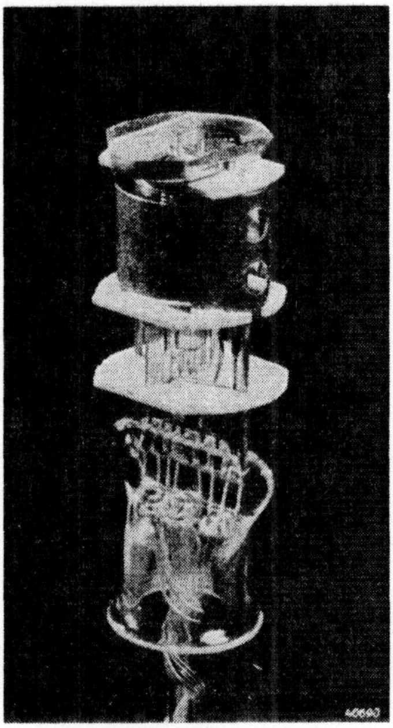

Fig. 38
Electrode assembly of the triode-heptode type
ECH 4, with ceramic discs. The heptode part is
above the centre disc and the triode part below.

[1]) The meaning of the terms "radio-frequency" or "high-frequency", "intermediate-
frequency" and "frequency-changing valves" is explained in Chapter IX.

Fig. 39
Electrode assembly of the indirectly-heated rectifier type UY 21, with ceramic discs.

In the octode used as the example for Fig. 33 it is seen that the electrode assembly is mounted between metal supporting discs with mica insulation. During the war mica was replaced by ceramic plates, which were found to be satisfactory. Fig. 38 is the electrode assembly and pinch of the triode-heptode, type ECH 4, with three ceramic plates between which the heptode portion (top) and the triode portion (bottom) are mounted. Fig. 39 shows another ceramic construction, that of a modern indirectly-heated rectifier valve, type UY 21, while in Fig. 40 can be seen the somewhat similar construction of the high-frequency pentode EF 22.

The mica disc in the top part of the dome-shaped bulb (8) in Fig. 33, which prevents lateral displacement of the electrode system as the result of mechanical shock, is sometimes replaced by small steel springs.

35. Glass valves with Pressed-glass Base

In 1938, to meet the needs of television and ultra shortwave reception, a new construction of valves was developed to supersede the pinch construction. The new method has important electrical, mechanical and manu-

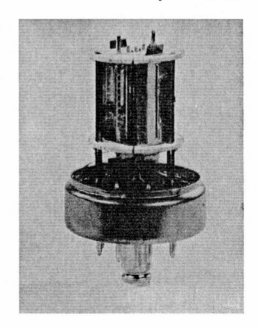

Fig. 40
Electrode assembly of a modern R.F. pentode, type EF 22, with ceramic insulating discs. The base guard plate with central guide pin and retaining ring are also shown. The bulb and screening cage have been removed.

48576

Fig. 41
Three forms of the modern glass construction of electronic valves. From left to right the A-technique, the B-technique and the C-technique. The A- and B-techniques are mainly applied for broadcasting valves, the C-technique for high-transconductance television valves and vhf receiving valves. There is a tendency to apply the A-technique also to special valves.

facturing advantages; for instance it affords a better method of sealing-on the valve base.

The characteristic feature of the new construction is the absence of the pinch. The base of the valve is a flat glass plate, with or without upturned rim, in which are connecting pins to which the system of electrodes is attached and which serve at the same time as contacts in a suitable valve-holder. This construction was next applied also to valves for broadcast receivers. These valves are designed somewhat smaller than those first made for television and ultra shortwave reception, so that we have two types in different sizes, one with 9 contact pins and one with 8.

Technical progress led at the same time to a systematic reduction of the dimensions of the electrode systems while retaining the same electrical properties as inherent in the larger-size valves. As a consequence there arose a need for smaller glass envelopes for the electrode systems, and this led to a third construction for broadcasting and special valves, much smaller than those of the second construction. Thus three designs or "techniques" of pinchless valves have arisen, all characterised by their dimensions and number of contact pins. Philips call the largest design the C-technique—the dimensions of valves according to this technique are 62 × 38 mm (excl. pins) and these valves have 9 contact pins. Those of the second design, the B-technique, have dimensions of 65 × 32 mm and the output valves and rectifying valves according to this technique have dimensions of 80 × 32 mm, and 8 pins. The smallest ones, the A-technique amplifying valves, have a size of 52 × 23 mm, whilst the output valves measure 70 × 23 mm and the rectifying valves 61 × 23 mm (all excl. pins); these A-technique valves all have 8 pins.

Fig. 41 gives an idea of the three designs of valves with pressed-glass base. A typical feature of the new construction is the fact that all the electrodes are carried through the base. Particularly the A-technique valves are so much smaller than those of the pinch construction that very small receivers or other electronic apparatus can be built with them without the number of valves having to be limited on account of lack of space. It is thus often possible to design more reliable and more easily adjusted receiving sets.

From the manufacturing point of view the A-technique possesses a considerable advantage over the B-technique, which we shall revert to in Chapter VIII when describing the manufacture of radio valves. In the following sections we shall confine our description mainly to the B- and A-techniques, but the characteristic properties of these pinchless glass valves hold also for those of the C-technique.

(a) Constructional Features of Valves with Pressed-glass Base

(a) B-Technique

A valve with pressed-glass base and without a "pinch" built according to the B-technique is illustrated in section in Fig. 42. It is the radio-frequency pentode type EF 22. The pressed-glass base (5) is dished, having an upturned rim. There are eight chromium-steel connecting pins, hermetically moulded into the base, as shown at (8) in Fig. 42. They are either 1.1 or 1.27 mm in diameter and are spaced round a circle 17.5 mm in diameter. The vertically-mounted electrode assembly is welded to the eight connecting pins and thus secured very rigidly. In the EF 22, as in fact in most valves of this construction, the electrode assembly is further supported by three U-section supports (4). In the centre of the glass base is the exhaust tube (11). When the electrode assembly has been mounted on the base, and the various electrodes connected to their respective pins, the bulb is sealed onto

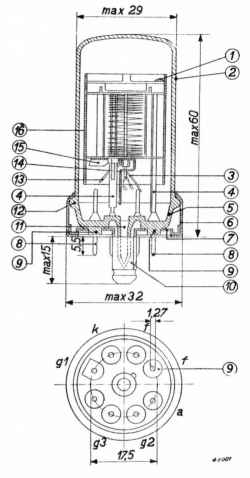

Fig. 42
Section of R.F. pentode, type EF 22
(actual size)
(1) Getter plate
(2) Bulb
(3) Connecting lead from an electrode
     (screen grid) to its contact pin
(4) U-section supports for the electrode
     assembly
(5) Pressed-glass base
(6) Metal retaining ring
(7) Guard plate fitted beneath the glass
     base
(8) Contact pin, 1.1 mm or 1.27 mm in
     diameter
(9) Circular bosses on the pressed-glass
     base which increase the leakage path
     between the contact pins and the
     guard plate
(10) Guide pin with cam
(11) Sealed pumping tube
(12) Fused joint between bulb and pressed-
     glass base
(13) Grid connecting lead
(14) Screening plate between grid lead
     and the remaining leads
(15) Extension of (14) between grid lead
     and anode
(16) Perforated   sheet-metal   screening
     cage

the upturned rim of the base, the fused joint being shown at (12). The valve is then exhausted by pumping and the getter is volatilised in the normal way. The plate for the getter (1) is mounted above the electrode assembly so that the mirror formed by deposition of the getter vapour is confined to the top portion of the bulb. After pumping, and the sealing of the exhaust tube by fusion, a guard plate with central guide pin (10) is fixed to the glass base by applying a metal ring (6) around the base and spinning it over the rim of the guard plate. The ring is of reduced diameter at the top and fits exactly round the bulb, resting on the protruding joint (12) where the base joins the bulb.

The electrode assembly for these pinchless valves is built up in the same way as that of valves with pinches. The electrodes are mounted between two discs of insulating material or of metal and insulating material, ceramic material being sometimes substituted for mica as shown in Figs 38, 39 and 40. The discs are held in position by the

59

Fig. 43
Cross-section of the diode-pentode UAF 41 in A-technique (actual size).

(1) Getter plate

(2) Wall of bulb

(3) Connections between the electrodes and the lead pins

(4) Supporting rods for the electrode assembly

(5) Valve base of pressed glass

(6) Metal rim cemented onto the envelope

(7) Screening cup in the glass base

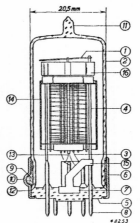

(8) Connecting pins of 1 mm diameter

(9) Cement securing the metal rim to the bulb

(10) Hemispherical projection of the metal rim serving to guide the valve into its holder

(11) Sealed off pumping tube

(12) Glazing joint between base and envelope

(13) Heater leads

(14) Screening cage

(15) Connection between the cathode pin and the screen capping in the base

(16) Diode system

above mentioned three metal supports of U-section instead of the two support rods shown at (10) in Fig. 33.

By using chromium steel, which has the same coefficient of expansion as glass, the connecting leads may be made relatively stout (1.1 or 1.27 mm diameter) so that they are strong enough to serve at the same time as contact pins to be inserted directly into the valve-holder.

The metallic coating which is applied to high-frequency valves of pinch construction is replaced in the pinchless valves by an internal screening cage surrounding the electrode assembly as described in Chapter VI, Section 33, and illustrated in Figs 42 and 43.

The method of building a valve with pressed-glass base makes it possible to bring all the electrode connections out through the base, including the grid connection, which for the octode type AK 2 shown in Fig. 33 appears at the top. This is permissible because the screening plates (14) in Fig. 42 maintain the anode-grid capacitance at a sufficiently low value. Moreover, the use of these screens and judicious arrangement of the sequence of connections to the various pins minimises the induction of hum on the grid from the alternating-current supply to the heater, as explained in Chapter XXVIII.

(β) A-Technique

Fig. 43 shows the construction of a pinchless valve in the A-technique (diode-H.F. pentode UAF 41). The base (see Fig. 44) consists of pressed glass with eight connecting pins. The electrode assembly is mounted on this base by welding supporting rods of the electrode system onto the pins, as is done also in the B-technique.

As seen from Figs 43 and 44, in the A-technique
the base (5 in Fig. 43) has no upturned rim.
Neither is there any exhaust tube fixed to the
base.

Eight chromium-steel connecting pins (8) are
moulded into the base spaced round a circle
of 11 mm diameter. These pins are 1 mm in
diameter. The pumping tube (11) is sealed off
at the top of the envelope. The electrode
assembly is built up as in the B-technique,
with the aid of mica discs and supporting rods,
which are round instead of U-shaped. Some

Fig 44
Base of pressed glass for
the A-technique with con-
tact pins and ring of
sintered glass.

of these rods extend
through the bottom mica disc.

At the top of the electrode system is a semicircular arm attached to
a supporting rod and with another rod-shaped piece of getter material
welded onto it.

The assembled electrode system is then attached to the connecting
pins of the glass base, after which the envelope is placed round it and
sealed air-tight to the base. With the B-technique this air-tight joint
is made by fusing the upturned rim of the base together with the rim
of the envelope, both having to be heated to about 800—900 °C to
flux the glass properly. The shorter the leads to the electrodes and the
narrower the envelope, the closer the components of the electrode
system come to lie to the sealing joint and the higher the temperature
to which they are raised in the sealing process.

If this fusing method were to be applied to valves of the dimensions of
the A-technique there would be a great risk of some of the parts being
oxidized or the cathode being poisoned, with the result
that there would be a higher percentage of rejects in
manufacture. The alternative would be to have a
stream of inert gas (e.g. nitrogen) passing through the
envelope while fusing, but that would of course make
the process of manufacture more complicated.

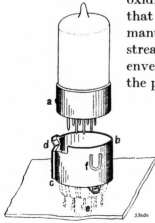

Fig. 45
A valve in A-technique with valve-holder to match.
(a) Hemispherical projection on the metal ring at the
    bottom of the valve
(b) Metal rim affixed to the "Philite" valve-holder.
(c) Groove in the metal rim into which the projection
    on the metal ring of the valve fits
(d) Spring for locking the valve in the holder
(e) Metal screening tube reaching to the base of the valve
(f) Punched springs for clamping the valve base rim
    in the metal can b

These difficulties have been overcome by devising for the A-technique valves an entirely different process for making an air-tight joint between the base and the envelope. This is what is called the **glazing technique,** which has several important advantages and is described in the following chapter on the manufacture of the valve. **In this process the temperature of the glass base and the envelope and consequently also of the components of the electrode system is kept appreciably lower than that in the fusing of the envelope onto the base.**

Contrary to the B-technique, in the A-technique the valves are evacuated through a pumping tube at the top of the envelope, after which the tube is sealed off airtight. Since the base of these valves has no guide pin another means is applied for ensuring that the valve is pressed into its holder in the right position. Fig. 45 shows a valve with its holder. A metal ring (6 in Fig. 43), cemented onto the bottom of the envelope where it is slightly recessed, has a semispherical projection (10 in Fig. 43 and a in Fig. 45) at a point lying halfway between the two filament pins. The "Philite" valve-holder has a metal rim b (Fig. 45) in which is a groove c with spring d. The valve can only be mounted in the holder when the projection on the valve rim is facing the groove in the holder rim.

When the valve has been pressed down into the holder the spring d slips in behind the projection a on the valve rim, thus ensuring that the valve does not get shaken out of the holder. In view of this locking arrangement the A-technique valves have been given the name of "Rimlock" Valves.

Thanks to the low temperature applied in the glazing process, the valve envelope retains its exact shape, thus making it possible to fit on the metal ring very closely around the envelope, so that the valve diameter of 23 mm is not exceeded. Since the envelope has considerable strength it is also possible to dispense with the metal rim and to provide for the guiding projection by blowing a small bulge in the wall of the envelope itself. This is of particular importance for valves working at very high frequencies, because then the metal rim would give rise to an undesired increase of capacities and of the damping.

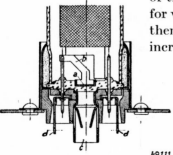

Fig. 46
Cross-section of the lower part of an A-technique valve with its appropriate holder.
(a) Screen cup in the glass base
(b) Strip connecting the screen cup with the cathode pin
(c) Metal screening tube in the Philite holder
(d) Contact springs

49111

Fig. 47
The electrode assembly of a triode-hexode in A-technique.
This shows how the triode part is mounted at the top
and the hexode part at the bottom of the valve.

Fig. 46 is a cross-section of the valve-holder with the valve in it. Pressed up against the bottom of the valve is a metal cup-shaped screen attached by a strip to the pin for the cathode connection. In the valve-holder is a metal tube c, which is normally earthed. This metal tube fits fairly well up against the cup-shaped screen in the valve base and together they form an effective screen between opposite pins, so that by this means, for instance, the grid-anode capacity is kept very low. In the A-technique valves also the electrode system is electrostatically screened off from outside influences by means of a screening cage (14 in Fig. 43).

The A-technique valves combining various functions in one, the multiple valves, such as the triode-hexode and the diode-pentode, differ from similar valves in the pinch or the B-technique construction in that the more complicated of the two electrode assemblies is at the bottom and the simpler one at the top. This simplifies construction of the supports and the mica plates holding the electrode assembly together. Furthermore there is less trouble from heat conduction through the connecting lead to the bottom end of the cathode. A short end of the cathode serves for the simpler electrode system and the rest for the more complicated one. The mean temperature of the longer part will presumably be less affected by heat conduction at the end than that of the shorter part. It is therefore preferable to have the longer part at the bottom. Fig. 47 shows the electrode system of a triode-hexode with the triode system uppermost and the hexode system underneath.

As far as the cathode temperature during the fusing-in process is concerned the A-technique valves could have been made with a diameter still smaller than 23 mm, were it not that the number of lead pins required in the base limits the minimum diameter to which one can go. Since at least eight of these pins are necessary for many types of valves, this number has been standardised for all valves. If the valve diameter were less than 23 mm the pins would have to be too

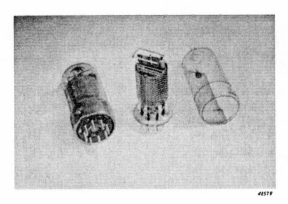

Fig. 48
From left to right: a complete valve in A-technique (diode-R.F. pentode), the electrode assembly of this valve mounted on the pressed-glass base, and the bulb.

close together and consequently there would be too much capacitance and too large dielectric losses between them. Moreover with a smaller diameter the temperature of the envelope wall would be too high for large power valves (e.g. output and rectifier valves), resulting in considerable dielectric losses and electrolytic action on the glass between the pins (see Chapter XXX, Section 186).

The diameter of 23 mm chosen for the A-technique valves is quite ample, also in respect to the voltages likely to be applied. It has even been found that in the case of a special television valve a peak voltage of 8000 V could be applied between two diametrically opposed contact pins without any fear of electrolysis of the glass or disruption.

(b) Advantages of Pinchless Construction

Stated briefly, the advantages of the pinchless construction, as compared with the construction employing a pinch, are as follows:

(1) **Compact dimensions of valves,** facilitating the design of smaller receivers with consequent reduction in the cost of materials.

(2) **Greater mechanical strength.** The mounting of the electrode assembly is much more rigid, so that the valve is better able to withstand mechanical shock and vibration. The use of a pressed-glass base with moulded-in connecting pins in place of the artificial-resin base eliminates the risk of parts working loose and of defective soldering between lead-in wires and base contacts.

(3) **All electrode connections emerge through the base of the valve,** making possible a more orderly chassis arrangement with more efficient screening.

(4) **Low glass temperatures.** In pinch constructions the glass at the pinch reaches a comparatively high temperature (in output valves it amounts to 200 °C), but in pinchless valves the temperature does

not exceed 90° C. The insulation is thus improved and the risk of electrolysis of the glass reduced.

(5) **Practical advantages of the new base.** With the elimination of the separate base and the use of eight pins moulded into the glass, the valve designer has much greater freedom in arranging the electrode connections. Further, the handy form of the base makes the replacement of valves easy, yet the valves are so firmly held in their holders that they cannot fall out even under severe conditions of mechanical shock or vibration.

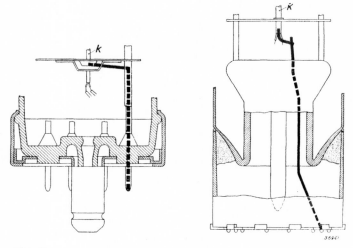

Fig. 49
Comparison between the lengths of the cathode lead in R.F. pentode type EF 22 with pressed-glass base (left) and in R.F. pentode type EF 9 with pinch (right).

(6) **Constant valve capacitances.** The dielectric constant of the "Philite" base was greatly affected by temperature variations. The new valves which have no such base do not suffer from changes in inter-electrode capacitances while warming up after switching on, this constancy of capacitance being enhanced by the low temperature of the glass base and also by the wider spacing of the connecting leads.

(7) **Closer tolerances in valve capacitances.** Because all electrode connections emerge at the base of the valve and the mirror formed by the volatilisation of the getter is confined to the top of the bulb at the maximum distance from the electrodes, the influence of the mirror on inter-electrode capacitances is much reduced, resulting in greater uniformity in capacitance as between valve and valve.

(8) **Improved performance on short waves.** The shortening of the connecting leads and their wider spacing results in improved valve characteristics, which confer substantial advantages for short-wave operation. Fig. 49, for example, gives a comparison between the length of the cathode connection in the EF 22 pinchless valve and in the EF 9 valve with pinch construction. The significance of short connections in short-wave operation is explained in Chapter XXV. Similar reductions occur in the length of the connections to other electrodes. It may be mentioned that in the earlier constructions the leads ran side by side through the pinch for a distance of 35 mm and spaced only 0.5 to 1.0 mm apart, thus greatly impairing the performance of the valve on shortwaves. With the pressed-glass-base construction each electrode can be connected to its base contact by the shortest route, and capacitative and inductive couplings to other electrodes are reduced to a minimum.

(c) Specific Merits of the A-Technique

In addition to the many advantages inherent in the A-technique in common with the B-technique, the former kind of valves possess the following specific merits:

(1) **Appreciably smaller dimensions, particularly in the diameter.** This means a still greater saving in space and cost of materials in the construction of radio sets where there is practically no sense in economising on the number of valves.

(2) **Greater safety margin in manufacture.** With the new method of manufacture any oxidization of parts of the electrode system while fusing the envelope onto the base is avoided, as is the poisoning of the cathode.

(3) **Very small tolerance in dimensions.** Thanks to this new method of manufacture the deviations of dimensions can be kept extremely small.

(4) **Better properties at very high frequencies.** The shorter electrode leads and the possibility of making valves without a metal ring round the base make the A-technique valves excellently suitable for application in amplifiers for veryhigh frequency.

(5) **The possibility of making battery-fed valves with very small filament current consumption.** The glazing technique is exceptionally favourable for valves to be used in battery radio sets. With these

sets it is of great importance to keep the power required for heating the cathode as low as possible. It is for this reason that the filament cathode, which is coated with a thin layer of oxide mixture mentioned in Chapter IV, is drawn as thin as possible. Philips use tungsten filament, this having a much greater tensile strength than nickel, so that the filament cathode can be drawn much

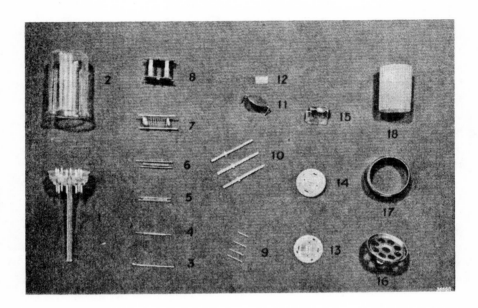

Fig. 50
Component parts of R.F. pentode type EF 22.
(1) Pressed-glass base with contact pins and pumping tube
(2) Bulb
(3) Bifilar heater winding
(4) Cathode
(5) Control grid
(6) Screen grid
(7) Suppressor grid
(8) Anode
(9) Metal strips for connecting the electrodes to the contact pins
(10) Three U-section supports for the electrode assembly
(11) Metal plate carrying the getter and forming a baffle to prevent deposition of getter material upon the electrode assembly
(12) Gauze for securing the getter material to (11)
(13) Lower insulating disc
(14) Upper insulating disc
(15) Screening plate in lower part of the valve for reducing anode-to-grid capacitance and grid hum.
(16) Guard plate for base with guide pin and circular openings for the contact pins
(17) Retaining ring for guard plate
(18) Screening cage

thinner. On the other hand, however, tungsten is more susceptible to oxidation than nickel wire.

The very low temperature of the glazing technique offers a great advantage, in that for some battery-fed valves of the A-technique the thickness of the filament cathode is reduced to 8 microns, for which a heating current of only 12.5 mA is required.

36. Metal Valves

About the year 1938 there appeared on the European market metal valves which became known as "steel valves". Their construction must not be confused with that of the American metal valves. Both constructions are pinchless, but whereas in the American valves the electrode assembly is mounted vertically as in glass valves with pinches or in valves with pressed-glass bases, in the "Steel Valves" the electrode assembly is mounted horizontally.

This is the main distinguishing feature of the steel valve, and it makes it possible for all the electrode connections to be brought out through the base in a simple manner. The detailed construction can be followed by reference to Fig. 51, which shows an internal view of an H.F. pentode, and Fig. 52, which is a sectional drawing of the same valve.

The electrode assembly is mounted horizontally between two channel-section supports with two mica plates (h) interposed. The supports are welded to the iron base-plate (a) which is 43 mm in diameter and has a downward-projecting rim. A number of holes about 3 mm in dia-

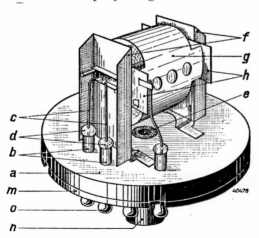

Fig. 51
Internal construction of the type EF 11 H.F.-pentode — a steel valve. The letter references are the same as in fig. 52.

meter and arranged in two groups on the circumference of a circle are punched in the base-plate and serve to carry the electrode connections through to the exterior of the valve. In each hole is soldered a small metal tube (b) and to each connecting lead (c) is attached an intermediate conductor of molybdenum which passes through one of the tubes, an air-tight seal being made by glass beads (d). The channel-section supports also act as screens for the electrode connections.

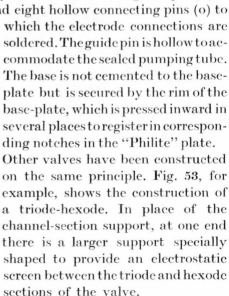

Fig. 52
Section of R.F. pentode, type EF 11.
(a) Steel base plate
(b) Metal tubes of special alloy having the same
coefficient of expansion as the glass beads (d)
(c) Electrode-connecting leads
(d) Glass beads
(e) Pumping tube
(f) Supports for the electrode assembly
(g) Electrode assembly
(h) Mica bridges
(i) Steel bulb
(k) Getter
(l) Sheet-metal baffle, for protecting the electrode
assembly from deposition of getter material
(m) Synthetic-resin base
(n) Guide pin
(o) Contact pin

In the centre of the base-plate a hole is
punched to accommodate the exhaust tube
(e). When the electrode assembly has been
welded to the base plate the iron envelope (i) is fitted over it and
likewise welded to the base-plate. The getter (k) is located in the top of
the envelope, and in order to avoid deposition of the getter material on
the electrode system a sheet-metal baffle (l) is interposed between the
getter and the electrode assembly; openings around the periphery give
free communication between the top and bottom sections of the bulb space.
The base (m) is a plate of "Philite" synthetic resin with a guide pin (n) of
the same material at the centre and eight hollow connecting pins (o) to

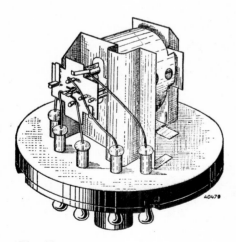

which the electrode connections are
soldered. The guide pin is hollow to accommodate the sealed pumping tube.
The base is not cemented to the base-
plate but is secured by the rim of the
base-plate, which is pressed inward in
several places to register in corresponding notches in the "Philite" plate.
Other valves have been constructed
on the same principle. Fig. 53, for
example, shows the construction of
a triode-hexode. In place of the
channel-section support, at one end
there is a larger support specially
shaped to provide an electrostatic
screen between the triode and hexode
sections of the valve.

Fig. 53
Construction of the triode-hexode frequency-changer type ECH 11.

69

It has been shown, therefore, that the horizontal construction permits the control-grid connection to be made to a base contact at one end of the electrode assembly and the anode connection to be taken from the other end of the assembly to a diametrically opposite pin, thus greatly reducing the coupling between these two electrodes. This construction, however, has the disadvantage that the valves have to be of relatively

Fig. 54
General appearance of a steel valve.

large diameter in order to accommodate the horizontal electrode assembly, and they therefore take up an unduly large area of the chassis of the receiver. High-power output-valves in this construction would be of such a large diameter as to be impracticable, and for this reason the construction is employed only for valves of small power. Comparison with a valve of the B-technique having a pressed-glass base shows that whereas the "steel-valve" is 43.5 mm in diameter and 43.5 mm high, the glass-base valve is only 29 mm in diameter and 60 mm high, which are much more convenient dimensions. The dimensions of the A-technique valves are even very much less and they render a considerably greater reduction of the size of apparatus possible. The glass-base construction with vertical electrode assembly can be used for all types of valves, including output valves and rectifiers.

The "steel" construction, however, does offer some important advantages over the pinch construction, and these may be summarised as below:

(1) **Very suitable to mass production** owing to the form of construction, the method of pumping (explained in Chapter VIII) and the fact that the getter material cannot be deposited on the electrode assembly.

(2) **Very robust construction.**

(3) **Short connections from electrodes to base pins,** rendering the valves very suitable for short-wave amplification.

(4) **All electrodes connected to the base**—and therefore at one end of the valve.

(5) **Excellent electrostatic and magnetic screening** and freedom from wall charges.

(6) **Close tolerances** on dimensions and on inter-electrode capacitances.

(7) **Valves may be used in any position.** Glass valves with pinch may only be used vertically with the base downwards or, if essential, horizontally; but not base upwards.

(8) **Unbreakable bulbs.**

37. Conclusion

The modern constructions, with pressed-glass base, and also valves with steel bulbs are now in general use and answer their purposes well. The pressed-glass-base constructions are particularly advantageous in valves for television and also in special valves for operation at very high frequencies (wavelengths from 10 metres down to 1 metre). For these purposes it is preferable, while employing the same general construction, to increase the diameter slightly and to fit nine pins instead of eight. On the other hand there is also a tendency to use A-technique valves for very-high-frequency purposes. The same constructions are also applied in groups of special valves and offer great possibilities for future development. In all probability the glass construction embodying the pinch will not be used for new designs of valves except, perhaps, for some large amplifying and rectifier valves.

CHAPTER VIII

The Manufacture of Radio Valves

The manufacture of radio valves is broadly divided into the following stages:

1. Production of component parts.

1a. Inspection and testing of components.

2. Mounting components to form complete electrode assemblies.

2a. Inspection and testing of electrode assemblies.

3. Mounting the electrode assembly in the bulb and evacuating the valve.

3a. Inspection and testing of evacuated valves.

4. External finishing and fitting base.

4a. Inspection and testing of external parts.

5. "Ageing"—or running the valve under normal working conditions.

5a. Final test.

Much of the production of component parts follows normal engineering practice and will therefore not be described in detail. With the help of a series of photographs, however, an attempt is made to give the reader some idea of the highly complicated processes employed in the manufacture of a valve—processes which involve the use of a large number of ingenious machines and carefully planned methods, the result of years of intensive research and labour on the part of specialist engineers and technicians, who are even now continuously improving and perfecting every detail.

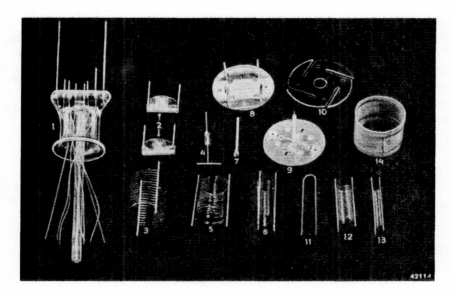

Fig. 55

Components of the electrode assembly of an octode, type CK 1.

(1) Pinch with pumping tube and leading-in wires fused in

(2) Getter holder

(3) Suppressor grid (grid No. 6)

(4) Control-grid connection with cross rod and glass bead for fusing onto bulb

(5) Outer screen grid (grid No. 5)

(6) Control grid (grid No. 4)

(7) Coiled-coil heater, coated with insulating material

(8) Upper plate with mica disc

(9) Cathode tube, and lower plate with mica disc

(10) Mica plate for centering assembly in the dome-shaped bulb

(11) Oscillator anode, consisting of two rods only

(12) Inner screen grid (grid No. 3)

(13) Oscillator control grid (grid No. 1)

(14) Anode

38. Glass Parts

These comprise the bulb and other glass parts, such as the glass tube for the pinch, glass insulators and so forth.

Fig. 56 shows one of the platforms in the glass works, with the glass-blowers in a circle around a central glass furnace. Here are blown bulbs of all shapes and sizes for valves, tubes and incandescent lamps. In Fig. 57 can be seen an automatic plant producing standard bulbs on mass-production lines, and thus releasing a large number of skilled glass-blowers for the manufacture of special bulbs.

Other glass parts such as pressed-glass bases of all-glass valves, pinches etc. are manufactured in the valve factory.

Fig. 56
Part of the glass factory. Glass-blowers, making bulbs of many shapes and sizes for valves and incandescent lamps, stand round the glass furnace in the centre of the platform.

39. The "Philite" Base

"Philite" is a synthetic-resin product which is mixed in powder form with sawdust or with some other body material, after which it is pressed to the desired shape in steel moulds under high temperature. Bases for valves are made from this material. Fig. 58 is a photograph taken in the Philite factory and shows some of the large presses.

40. Cathodes and Filaments

It has already been explained in Chapter VI that an indirectly-heated cathode consists of a small-diameter nickel tube coated, in the first instance, with a mixture of barium and strontium carbonate. The

Fig. 57
Machine for automatic production of bulbs in large quantities.

cathode tubes are obtained by drawing out the metal into a long continuous tube and then cutting it to the required lengths. Fig. 59
shows a machine designed for this operation.

After being cut to length, the cathode tubes are rolled near one end to
form a protruding ring. It is with the help of this ring that the cathode
rests on the lower mica disc of the electrode assembly. Fig. 60 shows a
machine on which the protruding ring is rolled into the cathode tubes.

The cathode tubes are next passed to another department where the
mixture of barium and strontium carbonates is applied by a process
which ensures a perfectly uniform coating.

Tungsten filament wire is drawn down to the required gauge in specially
constructed machines, and is then coiled on other machines. As indicated in Chapter VI, Section 26(b), the modern filament consists of a
straight wire or a coiled wire folded at several places. Certain heater-
voltage and current ratings necessitate a very long and thin wire, which
has to be accommodated in the limited space of the interior of the
cathode tube.

Fig. 58
Part of the "Philite" factory where, among other products, plastic bases for valves are moulded.

The filament must also have an insulating and heat-resisting coating. For this purpose a coating of aluminium oxide is applied. This is done by dipping the filaments in a bath in which the heat-resisting material is in suspension. The filaments are then heated to a high temperature in a furnace such as that illustrated in Fig. 61, in order to bake the insulation coating onto the wire.

Filaments for directly-heated cathodes are coated with the carbonates of barium and strontium by electric precipitation (cathaphoresis). The emitting material is applied directly to the wires, which are then stretched in zig-zag form in accordance with the length of wire required for the particular valve.

41. Grids

Valve grids usually consist of a spiral of very fine wire surrounding the cathode at a fixed distance. The dimensions of the grid must be accurate to within very close tolerances, as will be understood when it is stated that in some types of valve the distance between the grid and the cathode is only a few tenths of a millimeter. The grid must on no account come into

Fig. 59
Machine for cutting thin nickel tubing to exact length for cathodes.

Fig. 60 Machine on which a circular protuberance is rolled in the cathode tube near one of its ends. By means of this protuberance the cathode rests on the lower insulating mica plate of the electrode system.

Fig. 61a Furnace in which coiled heater filaments for indirectly-heated cathodes are placed to bake the insulating coating onto the wires.

Fig. 61b
Group of workers examin-
ing the coiled heater
filaments coated by heat-
resisting insulating ma-
terial after their removal
from the furnace wherein
they were baked.

contact with the cathode. This might very easily occur if the grid spiral became distorted, for example owing to expansion due to heat radiated from the cathode. In order to give the grid suf-ficient rigidity and thus prevent distortion, two or more thicker wires, termed grid rods, are placed longitudinally on the circumference of

Fig. 62 A grid-winding machine which will produce grids of any desired dimensions.

Fig. 63   Close-up view of a grid-winding machine.

the coil, and each turn of the grid is firmly attached to these rods.
The winding of the grids is done by automatic machines. The grid
rods, which are about $\frac{3}{4}$ mm in diameter and notched at regular intervals

to receive the grid
windings, are fed
side by side through
the machine which
winds a coil, the
turns of which are
laid accurately in
the notches of the
grid rods. The coil is
then pressed tight-
ly into the notches
in order to secure it.
Grids leave the grid-
winding machine
in long lengths as
shown in Fig. 63,

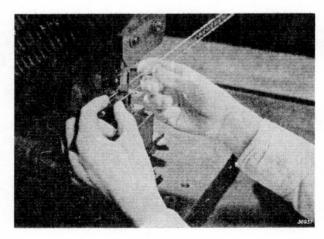

Fig. 64 The grids, wound in long lengths, are cut to exact
length on this machine.

Fig. 65   Finishing the separated grids.

Fig. 66   Testing grids with gauges.

each grid being separated from the next by a few turns wound on at greater speed. The windings, on their supports, are later cut off to the exact lengths required.

The grids as they leave the machine are first processed to stretch and straighten the supports. This is done by placing the grids in a

Fig. 67   Checking the dimensions of grids.

gas-filled tube and then heating by passing an electric current through them by means of two contacts. While the grids are hot they are slightly stretched, thereby straightening the supporting rods. After this process the grids are separated as shown in Fig. 64. Fig. 65 shows the sectional shape of the grids being corrected with special instruments, and in Figs 66 and 67 the grids are being carefully tested to a maximum tolerance of 0.03 mm. The spacing of the windings is also tested and, if necessary, corrected with tweezers.

42. Anodes

The most usual form of anode is a metal cylinder which surrounds the cathode

Fig. 68
Sheet metal wound on rollers, destined to become anodes, is cut to the correct width in this machine.

Fig. 69
Spot-welding the supports to anodes and screen plates.

and the various grids. Material for anodes reaches the valve factory as rolls of sheet metal, which are cut into strips of the correct width to suit the shape of anode required. This process is illustrated in Fig. 68. Anode plates are then stamped out of the strips and given the desired shape in a forming machine.

Screening plates are made in the same way as anode plates. Fig. 70 shows how the perforated plates for anodes and screening cages are made from sheet metal.

43. Miscellaneous Components

It is unnecessary to go into great detail concerning the manufacture of the numerous smaller components, but a few additional illustrations show some of the more interesting processes. Fig. 71, for instance, shows a machine for producing the pinch complete with leading-in wires. The glass for the pinch comes from the Philips glass factory in the form of long tubes. These are first cut to the required length and, after being heated and softened in a gas flame, are pinched by the machine, i.e. one end of the tube, while still hot and plastic, is pressed round the leading-in wires and supporting rods, and the other end is opened out into a flange. During this process also the pumping tube is fused in.

Small mica discs used in certain parts of the valve are stamped out by machines as shown in Fig. 72.

Other machines similar to that illustrated in Fig. 73 cut and shape fine wires. Fig. 74 shows how a small glass tube is fused onto the top of the bulb in order to form the funnel-shaped hole in which the top connecting wire for the grid contact is to be accommodated.

Machine for producing perforated plates from sheet metal.

44. Assembling the Electrode System

The various components,
having been made in their
special departments, are
delivered to the assembly
shop where the valves
are built and completed.
Mica and metal parts are
first thoroughly cleaned
and heated in ovens to
free them from gas. By
treating them in an at-
mosphere of nitrogen, the
gases contained in the
parts are replaced by
nitrogen, which is com-
paratively inert and, when released from the parts subsequent to
pumping, do not cause poisoning of the cathode. The electrode system
is then assembled. For glass valves with pinch, the assembly is built up
on the pinch itself, in which the electrode supports and leading-in

wires have already
been fixed. Fig. 55
shows the parts re-
quired for building
up a typical elec-
trode system, and
in Fig. 78 is shown
the final stage of
assembly with all
these parts correct-
ly mounted.

Fig. 71
Making the pinch, com-
plete with leading-in
wires, supports and
pumping tube.

83

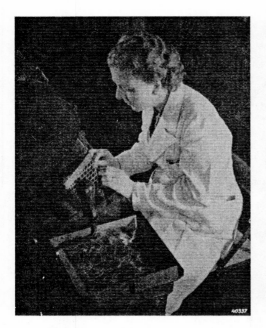

Fig. 72 Punching out mica discs.

Fig. 74 A thin glass tube is fused to the top of the bulb where the top grid connection is to emerge.

Fig. 73 Machine for cutting and bending the fine wires used for securing the electrodes.

Fig. 75 The bulb is inspected by viewing it against a strongly illuminated frosted glass.

Fig. 76
Mica discs in glass bulbs are raised to a high temperature in a special oven to free them from gas. During this process the bulbs are evacuated by pumping and then sealed by fusing.

Metal parts are spot-welded together and to the supports and leading-in wires. Fig. 79 shows the heater winding being welded to the connecting leads in the pinch, while further stages in assembly are illustrated in Figs 80 and 81, indicating the high degree of accuracy essential in producing modern valves.

For valves with pressed-glass bases constructed according to the A-, B- and C-techniques, and also for steel valves, the electrode system is built up separately. In the case of valves with pressed-glass bases the supports of the system

Fig. 77 Electric ovens for cleaning by reduction metal components before assembling.

Fig. 78
Final stage in building the electrode assembly for the octode type CK 1.

are then spot-welded to the connecting leads in the glass base, and in the case of steel valves these supports are fixed to the steel base itself.

45. The Sealing of the Bulbs and Pumping (Evacuating)

(a) Glass Valves with Pinch

When the internal components of a pinch valve have been assembled the whole mount is placed inside the glass bulb. Then the flange of the pinch is fused onto the bulb on a special machine, by placing the valve on its pumping tube in the machine and rotating it between a series of gas flames directed onto the spot where the joint has been made.

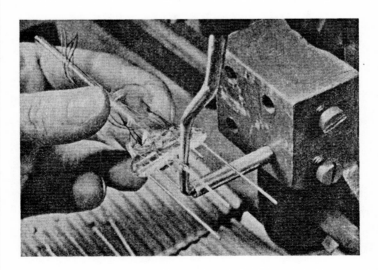

Fig 79
Welding the insulated heater winding to the leading-in wires in the pinch.

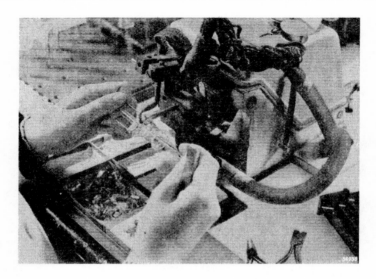

Fig. 80
Mounting the lower mica plate, with cathode and grid already
assembled, onto the pinch.

For this fusing of the glass
the rims of the envelope
and the pinch have to be
heated to a temperature
of 800—900 °C, i.e. above
the softening point of the
glass. Where a glass valve
has the control grid con-
nection at the top of the
bulb, the grid connecting
wire is threaded through
the glass funnel at the
top of the bulb and the
glass is then fused around
the wire at the same time
as the bulb is fused onto
the pinch. This completes

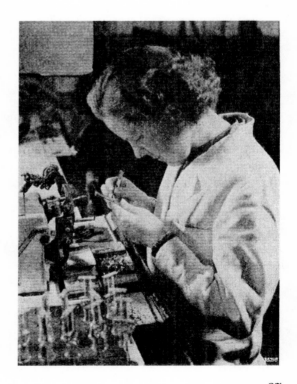

Fig. 81
Mounting the upper mica plate.
In the foreground can be seen
electrode assemblies complete
except for the anode.

Fig. 82 In this machine valves with pinch are pumped and the internal parts heated by high-frequency alternating currents.

the assembly of the valve with the leading-in wires emerging at the bottom—and possibly the grid connection at the top and with olny the pumping tube at the bottom giving communication to atmosphere.

Pumping is performed on the same machine as that which seals the electrode system onto the bulb, the air within the bulb being withdrawn through the pumping tube. Simultaneously with the pumping the electrodes and other metal parts inside the valve are raised to a high temperature in order to release any gases remaining in those parts. The raising to a high temperature is done by means of a coil automatically placed around the bulb and traversed by a high-frequency alternating current, which induces eddy currents in the metal parts of the valve, thereby heating them to a high temperature. After this degassing of the metal parts the filament is brought under tension, thereby converting the carbonates on the filament or on the cathode into oxides (see Chapter IV, Section 21a). When the pumping is completed the pumping tube is sealed off by fusing it close to the bulb, but just before this is done a high-frequency field is generated at the

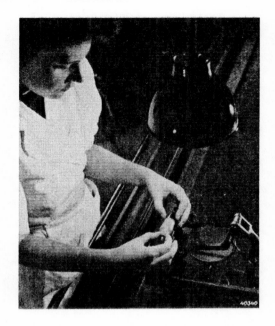

spot where the getter is situated, so as to cause the getter
plate to glow and evaporate
the barium or magnesium
material. As the metal cools
down the vapour is then
deposited chiefly on the inner
surface of the glass bulb opposite the getter, at the same
time absorbing any residual
gases, which are thus "bound",
leaving a practically perfect
vacuum in the bulb. Fig. 82
shows a pumping machine
for evacuating glass valves
with a pinch. The valves are then passed on to another machine
where they are fitted with a base and, as the case may be, a connecting cap, which are cemented onto the bulb. At the same time the
electrode leads are soldered onto the base and cap contacts. Finally,
in the case of valves that have to be given a metal coating, these are
metallized. This metallizing is done in a machine in such a way that
the parts that are to be left bare are protected while the metallic
covering is applied automatically to the remaining surfaces. In many cases the metalizing is given a coat of enamel upon which the valve type number, and possibly other data, is stamped by the process shown in Fig. 83.

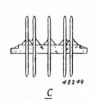

a   *b*   *c*

Fig. 84

Cross-section of the base plate of A-technique valves.

(a) base plate without glazing ring,

(b) ditto glazing ring just laid on,

(c) ditto after the glazing ring has been heated and set;
    owing to its surface tension it assumes a conical shape.

89

Fig. 85
Fusing machine for A-technique glass valves. The valves are placed on their pumping tubes and metal weighting caps are placed over the glass bases (uppermost), these caps being heated by rows of gas flames as the machine rotates.

(b) Glass Valves without Pinch, B-Technique

The electrode system mounted on the glass base is inserted in the glass envelope and the rims of the base and bulb are fused together, in the same way as under a), whilst also the pumping and degassing process is the same as with the pinch glass valves. But with the B-technique valves there is no "Philite" base to be fitted on, no leads to be soldered to the base contacts and no grid-contact cap to be placed on the top of the bulb.

After this pumping, degassing and fusing of the pumping tube, the metal base plate with spigot and guiding cam is affixed to the valve by means of a metal rim as described in Chapter VII, which is machine-flanged around the base plate and the fused joint of the bulb and base. Only the type number then has to be stamped on, after which the valve is ready for testing.

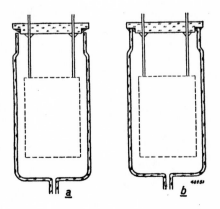

Fig. 86

Diagrams showing the bulb and base,

(a) before and

(b) after fusing together.

(c) Glass Valves without Pinch, A-Technique

The A-technique valves are not distinguished only by their small dimensions. A great technical advance lies in the new method of making an airtight joint between the bulb and the base without having to heat the glass and the inner components of the valve to such a high temperature as hitherto was required. In the A-technique the bulb and base are joined together by glazing with sintered glass, the softening point of which is much lower than that of the glass of the bulb and base. After the glass base with eight leading-in pins has been moulded in a special machine to the shape shown in Fig. 84a, a ring of sintered glass moulded with a binder is laid around the edge of the base plate (see Fig. 84b), which is then heated and gradually cooled off, thus neutralizing mechanical tensions in the glass, melting the powder in the glazing ring and causing it to adhere to the glass. Owing to the surface tension, the right value of which is obtained by adding certain admixtures in the glazing, the glazing rim assumes the shape shown in Fig. 84c, thus more or less convexly conical. The electrode system is welded onto the leading-in pins. Such a base with electrode assembly mounted can be

Fig. 87
Machine for pumping A-technique valves.

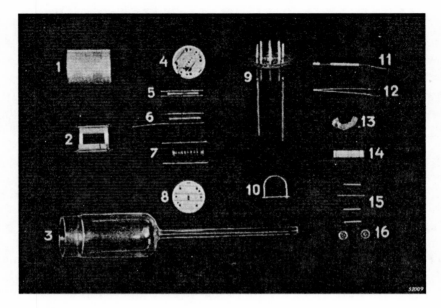

Fig. 88
The various components forming the double-diode-pentode UAF 41.
 (1) Screening cage
 (2) Anode of the pentode section
 (3) Bulb with pumping tube
 (4) Upper mica plate or disc
 (5) Control grid
 (6) Screen grid
 (7) Suppressor grid
 (8) Lower mica plate or disc
 (9) Pressed-glass bottom with connecting pins, supporting rods for the electrode
     assembly and ring of sintered glass
(10) Getter supporting stirrup and hollow rod which contains the getter material.
(11) Cathode
(12) Coiled filament folded to a V
(13) Screening cup with connecting strip
(14) Anode of the diode section
(15) Various connecting strips
(16) Mica discs.

seen in Fig. 89. As already indicated, in the A-technique the pumping
tube is not located at the base end but at the top end of the bulb
(Fig. 88).
The bulbs are then placed on the fusing machine with their pumping
tubes downwards (see Fig. 85) and the electrode systems with base
are placed inside the bulbs (see Fig. 86a). The bases are automatically
centered, so that the axes of the electrode systems coincide with the
axes of the bulbs. Small metal weighting caps are then placed loosely
over the bases.

The valves thus set up on the rotating part of the fusing machine are passed between two rows of gas flames directed radially towards the weighting caps, which are thereby heated. Owing to the heat irradiated inside the metal cap the layer of glazing is softened and the edge of the bulb penetrates into the glazing as a result of the weight of the base plate and that of the weighting cap. In this way both the edge of the bulb and the base plate are heated very gradually and evenly without causing any stresses in the glass. During the cooling down process that follows, on the same machine, the glazing sets and the edge of the bulb is secured tightly to the base (see Fig. 86b). In this method of fusing only the glazing compound has to be softened and not the glass parts of the valve, so that the temperature required is very much lower, about 450 °C being sufficient for the valves now being manufactured according

Fig. 89
The electrode system of the double-diode-pentode UAF 41 without screening cage.

to the A-technique. The temperature to which the cathode is thereby heated is at most 230 °C, whereas in the direct fusing of the bulb on the base it is as high as 500—600 °C. In order to avoid dangerous stresses arising in the glass through the cooling down following on the heating, the thermal expansion coefficient of the glazing compound has to be practically equal to that of the glass used for the bulb and base plate. After the joint has been made between the bulb and base on the fusing machine the valves are evacuated on a pumping machine (see Fig. 87) and the getter material is atomised in the same way as is done in the case of valves with a pinch. The pumping tube is then sealed off, leaving a small pointed end on top of the bulb.

The valves are then automatically lifted off the machine and transported by a conveyer belt to a machine where the metal rim is cemented onto the base.

Apart from the advantages of smaller dimensions, with this glazing technique the bulb can be made to a certain shape accurately to within say 0.1 mm and retains that shape exactly during the fusing process. In Fig. 43, for instance, one sees clearly that there is a recess in the

Fig. 90
Checking of the electrode system mounted on the pressed-glass base of an A-technique valve before inserting it into its bulb.

wall of the bulb close to the base plate, this being necessary for cementing on the metal rim, and if the bulb and base were fused together direct there is a great risk that the recessed part would lose its shape owing to the softening of the glass. Another consequence might be that with the small diameter of the valve the whole of the base plate may become soft when fusing, with the result that the contact pins no longer stand up straight or even become displaced. In such an event the pins would have to be re-aligned, and to avoid cracking the glass they would therefore have to be made of a soft material, as a result of which they might easily get bent and the valve base would then no longer fit into the socket. All these complications are avoided by applying the glazing technique, for the temperature of the base is kept very low and consequently it retains its shape, so that for the small diameters of the A-technique the pins can be made of hard material just as well as those for the larger diameters of the valves of the B- and C-techniques. A particularly important point is that manufacture can be speeded up enormously and that there are fewer rejects. Finally, thanks to the low fusing temperature there is no fear of poisoning the cathode.

(d) Metal Valves

In the case of metal valves the envelope and the base are electrically welded together in a special welding machine (about 2000 amps). The welding current is applied to the valve for only a very short time,

Fig. 91
Burning out valves on large ageing racks at a higher heater voltage than the normal rating.

so that the temperature inside the valve is kept low and there is therefore no fear of the cathode being poisoned. On another machine the valve is then evacuated through the metal pumping tube on the base plate, while at the same time the electrodes and other internal metal parts are degassed. With metal valves this degassing cannot be done by means of H.F. induction because of the screening formed by the metal envelope. It therefore has to be done by heating the iron bulb with gas flames to yellow-hot temperature; the internal parts are thereby raised to such a high temperature as to release the gases. Care is taken, however, not to heat the bulb to such a high temperature as to cause premature evaporation of the getter in the top of the bulb. After degassing of the internal parts the getter is evaporated by applying a jet flame to the top of the bulb. Finally the metal pumping tube is sealed by pinching, welding and shortening. The "Philite" plate with spigot and contact pins is then affixed to the valve by pressing in the collar of the base plate at three places, after which the electrode leads drawn through the hollow contact pins are soldered to the latter.

46. Burning Out ("Ageing"); Mechanical and Electrical Testing

The manufacture of the valves being thus completed, the valves are subjected to the "burning out" or ageing process. This consists of running the filaments or heaters for a few hours with a higher current than the normal rating, as explained in Chapter IV, Section 21a. For this purpose the valves are plugged into large frames, called ageing racks, one of which is illustrated in Fig. 91. Here all the filaments or heaters are connected in parallel to a supply source at rather more than normal voltage.

Fig. 92
View of the testing of valves. First they are preheated
by placing them in sockets (marked by A in the picture)
and afterwards they are checked at the place marked by B.
With the handle C the various commutations required
are produced for the consecutive measurements.

Before leaving the factory each valve is submitted to a large number of mechanical and electrical tests. Some of the test-desks are semi-automatic. First the valves are preheated by inserting them in sockets provided on the test-desks for this purpose (see A in Fig. 92),so that the cathodes are already warmed when they are inserted in the sockets provided for the testing operations (B in Fig. 92) where also their electrical characteristics are checked. On each test-desk is a roller commutator which can be moved to successive positions by means of a handle (C in Fig. 92), thus enabling male or female operators to check consecutively all the electrical properties of the valve, including the heater ratings, emission, characteristics and so on.

Any valves which do not come up to the specified requirements are at once rejected and destroyed. These electrical tests, however, are not in themselves sufficient because, although a valve may not exhibit any deviations from the electrical specification, it may still have some hidden defect which may become apparent when the valve is operated in a receiving set. Such defects include irregularities in manufacture, which may result in disturbing noises in the loudspeaker. They may be due to various causes, such as faulty insulation between the heater and cathode.

Fig. 93
Testing valves for mechanical defects which might give rise to disturbing noises. This is done by tapping with a light hammer.

In order to reveal any latent defects, therefore, each valve is also subjected to a series of mechanical tests. One such test consists of striking the valves with a light rubber hammer, as shown in Fig. 93, when a skilled operator can detect the existence and, by means of instruments connected to the valve, ascertain the cause of any hidden defect.

Fig. 94 Testing desk at which all essential voltages and currents of the valves are measured. The valve under test is in front of the left-hand panel. In the centre other valves are being warmed up.

Fig. 95
Shock-testing machine to check
mechanical rigidity of valves.

Final tests are then made
on a proportion of valves
selected at random from
every batch manufactured.
In these tests careful checks
are taken on all electrical
data, capacitances, microphonic effects, working life
and other characteristics
upon which so much may
depend in practical oper-
ation.

Thus it is ensured that
every valve placed on the
market fully meets all requirements, complies with

the close tolerances laid down
in the specification, and gives
the performance aimed at. In
this manner tens and hun-
dreds of thousands of mass-
produced valves of a single
type alone can be turned out
with astonishing uniformity
of characteristics. Figs 94,
95 and 96 give some idea of
the elaborate installations set
up for the electrical and
mechanical testing of valves.

Fig. 96
Tropical cabinet in which the temperatures and humidity of the tropics
are reproduced. After being left in this
cabinet for a period the valves are
examined for defects in insulation
and for corrosion.

Fig. 94, for instance, shows a test-bench on which all voltages and currents in the valve are measured rapidly. Fig. 95 shows a shock-testing machine in which an adjustable force vibrates the valves in both a horizontal and a vertical plane to test whether the valves are able to withstand the mechanical shocks and knocks to which they may be subjected both in transit and in use, particularly in automobile radio sets.

In the tropical cabinet, Fig. 96, atmospheric conditions such as obtain in the tropics are simulated, particularly high temperatures and humidity. Valves are placed in this cabinet to ascertain the effects of these exacting conditions upon the external insulating materials and the cement securing the bases to the valves.

CHAPTER IX

The Functions of Radio Valves

Every transmitting station radiates its own programme, and every programme induces in the receiving aerial an alternating voltage. Some of these voltages, especially those from low-power or distant transmitters, are extremely feeble. It is the function of the receiving set to select the voltage corresponding to the programme it is desired to hear; to amplify it; to change its form; and to amplify the modified wave again until the desired volume of sound is obtained in the loudspeaker or telephone. The electromagnetic wave radiated by the transmitter induces a voltage in the aerial which, plotted as a function of time, is not as a rule sinusoïdal but of a more complex form.

Before describing the various functions performed by valves in a receiving set, it is advisable to examine more closely the nature of the alternating voltages induced in the aerial circuit.

47. Nature of the Signals in the Aerial Circuit

Music, speech, television pictures and telegraph signals are transmitted in such a way that a wave of high-frequency electromagnetic vibration is radiated into space from the transmitting aerial. This wave is termed the **carrier wave** and its frequency must be sufficiently high to allow it to be radiated easily from the transmitting aerial. The carrier frequency is therefore always greater than 100,000 c/s, the actual value depending upon a variety of considerations. For normal broadcasting the carrier frequency is between 150 kc/s and 30 Mc/s, while for special purposes still higher frequencies are employed.

The relation between frequency and wavelength is:

$$f = \frac{3 \times 10^8}{\lambda},$$

where: f = frequency in cycles per second,
λ = wavelength in metres.

In the transmitter the high-frequency carrier is modified by the music, speech or other signal it is desired to transmit. This process is called **modulation.** In the receiving set it is possible to resolve this modulation and, by means of suitable apparatus such as, for example, a loudspeaker, to reproduce the original programme.

The acoustic vibrations of a programme of music or speech are transformed by a microphone in the studio into electrical vibrations, the amplitudes of which are proportional to the volume of sound, and the frequencies equal to the fundamental frequencies and harmonics present in the original sound vibrations. The audible range of frequencies extends from approximately 40 c/s to 10,000 c/s and the low-frequency vibrations which modulate the high-frequency carrier wave therefore lie within this range. The low-frequency (audio-frequency) vibrations are separated from the high-frequency vibrations in the receiver by a process known as **detection**, or rectification of the high-frequency wave.

In the transmission of pictures, as in picture telegraphy and television, the picture is "scanned" point by point, and a pick-up device transforms the brilliance of each point or picture element into a voltage proportional to the brilliance of that point. This voltage is employed to modulate a high-frequency [1]) carrier wave and is resolved again in the television receiver as an electrical voltage, which is then reproduced on the screen of a cathode-ray tube as a point of light of corresponding brilliance.

A high-frequency voltage of sinusoidal wave form may be represented as a function of time by the formula:

$$v = V_0 \cos (\omega t + \varphi), \tag{24}$$

where: v = instantaneous value of the voltage;
V_0= amplitude of the voltage;
ω = angular frequency of the vibration = $2 \pi f$, where f = frequency in cycles per second;
t = time in seconds;
φ = phase angle.

If at the commencement of the cycle $v = V_0$ the formula can be written in the simplified form:

$$v = V_0 \cos \omega t. \tag{25}$$

From this formula it is obvious that an H.F. voltage can be modulated by an L.F. voltage in two ways, that is to say by varying the amplitude V_0, or by varying the frequency ω. The first method is known as **amplitude modulation,** and the second as **frequency** or **phase modulation.**

---

[1]) The abbreviations H.F., I.F., and L.F. signify High Frequency, Intermediate Frequency and Low Frequency respectively. Instead of high frequency and low frequency the terms radio frequency and audio frequency are commonly used.

(a) Amplitude-modulated Signals

It results from the foregoing that in amplitude modulation the voltage in the transmitting aerial is an H.F. alternating voltage of a specified and constant frequency (the carrier frequency), the amplitude of which varies at low frequency. In the studio the modulation voltage is produced in the microphone by the effect of sound waves on the moving system of the microphone. The modulation voltage, after a suitable degree of amplification, is passed to the transmitter, where the carrier-frequency voltage, originally of constant amplitude, is influenced in such a way that its amplitude now varies in rhythm with the modulation frequency and in proportion to the amplitude of the modulating voltage. This process is illustrated in Fig. 97. The dot-and-dash line in diagram Fig. 97c is identical in shape with the wave form of the modulating voltage and is sometimes referred to as the **"envelope"**.

By **modulation depth** is understood the ratio between the amplitude of a sinusoidal envelope of a modulated high-frequency carrier wave and the amplitude of the unmodulated carrier voltage; thus:

$$m = \frac{V_M}{V_o} \times 100\%, \qquad (26)$$

where:

m = percentage modulation depth;

V_M = amplitude of the voltage represented by the sinusoidal envelope;

V_o = amplitude of the unmodulated carrier voltage.

The definition of modulation depth is represented graphically in Fig. 98, and the oscillogram reproduced in Fig. 99 represents a modulated H.F. voltage as recorded with a Philips cathode-ray oscilloscope.

With a modulation depth of 100% the amplitude of the envelope equals that of the carrier-wave voltage, which thus varies in rhythm with the modulation frequency and in amplitude between zero and twice the unmodulated carrier-wave voltage.

From Fig. 98 it is clear that the amplitude of the modulated carrier wave fluctu-

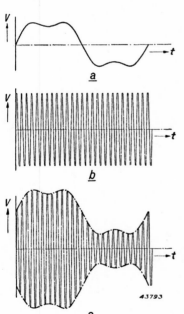

Fig. 97
Top: L.F. modulating voltage plotted as a function of time.
Centre: H.F. voltage (carrier wave) without L.F. modulation.
Bottom: Modulated H.F. voltage.

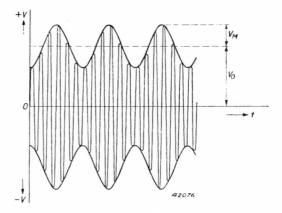

ates between $V_o + V_M$ and $V_o - V_M$. Further it is clear that, as the result of modulating an originally sinusoidal H. F. carrier-wave voltage (H. F. signal) by an L. F. alternating voltage (L. F. signal), the wave-form of the modulated carrier voltage is no longer a single sinusoidal wave. It can, in fact, be resolved into three separate sinusoidal oscillations, namely, one corresponding in frequency and amplitude to

Fig. 98
H.F. voltage modulated by a single low frequency. The modulation depth in this example is $33^1/_3\%$.

the unmodulated carrier wave; one with a frequency equal to that of the unmodulated carrier wave plus the modulation frequency; and the third with a frequency equal to that of the unmodulated carrier wave minus the modulation frequency, the amplitude of the two last mentioned being equal to $\frac{1}{2}$ m $\times$ the amplitude of the unmodulated carrier-wave voltage.

This phenomenon can also be demonstrated mathematically. Modulation causes the carrier wave to be attenuated or intensified in rhythm with the L.F. modulation voltage as represented in Fig. 98 and by the following equation:

$$v = \underline{V_o \ (1 + m \ \cos \ pt)} \ \cos \ \omega_0 t, \tag{27}$$

Fig. 99
Oscillogram of a modulated H.F. voltage comprising a 300 kc/s carrier and an audio frequency of 400 c/s, as recorded with a Philips cathode-ray oscillograph.

where

V_o = amplitude of the carrier-wave voltage;

m = modulation depth;

p = angular frequency of the modulation voltage = $2 \ \pi \ f_M$;

ω_0 = angular frequency of the carrier wave voltage = $2 \ \pi \ f_o$.

The term underlined in this equation is termed the **instantaneous value of the amplitude** of the carrier wave.

Equation (27) may also be written in the form:

103

$$v = V_0 \cos \omega_0 t + m\ V_0 \cos pt \cos \omega_0 t =$$

$$= V_0 \cos \omega_0 t + \tfrac{1}{2} m\ V_0 \cos (\omega_0 + p)\ t + \tfrac{1}{2} m\ V_0 \cos (\omega_0 - p)\ t. \quad (28)$$

Equation (28) clearly shows that modulation of an H.F. voltage by an L.F. voltage yields the three alternating voltages of different frequencies already referred to.

The two waves $\tfrac{1}{2} m\ V_0 \cos (\omega_0 + p)\ t$ and $\tfrac{1}{2} m\ V_0 \cos (\omega_0 - p)\ t$, one of which has a frequency higher than the carrier frequency by an amount, p, equal to the modulation frequency and the other a frequency lower than the modulation frequency by the same amount, are termed **side waves,** and the wave $V_0 \cos \omega_0 t$ is, of course, the original unmodulated carrier wave.

In radio transmissions of sound the modulation generally consists of frequencies between 40 c/s and 4,500 c/s [1]). Each of these frequencies, when combined with the carrier wave, forms two side waves, and all the modulation frequencies together form two symmetrical frequency spectra or **side bands,** one on either side of the carrier-wave frequency. One of these side bands extends from the carrier wave frequency + 40 c/s to the carrier wave frequency +, e.g. 4,500 c/s; and the other side band extends from the carrier wave frequency —40 c/s to the carrier wave frequency —, e.g. 4,500 c/s.

The whole frequency spectrum $(\omega_0 + p)$ to $(\omega_0 - p)$, where p represents the highest modulation frequency, is called the band width of the transmitter. It is thus clear that the band width is governed by and is proportional to the highest modulation frequency.

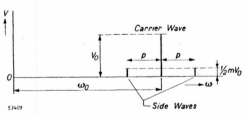

Fig. 100a
Carrier wave modulated by a single frequency, resolved into an unmodulated carrier and two side waves. The amplitude of the carrier is V_0, and of each side wave $\tfrac{1}{2} mV_0$.

Sometimes one sideband is suppressed in the transmitter, so that the frequency spectrum extends only from ω_0 to $(\omega_0 + p)$ or from $(\omega_0 - p)$ to ω_0. The suppression of one sideband is sometimes used in television transmission to limit the band width for the picture transmission because of the exceptionally high modu-

[1]) A frequency interval of 9000 c/s is usually maintained between the carrier frequencies of the various transmitters in a given broadcasting area. In these circumstances, in order to avoid interference between programmes transmitted on adjacent channels, the highest modulation frequency must not exceed 4500 c/s, for a side wave with a frequency of 6000 c/s transmitted from one station would appear as an interference in the programme of an adjacent channel, its frequency differing from that of the second station by 3000 c/s.

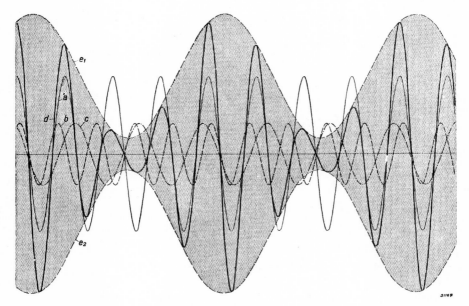

Fig. 100b Modulated H.F. oscillation d, comprising carrier a and two side waves b and c. The envelope is represented at e_1 and e_2.

————————————— carrier wave a
— — —·— — —·— — — side wave b
—·—·—,—·—·—,—·—·—·— side wave c
———————————— modulated carrier d
— — — — — — — — envelopes e_1 and e_2.

lation frequencies employed, which may be of the order of 2 Mc/s. Figs 100a and 100b indicate in yet another form the resolution of a modulated carrier wave into an unmodulated carrier wave with two side bands. In Fig. 100a the frequencies are plotted on the horizontal axis, the voltages shown for frequencies of $(\omega_0 + p)$, ω_0, and $(\omega_0 - p)$. Fig. 100b shows the voltages as functions of time (t) for the same oscillations, and also the resultant modulated H.F. voltage and the envelope.

The following table gives the ranges of frequencies actually employed in radio broadcasting and in television.

Type of Transmission		Carrier Frequency Range	Wavelength
Radio Broadcasting—Long wave		150— 300 kc/s	2000—1000 metres
do	Medium wave	500—1500 kc/s	600— 200 metres
do	Short wave	6— 30 Mc/s	50— 10 metres
Television		40— 100 Mc/s	7.5— 3 metres

It may also be helpful to give the following nomenclature which is more or less standardised to-day.

Carrier Frequency Range	Abbreviation	Wave Range	Unit
Very Low Frequency ($<$ 30 kc/s)	VLF	Myriameter waves	10,000 m
Low Frequency	LF	Kilometer waves	1,000 m
Medium Frequency	MF	Hectometer waves	100 m
High Frequency	HF	Decameter waves	10 m
Very High Frequency	VHF	Meter waves	1 m
Ultra High Frequency	UHF	Decimeter waves	10 cm
Super High Frequency	SHF	Centimeter waves	1 cm .
Extremely High Frequency	EHF	Millimeter waves	1 mm

(b) Frequency-modulated Signals [1])

In recent years considerable interest has been shown in the application of frequency modulation, an interest which has been further stimulated by the latest technical developments. The system holds great possibilities for the interference-free reception of radio transmissions. In frequency modulation the H.F. oscillation generated at the transmitting station is modulated in such a way that its frequency, and not its amplitude, varies in rhythm with the L.F. or programme modulation. The degree of modulation is determined by the extent to which the frequency varies from a constant mid-value of the high frequency, around which frequency the signal frequency varies symmetrically, the amplitude of the H.F. oscillation remaining constant.

If ω_0 ($= 2\pi f_0$) is the angular frequency of the unmodulated H.F. signal and p ($= 2\pi f_M$) the angular frequency of a sinusoidal modulation voltage, the foregoing description of frequency modulation shows that the frequency of a modulated signal will change p times per second, between the limits $\omega_0 + \Delta\omega_0$ and $\omega_0 - \Delta\omega_0$, $\Delta\omega_0$ being proportional to the amplitude of the L.F. modulation voltage.

The term $\Delta\omega_0$ is, of course, the amplitude of the frequency deviation about the mid-frequency ω_0. With a sinusoidal modulation the frequency can be expressed as a function of time by the formula:

$$\omega = \omega_0 + \Delta\omega_0 \cos pt. \tag{29}$$

In this formula ω is the **instantaneous value of the frequency** of the frequency-modulated signal. It can be shown that a wave of which the instantaneous frequency is represented by Equation (29) can itself be represented by the equation:

[1]) See also Th. J. Weijer's "Recent Developments concerning Frequency Modulation", Tijdschr. v. h. Ned. Radiogenootschap **8**, 1940, pp. 315—364, and also the bibliography given at page 122.

$$v = V_o \cos\left(\omega_o t + \frac{\Delta\omega_o}{p} \sin pt\right). \tag{30}$$

By substituting for $\dfrac{\Delta\omega_o}{p}$ the letter m, which is termed the **modulation index,** the formula becomes:

$$v = V_o \cos(\omega_o t + m \sin pt), \tag{31}$$

where:

v = instantaneous value of the voltage;

V_o = amplitude of the unmodulated H.F. signal;

ω_o = angular frequency of the unmodulated H.F. signal;

p = angular frequency of the modulating voltage;

m = modulation index =

$$= \frac{\text{amplitude of the frequency deviation}}{\text{frequency of the frequency deviation}} = \frac{\Delta\omega_o}{p}.$$

Equation (31) indicates that the phase, $\omega_o t$, of the unmodulated carrier wave is modulated by an L.F. oscillation m sin pt, and for this reason the system is also given the name **phase modulation.** The modulation index, m, can be determined in various ways. In the special case of phase modulation in which the amplitude $\Delta\omega_o$ is independent of the frequency p of the

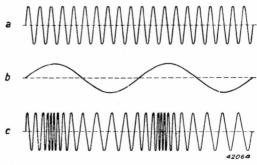

Fig. 101
(a) Unmodulated H.F. oscillation
(b) Modulating L.F. oscillation
(c) Frequency-modulated oscillation.

modulating L.F. signal but is proportional to the amplitude of the L.F. signal, the system is termed **frequency modulation.**

Fig. 101 shows a frequency-modulated wave plotted as a function of time. At (a) is shown the unmodulated wave of constant amplitude and frequency, (b) is the modulating L.F. oscillation and (c) the resultant frequency-modulated signal.

A frequency-modulated signal may occupy a considerable band width, for when Equation (31) is resolved it becomes:

107

$$v = V_0 \cos(\omega_0 t + m \sin pt)$$
$$= f_1 V_0 \cos \omega_0 t +$$
$$+ f_2 V_0 \{ \cos(\omega_0 + p)t - \cos(\omega_0 - p)t \} +$$
$$+ f_3 V_0 \{ \cos(\omega_0 + 2p)t + \cos(\omega_0 - 2p)t \} +$$
$$+ f_4 V_0 \{ \cos(\omega_0 + 3p)t - \cos(\omega_0 - 3p)t \} + \dots \text{etc.} \quad (32)$$

The factors f_1, f_2, f_3 etc. depend upon the value of m and can be determined mathematically. For small values of m (i.e. $m \ll 1$) $f_1 = 1$; $f_2 = \dfrac{m}{2}$; f_3 is proportional to m^2; f_4 is proportional to m^3 and so on. Thus, the factors f_3, f_4 and higher are extremely small. By disregarding these factors Equation (32) becomes:

$$v = V_0 \cos(\omega_0 t + m \sin pt)$$
$$= V_0 \cos \omega_0 t + \tfrac{1}{2} m V_0 \cos(\omega_0 + p)t - \tfrac{1}{2} m V_0 \cos(\omega_0 - p)t. \quad (33)$$

This equation is of the same form as Equation (28) for amplitude modulation and indicates that a frequency-modulated signal consists of a carrier wave with two side waves, the two side waves being, however, of opposite sign, whereas in amplitude modulation the side waves are of similar sign. Further, the band width is equal to twice the modulation frequency $\dfrac{p}{2\pi}$.

If the value of m is larger, the factors f_3, f_4 etc. in Equation (32) are also greater, and the side waves of frequencies $(\omega_0 + 2p)$; $(\omega_0 - 2p)$; $(\omega_0 + 3p)$; $(\omega_0 - 3p)$ etc. can no longer be disregarded. These frequencies are always greater or less than the carrier frequency by a multiple of the modulating frequency p. There is therefore a frequency spectrum of side waves of considerable width, in which the various frequencies are spaced at intervals equal to the modulation frequency p. In order to ensure interference-free reception of a frequency-modulated signal, $\Delta\omega_0$ must be large and, since $mp = \Delta\omega_0$, the modulation index m will also be large. Usually a value of modulation index greater than 10 is selected. In these circumstances, with a high modulation frequency, the side bands are of very considerable width.

Fig. 102 shows a frequency spectrum where $m = 10$, and another where $m = 20$, both being derived from a mathematical calculation of the factors f_1, f_2, f_3 etc. in Equation (32). The diagram shows that where the modulation index is high the band width is more than twice the frequency deviation $\Delta\omega_0$; but for practical purposes the band width may be taken as $2 \times \Delta\omega_0$, as the amplitudes of frequencies beyond this band are so small that they may be disregarded.

Thus, with a modulation index of 10 and a top modulation frequency of 10,000 c/s, the frequency spectrum occupies a band-width of

$$B = 2 \times 10 \times 10{,}000 \text{ c/s} = 100 \text{ kc/s.}$$

With amplitude-modulation, however, the band-width corresponding to a top modulation frequency of 10,000 c/s is only 20,000 c/s = 20 kc/s. Because of the considerable band-width necessary to obtain reasonably interference-free reception of frequency-modulated signals, high frequencies of the order of 40 Mc/s (wavelengths in the region of 7.5 m) are used for the carrier wave, for in the normal broadcasting band (500—1500 kc/s) there is no room for a 200 kc/s band-width channel, and in this frequency range, frequency modulation offers no advantage over amplitude modulation, which is therefore preferred on account of its simplicity. In the short-wave range (approximately 40 Mc/s), however, frequency modulation can offer the following valuable advantages:

(1) Less noise.

(2) Less susceptibility to atmospheric interference and other transients.

(3) Less interference by other programmes.

(4) Possibility of high-quality reproduction without sacrificing selectivity.

(5) Good reception at relatively low field strength at the receiving aerial.

All these advantages can be realised in the following circumstances:

(1) The frequency deviation $\Delta\omega_0$ must be much greater than the highest modulation frequency, p — say $\Delta\omega_0 = 10$ p.

(2) The interval between the carrier frequencies of two adjacent channels must be at least 2 $\Delta\omega_0$.

(3) Special provisions, to be described hereafter, must be made in the receiving set.

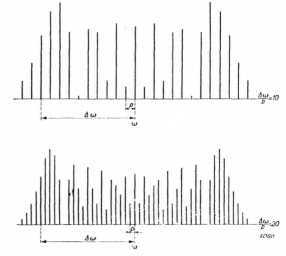

Fig. 102

Frequency-modulated signal analysed into components of various frequencies, for modulation indices of m = 10 and m = 20. The frequencies of the individual components are plotted horizontally and their amplitudes vertically.

48. Brief Explanation of the Operation of a Receiving Set

The reader has now been given a description of the nature of the signals received by the aerial, and some idea of their strength. The function of the receiving set is to amplify these signals, to "detect" them by separating voltages from the H.F. carrier voltage, and to re-amplify the modulation voltage to such an extent that the loudspeaker can emit sound waves of sufficient power.

The first process is to select the programme it is desired to hear, and to separate it from all other signals. This is achieved by the use of one or more tuned circuits consisting of coils and condensers. The selected signal is then amplified in one or more valves. These valves are coupled by other tuned circuits which also serve to keep the desired programme free from interference by other signals and side-bands picked up by the aerial.

Such amplification is termed **R.F. (radio-frequency) amplification** and takes place in **R.F. valves.** After it has received sufficient R.F. amplification the signal is passed to the **detector valve** for detection or rectification. In the detector stage the frequency and amplitude of the original modulation is separated from the carrier-wave frequency, which is suppressed.

The output of the detector consists, therefore, of L.F. (low-frequency) or, more correctly A.F. (audio-frequency) alternating voltages. These, in turn, are amplified in one or more **A.F. amplifying valves,** to an extent at which, when they reach the final amplifying valve or **output valve,** they are of sufficient strength to produce the power required for the loudspeaker or other sound reproducer.

Fig. 103 is a block diagram illustrating the sequence of the various stages in a receiving set of the type described above. Such a set is termed a **tuned radio-frequency (T.R.F.) receiver** and incorporates what is known as **direct R.F. amplification.**

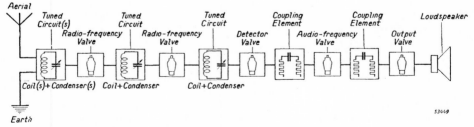

Fig. 103
Block diagram of a tuned radio-frequency receiver for amplitude-modulated signals.

This principle, however, is only used in the simpler types of receiver. Modern receivers, both large and small, are usually designed on the superheterodyne principle. **Superheterodyne reception** (often abbreviated to ,,superhet"), while employing all the essential features of T.R.F. reception, incorporates additional stages which permit more efficient amplification. An additional valve, termed the **oscillator valve,** generates an unmodulated H.F. oscillation the frequency of which differs slightly from the frequency of the carrier wave. This local oscillation is mixed with the tuned R.F. signal in one of the R.F. stages, by means of a valve which is called the **mixer valve, frequency changer** or **frequency converter.** In this stage, owing to the introduction of the local oscillation, new frequencies are set up, one of which is equal to the difference between the frequencies of the carrier and of the local oscillation. This "differential" frequency now becomes the carrier of the modulation of the original R.F. signal. Because the frequency of the local oscillation is always adjusted to a value higher or lower by a constant amount than the frequency of the signal to which the aerial circuit is tuned, the "differential" frequency is always the same and can therefore be selected or separated from other frequencies by means of circuits of fixed resonance, that is to say without the use of variable condensers. The signal thus selected can then be further amplified in valves. The "differential" frequency is termed the **intermediate frequency** (I.F.) and is amplified in the **I.F. amplifier** by means of **I.F. amplifying valves,** one or more in number, until the signal voltages are great enough to permit efficient detection in the detector stage. In European practice an intermediate frequency of approximately 125 kc/s or 475 kc/s is employed for broadcast reception, but for television receivers and in ultra-short-wave reception much higher intermediate frequencies are used.

After the detector stage of a superheterodyne receiver, the audio-frequencies are further amplified and then passed to the output valve in the same manner as in T.R.F. receivers.

It should be explained that the incoming signal may be amplified in one or more R.F. stages before being mixed with the locally-generated oscillation, or it may be fed to the mixer stage via one or more tuned circuits without amplification. The simplest superheterodyne receivers have no R.F. amplifier stage at all; more elaborate sets have one, or, in exceptional cases, more R.F. stages. The unmodulated H.F. voltage generated in the receiver is called the **oscillator alternating voltage,** and is produced by means of a valve termed the **oscillator valve,** a tuned circuit consisting of a variable condenser and a coil, and a feedback coil.

111

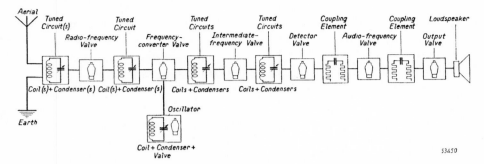

Fig. 104 Block diagram of a superheterodyne receiver for amplitude-modulated signals.

Fig. 104 is a block diagram illustrating the sequence of the various stages of a superhetercdyne receiver.

To summarise, the R.F. signal in the aerial of a superheterodyne set may or may not be amplified. If it is amplified this takes place in one or more R.F. amplifying valves and the amplified R.F. signal is mixed in the mixer valve with the oscillator voltage generated in the oscillator valve. The output of the mixer valve is a modulated I.F. voltage which is amplified in one or, in exceptional cases, more I.F. valves and is then passed to the detector valve. The output of the detector stage is an A.F. signal which is further amplified in the A.F. amplifying valves and in the output valve, which provides the power for operating the loudspeaker.

There are, of course, many variants of these basic types of receiver. There are, for instance, sets which operate without any R.F. amplification, the aerial signal being passed from the tuning circuit direct to the detector valve. The brief descriptions of methods of radio reception given above are applicable both to amplitude-modulated signals and to frequency-modulated signals, but in the case of frequency-modulation the detector stage is more complex, because the frequency-modulated signal must be converted into an amplitude-modulated signal before it can be rectified by the detector. This conversion takes place in the "frequency detector" or **converter stage,** which con‑ sists in essence of a resistance, an inductance, and a capacitance connected in series as shown in Fig. 105.

If the correct values are chosen for the resistance R, the inductance L and the capacitance C, an

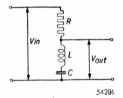

Fig. 105
Converter stage (frequency detector) for converting a frequency-modulated signal into an amplitude-modulated signal. If the correct values of R, C and L are chosen the voltage V_{out} varies linearly with the frequency of V_{in}, provided the amplitude of V_{in} is constant.

output voltage V_{out} is obtained, the value of which varies linearly, within close limits, with the frequency of the input voltage V_{in}, as shown in the graph reproduced in Fig. 106, provided the amplitude of V_{in} is the same at all frequencies. Fig. 106 shows the linear relation between the amplitude of the output alternating voltage V_{out} and the frequency ω of the input alternating voltage V_{in} under the stated conditions.

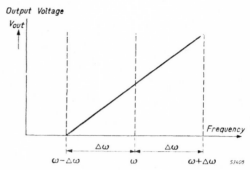

Fig. 106
Characteristic of the converter stage illustrated in Fig. 105. The output voltage is plotted as a function of the frequency ω, the amplitude V_{in} of the input signal being constant.

It has already been stated that the converter stage will function linearly only if the input voltage is constant at all frequencies. In order to achieve this condition, and thus to utilise fully the advantages of frequency modulation with respect to freedom from atmospheric and other interferences, it is necessary either to interpose what is known as a **limiter stage** prior to the converter stage, or to apply frequency feedback. Consideration here of the theory of limiting interference by the application of these measures would involve too great a digression, and it must suffice to remark that a limiter stage is frequently applied with the object of feeding to the frequency detector signals which are of constant amplitude in spite of all possible interferences. The characteristic of the limiter stage is as shown in Fig. 107. Usually, owing to interferences, both the amplitude and the frequency of the carrier wave are modulated. By avoiding amplitude modulation through limiting of the amplitude, the output of the amplitude limiter will contain only the frequency modulation resulting from the interference. In modern frequency-modulation transmissions the amplitudes of the frequency variations are

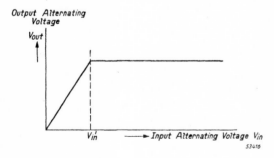

Fig. 107
Characteristics of an amplitude-limiting stage for frequency-modulated signals. The amplitude of the output signal voltage, V_{out} is shown as a function of the amplitude of the input signal V_{in}.

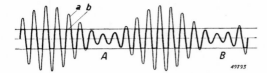

Fig. 108
Thin Line a: Amplitude-modulated signal.
Heavy Line b: Signal after passing through the amplitude limiter.

very large, corresponding to modulation indices m greater than 10, so that the frequency variations caused by interferences are extremely small compared with those corresponding to the A.F. modulation. Fig. 108 shows that portion of an amplitude-modulated signal which remains after it has passed through an amplitude limiter. At A and B the amplitudes are smaller than the threshold value V'_{in} of the amplitude limiter (see Fig. 107). In order to ensure that the signal contains no interferences due to amplitude modulation, the amplitude must not fall below this threshold value. The amplitude limiter must, therefore, be fed with a signal which has already been amplified sufficiently to ensure that no part of the interference modulation can appear in the amplitude limiter output.

Fig. 109 is a block diagram of the various stages of a superheterodyne receiver for frequency-modulated signals. A bibliography concerning frequency modulation is given on page 122.

49. Current Supplies for Radio Sets

All receiving sets must be supplied with electric current to enable the valves to operate. Usually this current is drawn from the direct-current or alternating-current mains, but in some sets it is obtained from accumulators or from batteries of dry cells.

Most mains-operated receivers nowadays are fed from alternating-current mains. As, however, with the exception of the cathode-heater supply, for which alternating current is quite suitable, valves are

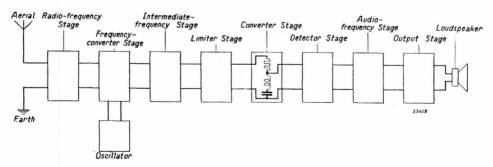

Fig. 109
Block diagram of a superheterodyne receiver for frequency-modulated signals.

operated entirely on direct voltages, provision must be made for converting the alternating supply from the mains into a direct supply.

Usually the mains alternating supply is transformed by the **mains transformer** into an alternating supply at a higher voltage, say 300 V, after which it is converted into a uni-directional supply by a **rectifier valve.** The uni-directional output of the rectifier cannot, however, be used in the receiver in that form because it is pulsating, i.e. it carries an alternating-current component. This component is suppressed in the **smoothing circuit** which follows the rectifier valve and consists of an arrangement of choke coils or resistances and condensers. The current supplies for radio valves are dealt with more fully in Chapter XXXII, and reference can also be made to Chapter XIX, where mains voltage rectifiers are discussed.

50. Summary of the Functions of Radio Valves

In addition to the functions already mentioned, valves are used for various subsidiary purposes, particularly in receivers designed for high-fidelity reproduction. They are employed, for example, in amplifying automatic-volume-control voltages and the like. It is seen, therefore, that valves have to perform a large number of widely divergent duties, each with different requirements by way of valve characteristics. For example, the operating conditions and required characteristics of an R.F. amplifier differ from those of an A.F. output valve.

The principal types of receiving valves include:

(1) R.F. amplifiers.
(2) I.F. amplifiers.
(3) Frequency changers (mixer valves).
(4) Oscillators.
(5) Detectors.
(6) A.F. amplifiers.
(7) A.F. output valves.
(8) Rectifiers.
(9) Electron-ray (fluorescent-screen) tuning indicators, the nature and functions of which are described in Chapter XXVI.

Sometimes valve systems designed for different functions are combined in a single bulb, and thus consitute a single valve, which is then termed a **multiple valve.** Thus, output pentodes or A.F. amplifiers often contain a detector system. The oscillator valve is also generally combined with the mixer valve.

CHAPTER X

Conventional Method of Representing Valve Electrodes in Circuit Diagrams

Before leaving the subject of the construction of radio valves it is necessary to indicate how valves are represented in circuit diagrams.

Fig. 110 is the conventional representation of a valve having an anode and an indirectly-heated cathode. If the valve has a control grid this is shown as a broken line, midway between the cathode and

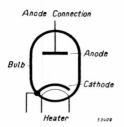

Fig. 110
Conventional representation of a diode.

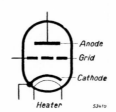

Fig. 111
Conventional representation of a triode.

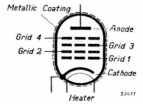

Fig. 112
Conventional representation of a multi-grid valve.

anode (Fig. 111). If the valve has several grids, arranged concentrically around the cathode, these are indicated by other broken lines and are numbered in their order starting from the innermost. Thus in Fig. 112 grid No. 1 is that nearest the cathode.

The various electrodes are indicated by letters: f for filament, h for heater, k for the cathode, g_1 for the first grid, g_2, g_3 etc. for successive grids, a for the anode and so on.

It is sometimes necessary to refer to the voltages of the various electrodes or the currents flowing to them. This is done by using the electrode letters as subscripts. Thus, V_a stands for anode voltage, V_{g2} for the voltage at the second grid, and I_k for the cathode current. A detailed list of symbols used is given at the end of this book.

CHAPTER XI

Classification and functions of Radio Valves

The valves employed in radio receivers are usually classified according to the number of electrodes they contain, the heater of an indirectly heated valve not being included in that number. The number of electrodes is intimately related to the functions which a particular valve performs in a receiver and this function is indicated in addition to the number of electrodes in the valve nomenclature.

51. Diodes

A valve having only two electrodes, that is to say one cathode and one anode and no control electrode, is termed a diode. Sometimes two separate diodes are contained in one and the same envelope, and the valve is then called a **double diode.** Usually a double diode has only one cathode, arranged either horizontally or vertically, and the two anodes take the form of two superposed cylinders surrounding the cathode as illustrated in Fig. 113. In other cases, as for example in diodes used for rectifying the mains alternating current, different forms of construction are employed (see Fig. 114).

Because a diode has no control grid it can only be used for rectification, that is to say for changing an alternating current into a uni-directional current. This rectifying action is applied in receiving sets for rectifying mains alternating current, thus providing the necessary high-tension supply; for separating the audio-frequency or modulation-frequency component from the R.F. or I.F. signal; and also for producing the direct voltages required for such processes as automatic volume control. Diodes used for rectifying mains voltage are referred to as "rectifiers", diodes used for separating the A.F. signal from the R.F. or I.F. signal are termed detector diodes or merely "diodes".

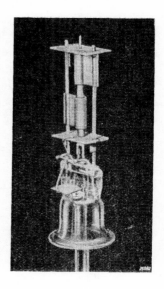

Fig. 113
Electrode system of double diode type CB 2.

117

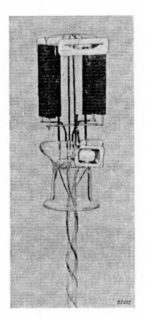

Fig. 114
Internal construction of full-wave rectifier, type AZ 1.

52. Triodes

A valve having one grid interposed between the cathode and anode is called a triode. By virtue of the control action produced by the grid, triodes can be used as amplifiers. In receivers their use is generally confined to lowfrequency amplification or to generating the local-oscillator voltage required in super-heterodyne receivers; other triodes are designed for use as power output valves.

53. Tetrodes

If a second grid is fitted between the control grid and the anode of a triode and is maintained at a positive potential with respect to the cathode, the valve becomes a tetrode, that is to say a four-electrode valve. The second grid is termed a **screen grid** and the first is employed as control grid. Tetrodes have greater possibilities as far as amplification is concerned than triodes, and possess the added advantage that feedback from the anode circuit to the control-grid circuit is substantially reduced by the screen grid. Tetrodes or screen-grid valves find their main application as R.F. or I.F. amplifiers but have now been largely superseded for this application by newer types of valves. Tetrodes are also available as output valves and then are so designed that the electron current is concentrated in the space between the grid and the anode. The resultant high density of the space charge minimises the secondary emission from the anode, as will be explained in Chapters XII and XIII. At one time yet another type of tetrode was used, in which the two grids consisted of a control grid and a space-charge grid located between the control grid and the cathode maintained at a slightly positive potential with respect to the cathode. This arrangement resulted in a larger anode current than would be possible under conditions of negative control-grid voltage and low anode voltage, and the valve gave a useful value of transconductance at very low anode voltages.

54. Pentodes

A pentode is a valve possessing five electrodes, of which three are grids. The first and second grids correspond to the control grid and screen grid of a tetrode, and the third grid, which is located between the screen grid and the anode, is termed the **suppressor grid**; it is usually connected either to the cathode within the bulb or in the base, or direct to earth.

The pentode is one of the most versatile types of valve and is employed in receivers for a large number of processes. There are pentodes specially designed for R.F. or I.F. amplification and these are generally known as R.F. pentodes. Many of these, however, are quite suitable for use as A.F. amplifiers and will give a considerrably greater degree of amplification than can be obtained with triodes, a stage gain of 100 to 200 being possible.

The pentode principle is also applied to the design of power output valves, and output pentodes are now the standard form of output valve for domestic receivers.

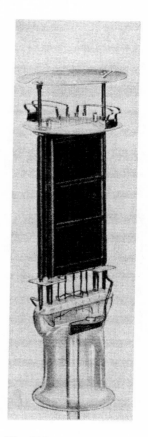

55. Hexodes

A hexode is a valve with six electrodes, four of which are grids. The first and third grids are control grids and the second and fourth are screen grids. Hexodes are employed as mixer valves in superheterodyne receivers, the R.F. signal being applied to the first grid and the local-oscillator voltage, generated in the receiver, being applied to the third grid. Both the R.F.

Fig. 115
Electrode system of directly-heated output triode, type AD 1.

signal (modulated) and local oscillation modulate the electron current flowing to the anode, with the result that the desired modulated I.F. signal is obtained. In suitable circuits hexodes can also be employed as R.F. and I.F. amplifiers.

56. Heptodes

A heptode, or seven-electrode valve, has five grids, the first four corresponding to the four grids of a hexode and the fifth, located between the fourth or screen grid and the anode, being the suppressor grid. Such valves are mainly used as mixing valves in the same way as hexodes but can also be employed as R.F. or I.F. amplifiers.

119

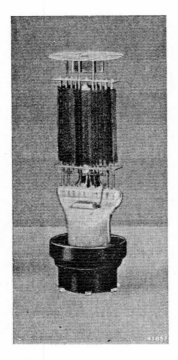

Fig. 116
Electrode system of typical 18-watt output pentode of high transconductance.

A more recent type of heptode (developed just before the war), which may be used for R.F. and A.F. amplification at low anode voltages, has a somewhat different electrode arrangement. The first grid serves to concentrate the electrons, the second grid acts as a space-charge grid, the third as a control electrode, the fourth is a screen grid, and the fifth a suppressor grid. This valve is equivalent to a pentode constructed around a space-charge grid.

57. Octodes

This type of valve has six grids in addition to cathode and anode, and at present is only employed as a frequency converter. The first and second grids function as the control grid and anode of a triode for generating the local-oscillator voltage in a superheterodyne receiver, the first grid being connected to a tuned circuit and the second to a feedback coil. The third and fifth grids are screen grids and the fourth grid is the control electrode to which the incoming R.F. signal is applied. The sixth grid acts as the suppressor.

58. Multiple Valves

It is common practice to combine two or more electrode systems in one envelope. Such valves are termed multiple valves. For example, a double diode can be combined with a pentode or triode, or a triode can be combined with a hexode. In most cases these combined valves have a common cathode and heater or, in the case of directly-heated valves, a common filament or two filaments either series or parallel connected. In valves where the cathode is mounted vertically the two systems are assembled around it, one beneath the other. For example, the bottom half of the cathode forms the cathode of one system and the top half the cathode of the second system. In directly-heated multiple valves the two electrode systems are often found mounted side by side, with a separate filament for each, the two filaments being connected inside the valve either in series or in parallel.

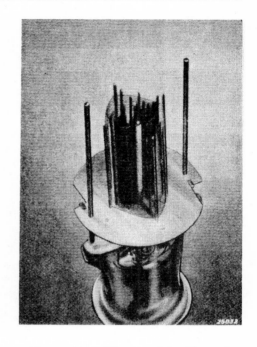

Fig. 117
Electrode system of an octode. The sixth grid and anode have been removed to show the oscillator grid, the two anode plates and the shielding plate of the oscillator section, grid number four and the diamond-shaped grid number five.

In standard practice multiple valves do not at present contain more than two systems, a double diode being considered as forming one system. The combination of more than two systems in one valve would involve great difficulties in manufacture and would introduce the risk of undesirable coupling between the various systems within the valve. Moreover, in the event of the failure of one system the complete valve would have to be replaced and this would be a costly matter. The combination of a double diode with an R.F. triode amplifier is called a **double-diode-triode.** The diodes are used as detector and for automatic volume control, while the triode is used as an A.F. amplifier with resistance coupling. Another common combination is the double-diode-output-pentode, the name of which is self-explanatory. There are also double-diode—R.F.-pentodes in which the pentode may be employed, for example, as I.F. amplifier in super-heterodyne receivers while the two diodes are used respectively for detection and automatic volume control.

The object of combining a triode with a hexode or heptode is to produce a valve

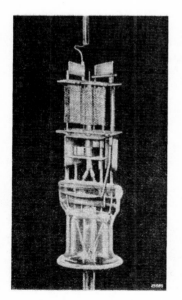

Fig. 118
Internal construction of a typical double diode-triode. The two cylindrical diode anodes are arranged below the triode assembly, from which they are separated by a shielding plate.

121

which performs the functions of local oscillator and mixer valve in superheterodyne receivers. The triode system forms the oscillator and the hexode or heptode forms the mixer. Triode-hexodes and triode-heptodes therefore perform the same functions as octodes as frequency-converters.

Another multiple valve sometimes encountered is a combination of A.F. triode and output tetrode for the output stage.

Bibliography Concerning Frequency Modulation

1) Carson: Notes on the Theory of Modulation, Proceedings of the Institute of Radio Engrs., *10*, (1922), pp. 57—83.

2) Balth. van der Pol: Frequency Modulation, Tijdschr. v. h. Ned. Radiogenootschap, *4*, (1929), pp. 57—70, in Dutch.

3) Th. J. Weijers: Recent Developments Concerning Frequency Modulation, Tijdschr. v. h. Ned. Radiogenootschap, *8*, (1940), pp. 315—364, in Dutch.

4) H. Roder: Ueber Frequenzmodulation, Telefunken Zeitung X, *53*, (1929), pp. 48—54, in German.

5) Armstrong: A Method of Reducing Disturbances in Radio Signaling by a System of Frequency Modulation, Proc. I.R.E., *24*, (1936), pp. 689—740.

6) Carson and Fry: Variable Frequency Electric Circuit Theory, Bell System Technical Journal, *16*, (1937), pp. 513—540.

7) Carson: Theory of the Feedback Receiving Circuit, Bell System Technical Journal, *18*, (1939), pp. 395—403.

8) J. L. Chaffee: The Application of Negative Feedback to Frequency Modulation Systems, Proc. I.R.E., *27*, (1939), pp. 317—331 and Bell System Techn. Journal, *18*, (1939), pp. 404—437.

9) Th. J. Weijers: Frequency Modulation, Philips Technical Review, *8*, (1946), p. 42.

10) Th. J. Weijers: Comparison of Frequency Modulation and Amplitude Modulation, Philips Technical Review, *8*, (1946), p. 89.

CHAPTER XII

Properties of the Valves

A valve has properties which can be defined partly by characteristics or a family of curves. If possible these characteristics should be such that the properties of the valve can be read from them for different working conditions. In the case of voltage-amplifying valves it is particularly of importance to know how great the amplification of the valve is in certain circumstances. For output valves it is important to know what power the valve can supply to the loudspeaker. Voltages, usually alternating, are applied to the input of a valve. In the case of **voltage amplification** it is desired to know what alternating voltage is generated at the anode with a certain anode impedance and grid alternating voltage, whereas for **output valves** it is desired to know the **power** that the valve can supply with the given grid alternating voltage. This power can be calculated from the alternating voltage across the anode load impedance of the output valve. The anode load resistance or impedance of the output valve of a radio receiver is formed by the loudspeaker, which is usually coupled to the output valve via an **output transformer.**

The characteristics of the valve have to be determined in such a way that the data of the valve are independent of the anode impedance, because the latter is a variable quantity, changing from one case to another. It must be possible to determine with the aid of the characteristics the properties and working of the valve for a certain anode impedance. By **working conditions** is understood, for instance, the **adjustment** of the valve, i.e., the various direct voltages applied to the electrodes, and the **anode impedance** that is employed.

59. Anode-Current-Grid-Voltage or Transfer Characteristic

An anode-current-grid-voltage curve is called the I_a/V_g or **transfer characteristic.** In order to obtain such a curve it is necessary to vary continuously the control-grid voltage for given, constant, direct voltages on the anode, on the screen grid (if present) and on any other grids that may be contained in the valve, and to measure the corresponding anode currents; by plotting these currents as a function of the grid voltage a curve is obtained. Usually when constructing such a curve

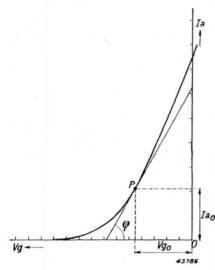

Fig. 119
I_a/V_g or transfer characteristic of a triode. The tangent of the angle φ formed by the tangent through the point P and the horizontal axis (V_g axis) indicates the transconductance of the valve at that point.

one starts from a grid voltage equal to zero and brings this down to the limiting negative value at which the anode current becomes sensibly zero. It is not usual to measure the anode current of receiver valves for positive values of the grid voltage.

Fig. 119 gives a transfer characteristic for a triode.

If, after the transfer curve of a triode has been constructed, a different anode voltage is chosen, then a new I_a/V_g or transfer curve is produced which runs alongside the first one. If measurements are taken for a number of anode voltages, then a **family of transfer characteristics** is obtained, from which a series of properties can be deduced.

Figs 126 and 128 show families of transfer characteristics of triodes.

As will be explained in the next chapter, high anode voltages have only little influence upon the anode current of valves with screen grids. The anode current is then mainly determined, apart from the control-grid voltage, by the screen-grid voltage. For that reason the family of transfer characteristics for screen-grid valves are set out for different values of the screen-grid voltage; these curves, however, hold good only as long as the anode voltage does not drop below a certain limit.

60. Transconductance of the Valve

One of the features of a valve that is of interest is the increase or decrease of the anode current per volt of the control-grid-voltage variation, as this tells us something about the controlling action of the grid. In determining this action it is necessary to keep the anode voltage and the voltages on any other grids constant, as otherwise no distinction can be made between the action of the control grid and that of the other electrodes whose voltages vary. As the I_a/V_g or transfer characteristic of a valve indicates the relation between the anode current and the grid voltage for constant voltages on the other electrodes, it is possible to read from that characteristic the anode-current variation per volt of the control-grid voltage. This quotient gives us a certain number. If, however, the quotient of the anode-current variation and the grid-voltage variation is determined for a grid-voltage variation smaller or greater than one volt, then a somewhat different figure is obtained. It is customary to use as characteristic figure the quotient of a very small anode-current variation dI_a and the corresponding very small grid-voltage variation dV_g. For a triode this gives us the expression $\left(\dfrac{dI_a}{dV_g}\right)_{V_a\,=\,\text{const.}}$.

If a tangent is drawn through a certain point of the I_a/V_g curve (see Fig. 119) then the angle contained within that tangent and the V_g axis gives the quotient $\dfrac{dI_a}{dV_g}$.

The greater the quotient, the steeper the tangent with respect to the horizontal axis; this quotient is called the **transconductance** or **mutual conductance** of the valve at the given point (indicated by the grid voltage or the anode current from which one departs

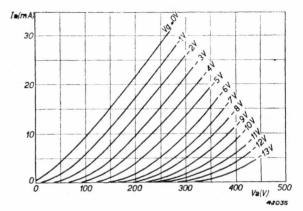

Fig. 120   Plate characteristics of a triode.

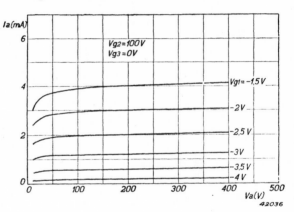

Fig. 121   Plate characteristics of a pentode.

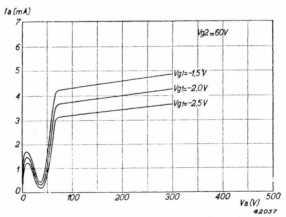

Fig. 122   Plate characteristics of a tetrode.

125

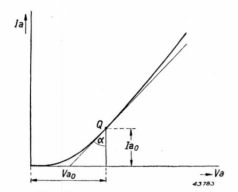

Fig. 123
Plate characteristic of a triode. The tangent of the angle α formed by the tangent to this curve through the point Q and the I_a axis indicates the internal resistance of the valve at this point of the characteristic.

for a certain anode voltage). From Fig. 119 it is further seen that the transconductance of the valve is not equal for all grid voltages, but generally increases as the negative grid voltage becomes smaller. This transconductance of the valve will be indicated by the symbol g_m, according to the definition

$$g_m = \left(\frac{dI_a}{dV_g}\right)_{V_a = const.} \quad (34)$$

Contrary to the foregoing, however, the transconductance of the valve is often measured by varying the grid potential 1 volt and determining the corresponding variation of the anode current. This gives the quotient $\dfrac{dI_a}{dV_g}$ by approximation. From the above it follows that the transconductance has the dimension of a **conductance** (the reciprocal of a resistance, whence the term transconductance). It is generally expressed in terms of mA/V or μA/V.

61. Anode-Current-Anode-Voltage or Plate Characteristics

An anode-current-anode-voltage or plate curve is termed an I_a/V_a or **plate characteristic** (plate was the original name for anode). Such a curve is constructed by measuring for every anode voltage the corresponding anode current at a given control-grid voltage and given constant voltages on any other grids present in the valve. By plotting such plate characteristics for various values of the control-grid voltage (any screen-grid voltage remaining constant) a **family of plate characteristics** is obtained as shown in Fig. 120 for a triode, in Fig. 121 for a pentode and in Fig. 122 for a tetrode. For screen-grid and pentode valves there is a family of characteristics for each screen-grid voltage.

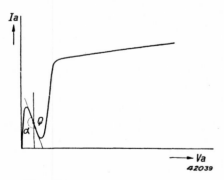

Fig. 124
Plate characteristic of a tetrode. Within a certain range of anode voltage the angle α is greater than 90° and the internal resistance negative.

126

62. Internal Resistance

While on the one hand it is of interest to know the anode-current variation for a certain control-grid voltage variation, on the other hand it is also of importance to know how great the anode-current variation is for a certain anode-voltage variation. This knowledge tells us something about the influence of the anode voltage upon the anode current at a constant control-grid voltage and with cons'ant voltages on any other grids. If, for instance, the control-grid voltage in a certain valve circuit is changed then, as a rule, in consequence of the resulting anode-current variation, the anode voltage also will change (in the case, for example, when the voltage drop across an anode series resistance is changed). The plate characteristic indicates the relation between the anode current and the anode voltage for constant voltages on the other electrodes; thus one can read from this characteristic the anode-current variation per volt of anode-voltage variation. This quotient, the anode-voltage variation divided by the corresponding anode-current variation, yields a certain figure. As in the case of transconductance, for a given point of the plate curve, the quotient of a very small anode-voltage variation dV_a divided by the corresponding very small anode-current variation dI_a is usually taken as the characteristic figure. This gives us the expression $\left(\dfrac{dV_a}{dI_a}\right)_{V_g \,=\, \text{const.}}$.

If a tangent is drawn through a certain point of the plate characteristic (see **Fig. 123**) then the angle contained within that tangent and the I_a axis yields the quotient $\dfrac{dV_a}{dI_a}$. This quotient has the dimension of a resistance; it is the **alternating-current resistance** between anode and cathode and is usually called the **internal resistance** of the valve, which is expressed in ohms, kilohms or megohms. This alternating-current resistance or internal resistance of the valve is not to be confused with the **direct-current resistance**, which is calculated from the quotient of the anode direct voltage corresponding to the working point and the anode direct current.

The larger the angle a (see **Fig. 123**), the greater the internal resistance. With $a = 90°$ (thus when the tangent is horizontal) the internal resistance is infinitely great, and when a is greater than $90°$ the internal resistance is negative. This is the case with tetrodes within a certain range of anode voltage (see **Fig. 124**). In the normal design of receiving sets this negative internal resistance—which is a result of the secondary emission of the anode (see Chapter XIII)—is not utilised.

Usually the internal resistance is indicated by R_a. Its definition is therefore:

$$R_a = \left(\frac{dV_a}{dI_a}\right)_{V_g = \text{const.}} .\tag{35}$$

From the flatness of the plate characteristics of a pentode in a certain anode-voltage range (see Fig. 121) it may be deduced that in that range the internal resistance of the pentode is very great. It is very difficult to determine this high internal resistance exactly from I_a/V_a curves. When the anode voltage of pentodes is reduced to low values the internal resistance diminishes rapidly.

The internal resistance of a valve can also be determined from its family of transfer characteristics. For the point P in Fig. 126 this amounts to:

$$R_a = \frac{V_{a3} - V_{a2}}{I_{ao1} - I_{ao}} .$$

Since in a pentode the internal resistance is very great, it follows that for a variation of the anode voltage from V_{a3} to V_{a2} the anode current variation I_{ao1}-I_{ao} must be small. This implies that the transfer characteristics of a pentode for different anode voltages lie very close together, and the anode-current deviations between the various characteristics are often so small as to fall within the line thickness of a transfer characteristic.

63. Amplification Factor

According to Formula (34) the anode-current variation of a valve when the grid voltage is changed is determined by $dI_a = g_m dV_g$. According to Formula (35) the anode-current variation when the anode voltage is changed is determined by $dI_a = \frac{dV_a}{R_a}$. If, now, there is simultaneously a grid-voltage variation and an anode-voltage variation, the influences of both variations on the anode current can be added and the resulting anode-current variation is to be represented by

$$dI_a = g_m dV_g + \frac{dV_a}{R_a} .\tag{36}$$

If dI_a is made zero, then we have:

$$g_m dV_g = -\frac{dV_a}{R_a},$$

or

$$-\frac{dV_a}{dV_g} = g_m R_a .\tag{37}$$

The ratio $-\dfrac{dV_a}{dV_g}$ between the small increase of the anode voltage and the small decrease of the grid voltage, which neutralizes the anode-current variation caused by the anode voltage, is called the **amplification factor** of the valve, and this is represented by the symbol μ.

The amplification factor is, therefore, the ratio of the anode-voltage variation to the grid-voltage variation which bring about one and the same anode-current variation. The anode-voltage variation required to change the anode current say 1 mA is very much greater than the requisite grid-voltage variation. The effect of the grid-voltage variation is consequently many times greater than that of the anode-voltage variation. From the given definition and Formula (37) it follows that the amplification factor is equal to the transconductance multiplied by the internal resistance:

$$\mu = g_m R_a = \left(\frac{dI_a}{dV_g}\right)_{V_a = const.} \times \left(\frac{dV_a}{dI_a}\right)_{V_g = const.} = -\left(\frac{dV_a}{dV_g}\right)_{I_a = const.} \tag{38}$$

The amplification factor may therefore be regarded as the ratio of an anode-voltage variation causing a certain anode-current variation to the grid-voltage variation required to restore the changed anode current to its original value. The amplification factor can be determined either from the family of plate characteristics or from the family of transfer characteristics. The data for a constant anode current lie on a horizontal line in the I_a/V_a diagram, and from Fig. 125 it appears that V_g may drop from —3 to —4 V if the anode voltage is raised simultaneously from 185 to 215 V. From this it follows that the amplification factor of this valve under the assumed conditions is equal to $\mu = \frac{30}{1} = 30$.

For a pentode it is difficult to determine exactly the amplification factor from the family of plate characteristics, because the curves are nearly horizontal. It follows from this, however, that the amplification factor must be very great.

The amplification factor

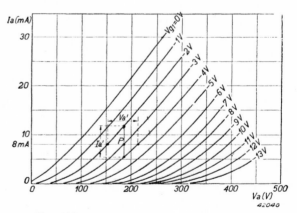

Fig. 125
Family of I_a/V_a or plate characteristics of a triode. When the negative grid bias at the working point ($I_a = 8$ mA and $V_a = 185$ volts) is increased from —3 V to —4 V the anode voltage has to be raised to 215 V in order to restore the anode current to its initial value.

of a valve indicates the theoretical limiting value of amplification (anode alternating voltage divided by the grid alternating voltage), but it is never possible to reach this value in practice (see Section 68). The amplification factor can also be determined from the family of transfer characteristics (see Fig. 126) by drawing a horizontal line through the point P. The amplification factor is then $\mu = \dfrac{V_{a3} - V_{a2}}{V_{go1} - V_{go}}$. The reciprocal value of the amplification factor is sometimes symbolized in radio-engineering literature by the letter D ("Durchgriff") and is expressed as a percentage. In the case of a valve with an amplification factor of 25 the reciprocal value is therefore $\frac{1}{25} \times 100\% = 4\%$. The formula from which this quantity can be found follows from Equation (38):

$$g_m R_a D = 1. \tag{39}$$

64. Relation between the Families of Transfer and of Plate Characteristics

The family of transfer characteristics and that of plate characteristics are closely related and the one may be deduced from the other. Let us take the family of plate characteristics in Fig. 127. By drawing a vertical line for the 250-V (constant) anode voltage it is possible to derive all the data required to plot a transfer characteristic. Each point of intersection of the vertical line and a plate characteristic gives a certain grid voltage and the corresponding anode current. By repeating this for other anode voltages one obtains a family of transfer characteristics.

Inversely it is also possible to construct a family of plate characteristics from a family of transfer characteristics. If a vertical line is drawn through the latter (in Fig. 128 through $V_g = -2$ V) a number of points of the plate characteristic will lie on that vertical line. A family of plate characteristics is obtained when this process is repeated for other negative grid voltages.

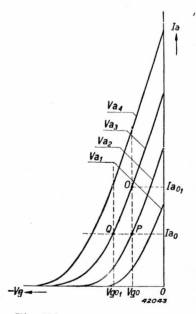

Fig. 126
I_a/V_g or transfer characteristics of a triode. From these curves it is possible to determine the amplification factor and the internal resistance.

65. Various Kinds of Couplings

As has already been mentioned in passing, a suitable **coupling element** has to be introduced in the anode circuit in order to convert the anode-current variations caused by the grid alternating voltage into voltage variations. For this purpose either resistances are used or else a more complicated arrangement is employed.

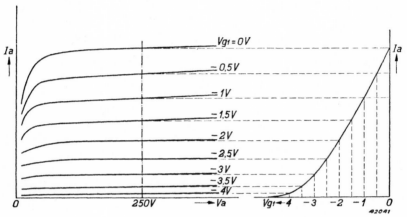

Fig. 127
Relation between the plate characteristics and the transfer characteristic for a pentode having $V_a = 250$ V.

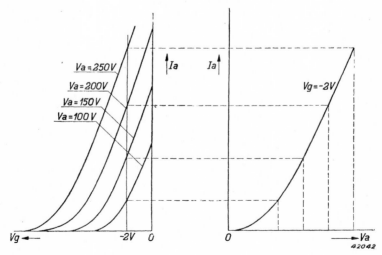

Fig. 128
Construction of a plate curve of a triode from the family of transfer characteristics.

131

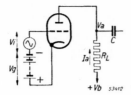 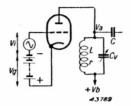

Fig. 129
Fundamental circuit dia-
gram of a resistance
coupling.

Fig. 130
Anode coupling circuit using
a composite impedance (tuned
circuit).

Fig. 129 gives a diagrammatic representation of a **resistance coupling.**
In the anode circuit a resistance R_L is introduced, across which voltage
variations are set up; these are applied via the condenser C to the grid
of the next amplifying valve. V_i is the grid alternating voltage and V_b
the voltage of the anode-power-supply source. The point $+V_b$ is
earthed [1]) for high frequency or low frequency, via the supply source.
As a consequence of the voltage drop in the anode resistance caused by
the anode direct current, the direct voltage V_a on the anode is lower
than the voltage V_b of the supply source. This voltage drop will
diminish when the grid voltage becomes more negative. The phase of
the alternating voltage across R_L is therefore shifted 180° with respect
to the phase of the alternating voltage at the valve input.

Fig. 130 shows how a composite impedance may be introduced into
an anode circuit. In place of the resistance R_L we have a coil
(inductance) and a condenser connected in parallel. If the inductance
of this tuned circuit is L and the capacity C_v, while the coil has an
ohmic resistance r, then the impedance of this circuit at its resonant
frequency is $\frac{L}{rC_v}$. Since the ohmic resistance of the coil is comparatively
low, the constant voltage drop across the coil is usually negligible, and
so $V_{ao} = V_b$. Due to the anode-current variations there occurs across
the impedance of this circuit an alternating voltage which is superimposed upon the mean value $V_{ao} = V_b$. Owing to the fact that for
frequencies other than the resonant frequency the impedance falls, these
frequencies will be amplified to a smaller extent. This accounts for the
selectivity of the amplifier.

Another method of coupling is shown in Fig. 131. In the anode circuit
an inductance is introduced which has a high reactance at the frequency
to be amplified. In order to ensure a sufficiently high reactance for

---

[1]) i.e. connected effectively to the cathode.

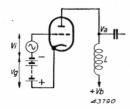

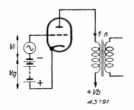

Fig. 131
Anode coupling circuit
using a choke coil.

Fig. 132
Coupling of an anode circuit by means
of a transformer (n is the ratio of the
transformer).

low frequencies this inductance has an iron core when used in a L.F.
amplifier. Such a coupling is called a **choke coupling.**
In the case of **transformer coupling,** sometimes applied in L.F. amplifiers, the inductance of the primary winding may be regarded as a
choke. Here is therefore another form of choke coupling in which the
alternating voltage occurring across the primary winding is stepped up
or down by transformer action, depending upon the ratio.

66. Representation of the Anode Resistance in a Family of Plate Charac- teristics

When a resistance is introduced in the anode circuit, as illustrated in
Fig. 129, the current flowing through it causes a voltage drop equal to
$I_a R_L$. The difference between the supply voltage V_b and the anode
voltage V_a is thus equal to the voltage drop $I_a R_L$, so that the following
equation is obtained:

$$V_b - V_a = I_a R_L \tag{40}$$

or

$$V_a = V_b - I_a R_L. \tag{41}$$

This equation may be written in the form $I_a = f(V_a)$ and is represented
on a family of plate characteristics by a straight line, usually called
the **load line** or **resistance line.** It must intersect the V_a axis in the point
$V_a = V_b$, since in that case $I_a = 0$, and the I_a axis in the point
$I_a = \dfrac{V_b}{R_L}$, because $V_a = 0$ (see Fig. 135). When R_L is of a very high
value the line is very flat, whereas for low values of R_L it is steep.
When $R_L = 0$ the line is vertical, because then the anode voltage is
always equal to V_b irrespective of the value of I_a.
With the aid of the plate curves and a load line it is easy to determine
the anode current I_a and the anode voltage V_a to which the valve
adjusts itself for given values of the supply voltage V_b, the anode

resistance R_L and the negative grid voltage V_g. These values correspond to the point where the resistance line intersects the curve corresponding to the given value of V_g. For every value of V_g a different value of V_a is found, and from the anode-voltage variation caused by a certain grid-voltage variation it is possible to determine the amplification.

67. Equivalent Circuit for an Amplifying Valve

When a valve is controlled by an alternating voltage on the grid it can be regarded as an alternating-current source or as an alternating-voltage source (on the anode side), with certain properties. This makes it possible to calculate the amplification and other values in a simple manner. Whether a current or voltage source is considered is purely a matter of convenience.

(a) The Valve as Current Source

In the case of a valve with resistance coupling as indicated in Fig. 129, as a consequence of the alternating voltage V_g on the grid, the anode current and the anode voltage undergo periodical changes. The anode alternating current which results from the grid alternating voltage V_g alone must, according to Equation (34), be equal to $g_m V_g$. The anode alternating voltage produces an anode alternating current which, as can be deduced from Equation (35), is equal to $\dfrac{V_a}{R_a}$. Adding together the influences of V_g and V_a we obtain a resulting anode current

$$I_a = g_m V_g + \frac{V_a}{R_a}. \tag{42}$$

Through the valve and the anode series resistance R_L an anode current I_a flows and as a result of the existence of R_L an alternating voltage V_a is produced on the anode which is phase shifted by 180° with respect to V_g and I_a. When the grid voltage increases (in positive direction) the anode voltage decreases. Consequently, V_a is equal to $-I_a R_L$. By substituting $-I_a R_L$ for V_a in Equation (42), we obtain

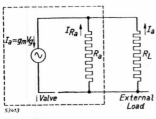

$$I_a = g_m V_g - \frac{I_a R_L}{R_a}, \tag{43}$$

or

$$I_a = g_m V_g \frac{R_a}{R_a + R_L}. \tag{44}$$

Fig. 133
Equivalent circuit diagram of the valve where this is regarded as a current source $g_m V_g$ with an internal resistance R_a connected parallel. The part within the broken-line rectangle represents the valve and the part outside it the external load.

This formula represents that part I_a of the current produced by a current source $g_m V_g$ which flows through the resistance R_L when two resistances R_a and R_L are joined in parallel and connected to that current source. Consequently, as represented in Fig. 133, a valve controlled by a grid alternating voltage V_g can be regarded as a current source, $g_m V_g$, with a resistance R_a connected parallel and feeding an external resistance R_L.

(b) The Valve as Voltage Source

In Equation (44) I_a may be replaced by $-\dfrac{V_a}{R_L}$ and we then obtain:

$$-\frac{V_a}{R_L} = g_m V_g \frac{R_a}{R_a + R_L} \tag{45}$$

or

$$V_a = -g_m R_a V_g \frac{R_L}{R_a + R_L}. \tag{46}$$

Now by subsituting μ for $g_m R_a$ we obtain

$$V_a = -\mu V_g \frac{R_L}{R_a + R_L}. \tag{47}$$

In this formula V_a is that part of the voltage supplied by a voltage source μV_g that obtains across the resistance R_L when there are two resistances R_a and R_L in series with this voltage source. Therefore, as indicated in Fig. 134, a valve controlled by the grid alternating voltage may be regarded as a voltage source μV_g with an internal resistance R_a connected in series and feeding an external resistance R_L.

The Formulae (44) and (47) given above may also be derived in the following way: If a valve controlled by a grid alternating voltage is considered as a system with two terminals we can determine the properties of that system by experimenting with the terminals. When the two terminals are not connected (open circuit), the voltage can be measured, and when they are short-circuited the current can be measured. Now the internal resistance of a system equals the quotient of the alternating voltage when not loaded ($R_L = \infty$) and the short-circuit current ($R_L = 0$). For $R_L = \infty$ the anode alternating current

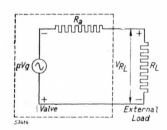

Fig. 134
Equivalent circuit diagram of a valve where this is regarded as a voltage source μV_g, with an internal resistance R_a connected in series. The part within the broken-line rectangle represents the valve and the part outside it the external load.

equals zero and apparently therefore the anode alternating voltage, according to Equation (42), equals $-\mu V_g$ (bearing in mind that $g_m R_a = \mu$). If the valve is short-circuited for alternating current on the anode side then the short-circuit current equals $g_m V_g$. The quotient of no-load voltage and short-circuit current is thus

$$\frac{\mu V_g}{g_m V_g} = R_a.$$

Regarded in this way, therefore, the valve represents a voltage source with an e.m.f. equal to μV_g and an internal resistance R_a connected in series. If such a voltage source is connected to a resistance R_L, then it is known that the part of the voltage supplied by the e.m.f. that exists across R_L equals $\mu V_g \dfrac{R_L}{R_a + R_L}$. Further, any voltage source with an e.m.f. equal to V_0 and an internal resistance R_a can be replaced by a current source $I_0 = \dfrac{V_0}{R_a}$ with a parallel connected resistance R_a [1]). Substituting for V_0 the voltage μV_g, we find the value of the current I_0 to be $\dfrac{\mu}{R_a} V_g = g_m V_g$. By this means we arrive at the Formula (47), since, as is known, that part of the current supplied by a current source $g_m V_g$ which flows through the resistance R_L is equal to $g_m V_g \dfrac{R_a}{R_a + R_L}$ [2]), when R_L and R_a are connected in parallel with each other and to the current source.

68. Dynamic Transconductance and Dynamic Characteristics

A grid-voltage variation results in a certain anode-current variation. The latter is governed by the product of the transconductance and the grid-voltage variation. This applies only so long as the anode voltage remains constant.

When a load resistance is introduced in the anode circuit the anode voltage also varies with the anode current. As a result the anode-current variation that would occur without a load resistance in the anode circuit is reduced. In this case, too, the quotient of the anode-current variation and the grid-voltage variation can be determined, and this quotient is called the **dynamic transconductance**. The curve representing the anode current for a certain anode load resistance as a function of the grid voltage is termed the **dynamic transfer characteristic**. In the case

---

[1]) See Appendix II, 5.
[2]) See Appendix II, 4.

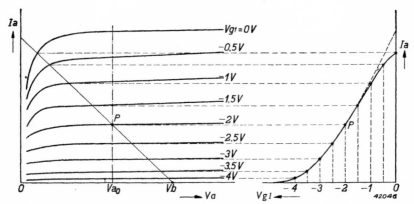

Fig. 135
Family of plate characteristics of a pentode with resistance line. This line represents
the relation between I_a and V_a due to the resistance in the anode circuit when the valve
is fed by the voltage source V_b via the resistance. In the right-hand part the full line
is the dynamic transfer characteristic constructed from the family of plate characteristics
with the aid of the resistance line; the broken-line curve represents the static transfer
characteristic for an anode voltage V_{a0} and the working point P.

of a pentode, above a certain minimum anode voltage there is only
a little difference between the dynamic transfer characteristic and the
static transfer characteristic previously discussed, because with this
type of valve the anode voltage has little influence upon the anode
current.

Fig. 135 shows that the bottom part of the dynamic transfer characteristic of a pentode coincides with that of the static characteristic,
and that the upper part of the dynamic characteristic is less steep;
only in the latter part the transconductance is less than that in the
case of the static curve. Over the corresponding parts of the anode
characteristics the internal resistance of the pentode is small compared
with the external resistance. In the range within which the internal
resistance is large in comparison with the external resistance, however,
the dynamic transconductance is equal to the static transconductance.

For triodes this is not the case, as seen from Fig. 136, because the
external resistance of these valves is normally much larger than the
internal resistance, and consequently the dynamic transconductance
is everywhere greater than the static transconductance. Fig. 136 shows,
however, that the dynamic transfer characteristic of a triode is much
straighter and that the top curvature found with pentodes is absent.

With the aid of Equation (47) in the previous section the dynamic
transconductance can be easily calculated from the data of a valve.
As defined above, the dynamic transconductance g_d is the quotient

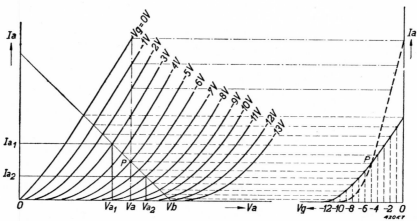

Fig. 136
Family of plate characteristics of a triode and resistance line. In the right-hand part the full line represents the dynamic transfer characteristic derived, whilst the broken line gives the static transfer characteristic for the anode voltage V_a corresponding to the working point P.

of the anode current flowing through the external resistance R_L divided by the grid alternating voltage which produces that current. From Equation (44) it follows that

$$g_d = g_m \frac{R_a}{R_a + R_L}. \tag{48}$$

The dynamic transconductance is thus less than the static transconductance g_m since $\dfrac{R_a}{R_a + R_L}$ will always be less than unity. However, when R_a is very great compared with the external resistance R_L, the fraction $\dfrac{R_a}{R_a + R_L}$ is almost equal to unity and it can then be assumed that the dynamic and static transconductances are the same. This is the case with pentodes that have a high internal resistance; Fig. 135 shows that in a certain range the dynamic and static transfer characteristics coincide. With triodes, since their internal resistance is low, the dynamic transconductance generally deviates considerably from the static transconductance; when, for instance, R_a and R_L are equal, the dynamic transconductance is half the static transconductance.

The calculation of valve amplification is simplified by using the dynamic transconductance. The anode alternating current passing through the load resistance produces across the latter an alternating voltage equal to $I_a R_L$. By definition the amplification A is equal to $\dfrac{V_a}{V_g}$ [1]).

Since $V_a = I_a R_L$ we find, with the aid of Equation (47),

$$V_a = V_g\, g_d R_L = V_g\, g_m \frac{R_a R_L}{R_a + R_L}. \tag{49}$$

From this we obtain for the amplification:

$$A = \frac{V_a}{V_g} = g_m \frac{R_a R_L}{R_a + R_L}. \tag{50}$$

When R_a is large compared with R_L this formula is simplified to $A = g_m R_L$ (this applies to pentodes) [2]). The amplification can, however, also be determined with the aid of Equation (47), according to which it must be equal to:

$$A = \frac{V_a}{V_g} = \mu \frac{R_L}{R_a + R_L} \tag{51}$$

(μ = amplification factor of the valve). Bearing in mind that $\mu = g_m R_a$ and replacing μ in Equation (51) by $g_m R_a$ we arrive again at Equation (50). If it is desired to determine from the normal valve data the amplification with resistance coupling, it must not be overlooked that the voltage drop in the anode resistance may be rather large and consequently the normal published static data do not hold good. The anode direct voltage is much lower than the anode voltage indicated in the published valve data when the anode is fed via a coupling resistance from a H.T. source [3]) with a voltage equal to the published static anode voltage of the valve. As a consequence the anode direct current and the transconductance at the working point are also much smaller. In such a case, therefore, suitable static valve data must be available in order to calculate the correct amplification. Generally speaking it is therefore better to use the families of plate or transfer characteristic, which give the characteristic values of the valve over a very wide range. From Equation (51) it also follows that when R_L is very large compared to R_a and the fraction $\dfrac{R_L}{R_a + R_L}$ is therefore practically equal to unity, the amplification of the valve is nearly equal to μ. Thus the amplification factor indicates the upper limit of amplification obtainable.

---

[1]) Here attention is no longer paid to the signs of V_g and V_a.

[2]) The fraction $\dfrac{R_a}{R_a + R_L}$ is then equal to unity, so that $\dfrac{R_a R_L}{R_a + R_L}$ equals R_L.

[3]) H.T. source stands for high-tension supply source.

69. Alternating-Current Resistance in the Anode Circuit

As already observed, in the anode circuit there is often an impedance, for example a tuned circuit, that has a very low d.c. resistance but a considerable a.c. resistance. At the resonant frequency a tuned circuit acts as a pure resistance and its impedance can therefore be represented by a straight line in a family of plate characteristics. Owing to the fact that the ohmic resistance of the circuit is low and thus can be disregarded, there is practically no voltage drop in the circuit. Consequently the a.c. resistance line must pass through the working point P, which is determined by the anode voltage, i.e. by the voltage of the supply source, and by the negative grid bias (see Fig. 137). The slope of the a.c. resistance line is determined by the value of the resonant resistance $\left(\dfrac{L}{rC}\right)$ of the tuned circuit.

For a given value of anode-load impedance, the instantaneous values of I_a and V_a corresponding to a certain V_g can be found by determining the points of intersection of the load line and the curves. A variable grid voltage V_g is represented by a point which moves along the load line.

In the case of a valve operated with maximum swing of the control voltage, i.e., where the grid carries an alternating voltage that varies between the starting point of grid current and the value of the grid voltage which corresponds to zero anode current, it is seen that the anode voltage varies between a minimum value and almost double the value of the anode direct voltage; thus the anode voltage may considerably exceed the value of the direct supply voltage, a fact that has to be taken into account when applying high anode direct voltage (breakdown voltage).

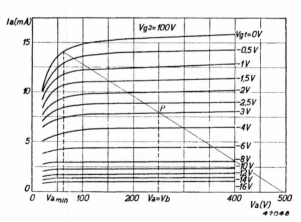

Fig. 137
Family of plate characteristics of a pentode; the resistance line through the working point P represents the resistance at resonance of the tuned circuit inserted in the anode circuit.

Given the impedance in the anode circuit, then, just as is the case for a resistance coupling, the

140

amplification can be calculated with the aid of the Formulae (50) and (51). In contrast to the resistance coupling, however, it is nearly always possible in this case to use directly the published static characteristic data, since these generally correspond to the working voltages applied.

If the anode impedance is not a pure resistance, thus if a tuned circuit is used as an impe-

Fig. 138
Family of plate characteristics of a pentode with ellipsoidal load lines for various grid-alternating-voltage amplitudes; these lines represent the current-voltage relation when an impedance is placed in the anode circuit which causes a phase shift between current and voltage. These curves have been recorded by means of a Philips cathode-ray oscillograph.

dance for a frequency other than the resonant frequency, the relation between anode current and anode voltage can no longer be represented by a simple resistance line, because phase shift takes place between current and voltage. Where there is a capacitive load the voltage lags behind. In that case, when a sinusoidal alternating voltage is applied the locus indicating the relation between anode current and anode voltage will be ellipsoidal (see Fig. 138). Such an ellipsoidal load line is obtained, for instance, when a L.F. transformer is used as the coupling element or in the case of the output valve. Where there is maximum control, however, an undistorted ellipse will never be formed, because then the voltages and currents occurring are not purely sinusoidal.

70. The Power in the Anode Circuit

If a resistance is introduced in the anode circuit then, in consequence of the anode alternating current, an alternating voltage will occur across it. And if the anode alternating current and voltage are sinusoidal and in phase the alternating-current power dissipated in the load resistance will be equal to half the product of the amplitudes of the anode alternating voltage and the anode alternating current. If the grid voltage in Fig. 139 is varied between the points Q and R then the anode-current amplitude will be $I_{a\,max}$ and the anode-voltage amplitude $V_{a\,max}$. The a.c. power supplied is then equal to $\dfrac{I_{a\,max} \cdot V_{a\,max}}{2}$.

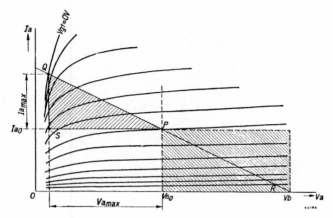

Fig. 139
Plate characteristics of a pentode with the straight line
representing the voltage-current relation resulting from
the introduction of a resistance in the anode circuit.
If the grid swing extends from Q to R the power delivered
to the external resistance is equal to the area of the
triangle PQS.

This power can be found directly from the family of plate curves; the area of the triangle PQS is a measure of the a.c. power delivered. This power varies for different values of the load resistance, so that there will be one particular resistance which gives maximum a.c. power transfer [1]). The resistance also absorbs d.c. power, which is equal to $I_{ao} \times (V_b - V_{ao})$ and is represented in Fig. 139 by a rectangle with $(V_b - V_{ao})$ as base and I_{ao} as height. The a.c. power supplied is of interest almost exclusively for output valves. The anode circuit of an output valve is connected to the loudspeaker via a transformer which steps down the voltage (see **Fig. 140**). If the loudspeaker is regarded as a resistance R_l then, disregarding the transformer resistance, the a.c. resistance in the anode circuit is equal to n^2R_l (n representing the ratio of the transformer).

However, as there is practically no direct voltage drop, $V_{ao} = V_b$ and the loss of d.c. power is negligible. The a.c. power drawn from the anode circuit via the transformer sets up a current in the loudspeaker resistance which causes the cone to vibrate. It is obvious that the a.c. power available in the anode circuit increases as the alternating-voltage amplitude at the control grid is made greater. There is, however, a limit to this increase, because the

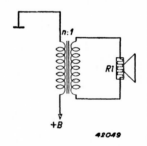

Fig. 140
Example of a loudspeaker connected to the output stage via a transformer.

[1]) These statements are only approximately correct. As will be explained in Chapter XV, distortion of the anode current takes place. In the case of pentodes this causes mainly generation of the third harmonic. When there is 10% third harmonic the fundamental component of the anode current will be about 10% greater than PQ in Fig. 139 (see also Fig. 168). In reality, therefore, the power supplied will be found to be 20% greater than that found by calculation of the area of the shaded triangle.

142

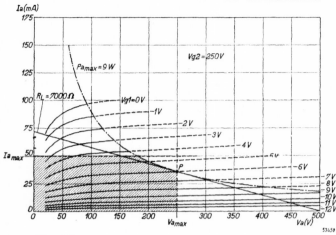

anode-voltage amplitude can never exceed the anode direct voltage available and the amplitude of the anode-current variations can never be greater than the anode d.c. (the anode current cannot be negative). At a certain load resistance the a.c. power that can be delivered is therefore limited by the anode direct voltage and anode direct

Fig. 141

Family of plate characteristics of a pentode in which the limits of the maximum admissible values of anode current $I_{a\,max}$, the anode voltage $V_{a\,max}$, and the anode dissipation $P_{a\,max}$. are indicated.

current, which are determined by the position of the point P.

71. Choice of the Working Point

For an output valve a working point will as a rule be chosen that allows of the highest possible a.c. power being obtained. Owing to various circumstances, however, the choice is subject to limitations. In the first place for every valve a maximum permissible anode direct voltage $V_{a\,max}$ is fixed, which may not be exceeded on account of the danger of breakdown of the insulating parts, due allowance being made for the amplitude of the maximum superposed alternating voltage. The working point must always lie on or to the left of a vertical line drawn through $V_{a\,max}$ (see Fig. 141). Similarly there is also a limiting value of the maximum anode current, which is related to the maximum cathode current $I_{k\,max}$ (emission current of the cathode). Usually the maximum value of the cathode direct current $I_{k\,max}$ is published, and in the fixing of this account has been taken of the direct current to the other electrodes and of the amplitudes of the superimposed maximum alternating currents occurring. Now in the case of a pentode, for instance, there is a certain ratio between the anode current and the screen-grid current which remains practically constant for the whole usable range of gridvoltage. Therefore, given the maximum cathode current, it is possible to determine the maximum admissible anode direct current by using the values of I_a

143

and I_{g2} published for the normal working point. This current value is then found from

$$I_{a\,max} = \frac{I_a}{I_a + I_{g2}} \times I_{k\,max}. \qquad (52)$$

The working point must lie on or below a horizontal line drawn through $I_{a\,max}$ in order not to exceed the maximum cathode current. A third limitation lies in the maximum admissible average anode dissipation, which is related to the heating of the anode ($P_a = I_{ao} V_{ao}$). This anode dissipation is likewise published for every valve. All the working points where this maximum anode dissipation is reached lie on a hyperbola, so the working point may not, therefore, lie to the right of that curve. The shaded area in Fig. 141 indicates the area outside which the working point must not lie.

The boundary lines apply only for the working point. The load line may run partly outside this area and consequently momentary values may have a product greater than $P_{a\,max}$, provided that the product of the average values lies within the limits indicated.

Further account has to be taken of the fact that if the negative grid bias is too small, grid current will begin to flow (in the case of indirectly-heated valves this generally takes place at about —1.3 volts).

The **grid swing**, i.e., the range of usable negative grid voltage, can therefore only be utilised up to this point. Now the working point and the load line of an output valve are chosen with due regard to these factors in such a way that the largest possible power is delivered. Therefore the triangle **PQS** of Fig. 139 has to be chosen as large as possible.

In the case of a valve for voltage amplification the choice of the anode resistance is determined by entirely different considerations. Here the resistance must be of such a value that the largest possible amplification is attained. In this case, therefore, the aim is to get the largest possible ratio of anode to grid alternating voltages. The data published for Philips valves always include the most favourable working point and anode load resistance.

CHAPTER XIII

Action of the Various Grids

An amplifier valve for receiving sets has at least one grid but generally more than one. The grids of an amplifier valve have various functions and consequently are designated by different names. They can be classified under three headings, viz.:

1) control grids,
2) screen grids,
3) suppressor grids.

Every amplifier valve has one control grid and some have more than one. In addition it may have one or more screen grids; when a valve has a screen grid it often has also a suppressor grid. The action of the various grids is briefly explained below.

72. The Control Grid

(a) The Control Voltage

The controlling action of the control grid has been briefly explained in Chapters III and IV, where it was shown that the anode current is related in a certain way to the control voltage or effective potential [in the space-charge range the anode current is approximately proportional to the $^3/_2$ power of the effective potential; cf. also Equation (12) on page 19]. In triodes the effective potential is

$$V_s = p \left(V_g + \frac{V_a}{\mu} \right),$$

where μ represents the amplification factor of the valve and p is a factor related to the dimensions of the valve and is usually only a little less than unity. If $\dfrac{V_a}{\mu}$ is greater than the absolute value of V_g then, notwithstanding the negative control-grid bias, the effective potential is positive and consequently there is an anode current, which is limited by the space charge, and if the control grid is sufficiently negative there will be no electron current (grid current) flowing to that grid.

The effective potential is the mean potential in the plane of the control grid. This potential, however, is not everywhere the same and equal to

145

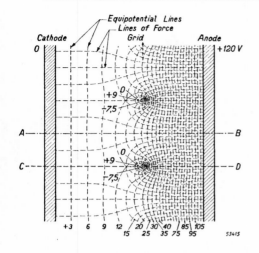

Fig. 142
Equipotential lines and lines of force in the middle part of a triode with flat, parallel electrodes. The field distribution at the extremities of the electrode system are not shown here.

the voltage V_g. In consequence of the potential of the positive anode behind the grid (assuming the anode has a considerably higher potential than that of the grid), the potential between the grid wires is higher than that of the grid wires themselves. In Fig. 142 equipotential lines obtaining for a triode with a negative grid (without space charge) are shown, whilst the curves a and b of Fig. 143 represent the potential, as a function of the distance from the cathode, of the sections A-B and C-D respectively of the field represented in Fig. 142. If in Fig. 142 the variation of the potential is determined in the grid plane from one grid wire to the other, a curve as shown in Fig. 144 is obtained. From this it appears that the potential in the centre between the grid wires is much higher, through the influence of the anode potential, than the potential of the grid wires themselves (in a range a the potential is actually positive). The field of the anode, so to speak, penetrates between the grid wires.

(b) The Island Effect

Fig. 142 shows how the field in the vicinity of the cathode and of the anode is almost perfectly homogeneous, the equipotential lines running parallel to these electrodes. Consequently the field strength at the surface of the cathode is everywhere constant, as is expressed by the angle α in Fig. 143. In valves

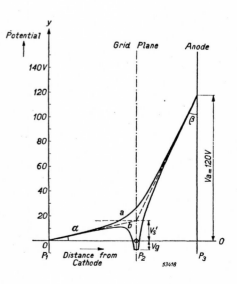

Fig. 143
Curve a: Potential distribution in the section A-B of the field of Fig. 142. Potential as a function of the distance to the cathode.
Curve b: Potential distribution in the section C-D of the field of Fig. 142.

146

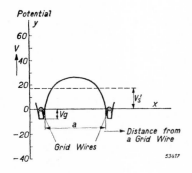

Fig. 144
Potential distribution in the grid
plane of the field of Fig. 142
(potential as a function of the
distance to a grid wire).

where the distance between two grid wires is of the same order of magnitude as or considerably greater than the distance between grid and cathode the field strength at the surface of the cathode will not, however, be everywhere equal. If the meshes of the grid are large then, owing to the anode potential influencing the field-strength distribution as far as the surface of the cathode, this field-strength distribution will not be homogeneous near the cathode, and thus the equipotential lines in the vicinity of the cathode surface will no longer be parallel. Fig. 145 gives the equipotential lines for the case where the distance between the grid wires is of the same order of magnitude as that between the grid plane and cathode. If now the potential distribution is plotted for such a field distribution in the sections A-B and C-D a diagram is obtained as shown in Fig. 146. Fig. 146 shows that the angle a_1 at which the potential-distribution curve of the section A-B (curve a) leaves the cathode surface is greater than the angle a_2 of the curve b for the section C-D through a grid wire. In this example the field strength in that part of the cathode surface opposite the centre of the grid openings is consequently greater than that in the part lying opposite a grid wire. If the grid is strongly negative the field strength in a part of the cathode lying opposite a grid wire may be inversely directed (a_2 negative).

The field strength at the cathode surface can be determined for different values of V_g from the point opposite one grid wire to that opposite the other (see also Fig. 147). From the configuration of the equipotential lines near the surface of the cathode (see Fig. 145) it appears that the field strength is not constant but is as shown, for instance, in Fig. 147b. For $V_g = 0$ the

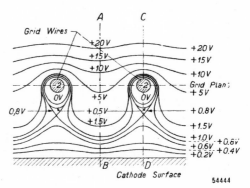

Fig. 145
Equipotential lines in the space between grid and cathode of a triode in which the distance between grid and cathode is of the same order of magnitude as the grid mesh.

147

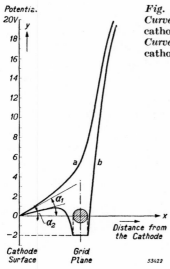

Fig. 146
Curve a: Potential as a function of the distance from the cathode in the section A-B of the field of Fig. 145.
Curve b: Potential as a function of the distance from the cathode in the section C-D of the field of Fig. 145.

field strength distribution is such that the electrons are drawn from the cathode everywhere along the cathode surface between two grid wires. As the negative grid bias V_g becomes greater (see also the curves for $V_g = -10$ V and -20 V) this will not, however, be any longer the case. Then the field strength will only in a certain area of the cathode surface be so directed as to cause the electrons to be attracted away from the cathode (the areas A-A′ and B-B′ in Fig. 147b). Outside that area the electrons are driven back towards the cathode. In the case of the curve for $V_g = -10$ V the average field strength is still negative, but in the case of the curve for $V_g = -20$ V it is positive.

In the small area B-B′, however, the field strength is negative and in that area the cathode still emits electrons. Above a certain negative grid bias the cathode of such a valve will no longer emit uniformly, the emission then taking place only at certain spots or "islands", and one may refer to this as the **"island effect"**. As the negative grid bias increases these islands become smaller and smaller, until finally suppressed. The result of this island effect therefore is that with a positive average field strength at the cathode an anode current may still flow, which would not be the case were it not for this particular effect. Another result of the island effect is that a large negative grid bias is required to reduce the anode current to zero. The island effect will manifest itself to a greater extent as the anode voltage rises.

The influence of the island effect upon the transfer characteristics is illustrated by Fig. 147c where the full lines represent the transfer characteristics for various anode voltages without island effect and the broken lines the corresponding characteristics with island effect.

Owing to the island effect the transconductance at the working point is less and the anode current is greater than would be the case without island effect. Furthermore the characteristics are more curved, so that, especially with large signals, more distortion takes place. For these

148

reasons this phenomenon has to be very scrupulously avoided in the construction of most valves. In some valves, on the other hand, the island effect is desired, as will be seen in Chapter XXII.

73. The Screen Grid

(a) Principle

If a second grid is introduced in a triode between the control grid and the anode and a direct voltage is applied to that second grid which is positive (about 60—250 V) with respect to the cathode, then that grid will act as a screen grid. In most cases this grid has a very small mesh. The screen grid acts as anode for a triode consisting of cathode, control grid and screen grid. Due to the screen grid a positive **potential plane** is formed in the valve between control grid and anode. This potential plane draws the electrons through the negative control grid away from the cathode, so that they reach the potential plane formed by the screen grid at a velocity corresponding to the po-tential of the screen grid. In doing so, some of the electrons collide with the screen-grid wires, but most of them shoot through the meshes of the screen grid and im-pinge on the anode. Hence the number of electrons impinging on the anode per time unit is de-termined mainly by the voltages on the control and screen grids,

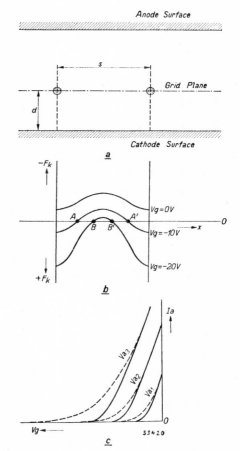

Fig. 147
a) Triode in which the spacing of the grid wires is greater than the distance d be-tween grid and cathode.
b) Curves representing the field strength F_k along the surface of the cathode (x axis) for various grid voltages (V_g).
c) *Full lines:* transfer characteristics at different anode voltages without island effect.
   Broken lines: transfer characteristics at different anode voltages with island effect.

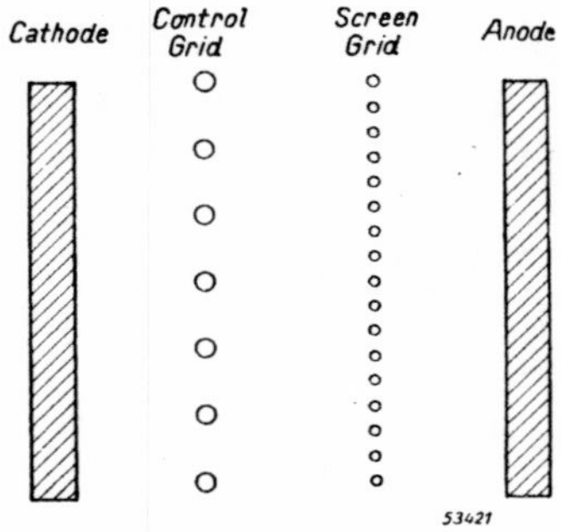

Cathode   Control   Screen   Anode
           Grid   Grid

Fig. 148
Section of an electrode system
of a screen-grid valve (tetrode)
with flat, parallel electrodes.

whilst the voltage on the anode has hardly any influence upon it. **Consequently the anode current is only in a limited degree dependent upon the anode voltage.** The anode-current-anode-voltage characteristics of screen-grid valves are therefore within a certain range of anode voltage practically horizontal (see also Fig. 121 on page 125). The name screen grid is due to the fact that a positive grid between control grid and anode screens the effect of the anode potential upon the electron stream flowing through the control grid. Whereas in a triode the anode performs the double function of creating a positive potential plane and intercepting the electrons, in a screen-grid valve these two functions are divided among two different electrodes.

Fig. 148 represents a section through a screen-grid valve (tetrode) in which the electrodes are mounted parallel to each other.

(b) Current Distribution

The action of the screen grid is governed by the distribution of the electron current between the screen grid and the anode. This distribution is influenced by the anode voltage. The total electron current passing from the cathode through the control grid, which distributes itself between the screen grid and the anode, actually remains independent of the anode voltage, since it is determined by the potential in the plane of the control grid, whilst the latter in turn is determined by the potential in the plane of the screen grid. Now the mean potential in the plane of the screen grid and thus also the electron current will depend only in a small degree upon the anode voltage, because this grid is as a rule very closely wound. The anode voltage, however, always continues to exercise a slight

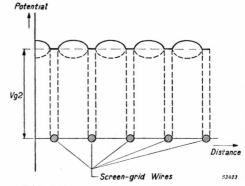

Fig. 149
Distribution of the potential in the screen-grid plane when the anode voltage is appreciably higher (full-line curve) or appreciably lower (broken-line curve) than the screen-grid voltage.

150

influence upon the mean potential
in the plane of the screen grid, as is
shown in Fig. 149, since the field
of the anode extends through the
meshes between the screen-grid
wires, though less so than is the case
with the control grid of a triode.
It is for this reason that the anode
voltage as a matter of fact exercises
little influence upon the electron
current.

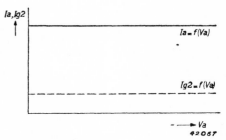

Fig. 150
Anode-current-anode-voltage characteristic (full line and screen-grid-
current-anode-voltage characteristic
(broken line) of a screen-grid valve
assuming the electron paths in the
valve to be parallel (disregarding phenomena related to the space charge).

The electrons leave the plane of
the screen grid at a velocity corresponding to the potential in the
screen-grid plane. Assuming that
the electrons describe parallel paths perpendicular to the screen-grid
plane, then a very small voltage on the anode would suffice to cause all
the electrons shooting through the meshes to arrive at the anode. The
current distribution would then be such that the electrons whose paths
end against the screen-grid wires form the **screen-grid current** and those
electrons whose paths pass between the screen-grid wires form the
anode current.
The electron current leaving the cathode would then be distributed
between the screen grid and the anode in the same proportion as that
between the area of the screen-grid wires and the area of the screen-grid
meshes. If a very low anode voltage sufficed to cause all electrons
shooting through screen-grid meshes to collide against the anode, then
the anode-current-anode-voltage characteristics and the screen-grid-
current-anode-voltage characteristic would take the form represented
in Fig. 150. Actually, however, if the effect of the secondary emission
of the screen grid and the anode is ignored (see under d), the anode-
current and screen-grid-current characteristics are as represented in
Fig. 151. In the current distribution shown in Fig. 151 two ranges of
anode voltage may be distinguished, viz.:

(a) $V_a \leqslant V_{g2}$; this is the condition under which electrons return from
    the space between screen grid and anode. This range is indicated
    in Fig. 151 by the letter A.

(β) $V_a > V_{g2}$; this is the condition under which all electrons shooting
    through the screen grid reach the anode. In Fig. 151 this range is
    indicated by the letter B.

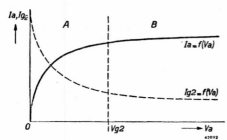

Fig. 151
Anode-current-anode-voltage character-istic (full-line curve) and screen-grid-current-anode-voltage characteristic (broken line) as actually measured in a screen-grid valve in the absence of the phenomena related to secondary emission (without space charge). In the current distribution shown two anode-voltage ranges can be distinguished, that in which V_a lies between 0 and approximately V_{g2} — the range of returning electrons (A), and that in which $V_a > V_{g2}$ — the range of continuing electrons (B).

(a) Current Distribution in the Range of Returning Electrons

The shape of the plate characteristics in this range (see Fig. 151) is dependent upon the fact that, over a certain range of anode voltage, the electrons are deflected by the field distribution in the vicinity of the screen grid. This deflection takes place in the direction of the positive screen-grid wires.

If an electron a (see Fig. 152) moves midway between two grid wires it will not be subject to any lateral force and will continue in the direction perpendicular to the screen-grid plane. The velocity at which the electron leaves that plane corresponds to the potential at the place where this crosses the screen-grid plane. This potential is roughly equal to the mean potential in the screen-grid plane. An electron b, however, is subjected to a lateral attractive force in the vicinity of a grid wire, in consequence of which its path turns away from the original direction at an angle a. The electron will then reach the anode only if the normal component of the velocity $v_n = v \cos a$ is greater than the velocity corresponding to the potential difference $V_{g2} - V_a$ between screen grid and anode. If the normal component v_n is smaller than that corresponding to $V_{g2} - V_a$ than the anode is not reached. The electron in that case turns back before the anode is reached, moving again in the direction of the screen grid, and either travels direct towards a screen-grid wire or arrives in the space between control grid and screen grid, there again turning back and con-

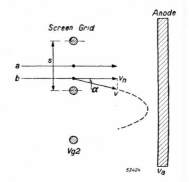

Fig. 152
Deflection of an electron passing through the screen grid. The deflection is greater when the electron passes closer to a grid wire (for instance b); there is no deflection if an electron passes through the screen-grid plane exactly midway between two wires (a).

tinuing once more in the direction of the screen grid, sometimes reaching the screen-grid wire after travelling to and fro several times around it.

On the basis of the above-mentioned requirement for the normal component v_n, the equation applying to an electron that reaches the anode is therefore

$$v_n \gtreqless \sqrt{\frac{2e}{m}(V_{g2} - V_a)}. \tag{53}$$

Considering that $v_n = v \cos \alpha$ and $v = \sqrt{\frac{2e}{m} V_{g2}}$ and substituting this in Equation (53) one obtains:

$$\sqrt{\frac{2e}{m} V_{g2}} \cos \alpha \gtreqless \sqrt{\frac{2e}{m}(V_{g2} - V_a)}, \tag{54}$$

or

$$V_{g2} \cos^2 \alpha \gtreqless V_{g2} - V_a, \tag{55}$$

or

$$V_a \gtreqless V_{g2} \sin^2 \alpha. \tag{56}$$

From Equation (56) it appears that an electron leaving the screen-grid plane at an angle α with respect to the perpendicular to that plane can reach the anode only if the anode voltage is greater than $V_{g2} \sin^2 \alpha$. If $\alpha = 0$ then, as is to be expected, V_a may be just greater than zero. Now upon passing through the screen-grid plane the electrons undergo a deflection which depends upon the place where their path intersects that plane, since the angle α depends upon the distance x between that point and the point midway between two grid wires, α being by rough approximation proportional to x.

When the anode voltage $V_a = 0$ the anode current $I_a = 0$. In this case, therefore, all electrons return to the screen grid. As the anode voltage increases from zero, first the electrons with a small deflection will reach the anode, or, as it is sometimes expressed, will be "taken over" by the anode, these being followed successively by the electrons with larger angles of deflection. As soon as the anode voltage is so high that even for the largest angle of deflection occurring $V_a > V_{g2} \sin^2 \alpha$, the anode current is maximum.

It can be proved that in the range of returning electrons the anode current I_a forms a certain part of the total current I_k emitted by the

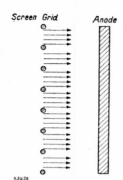

Fig. 153
Movement of the electrons through the screen-grid plane towards the anode.

cathode, the approximate relationship between anode and cathode currents being as follows:

$$I_a = I_k \, K \, \sqrt{\frac{V_a}{V_{g2}}}, \qquad (57)$$

in which K is a constant inversely proportional to the spacing s (see also Fig. 152) of the screen-grid wires and depending upon the construction of the valve.

(β) Current Distribution in the Range of Continuing Electrons

In this range all electrons shooting through the meshes of the screen grid reach the anode. Owing to the field distribution in the vicinity of the screen grid, which depends upon the potential; at the anode and at the screen grid, the electrons leaving the control-grid plane may be concentrated either on the wires or on the meshes of the screen grid, and in a particular case the electrons will move along parallel paths through the screen-grid plane (if the potential caused by the anode potential in the screen-grid plane is equal to the potential of the screen grid). As a result there is a certain relation between the current distribution between anode and screen grid and the potentials of these electrons.

In this range, as in the range of returning electrons, the shape of the plate characteristic is related to the fact that due to the field distribution in the vicinity of the screen grid the electrons are deflected.

(c) The Space Charge in the Valve Section and Screen Grid between Anode, the Virtual Cathode

The foregoing considerations regarding the current distribution hold with accuracy only if there is no space charge between screen grid and anode. In most cases, due to the high electron velocities occurring in practice, the space charge will in fact be so small that its influence need not be taken into account. There are many cases, however, where a space charge of great density is purposely produced in the space between screen grid and anode. For that reason the potential field in the space between screen grid and anode of a tetrode will now be considered.

154

For the sake of simplicity we will take as a basis the case of flat parallel electrodes, with the electrons passing through the screen-grid plane in a perpendicular direction (see also Fig. 153) at a velocity corresponding to the mean potential V_{g2} in the screen-grid plane. If the screen-grid potential is V_{g2} and the anode potential V_a, then, no space charge being present, the potential distribution will be as represented by the broken line a in Fig. 154. Increasing the electron current in the space between screen grid and anode, for instance, by raising the control-grid voltage,

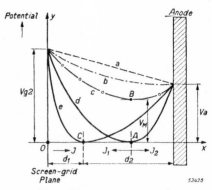

Fig. 154

Potential distribution between screen grid and anode at different current intensities.

produces a drop of the potential in the space between screen grid and anode.

The potential-distribution diagram for this section of the valve will then follow a curve similar to the dot-and-dash line b in Fig. 154. If the electron current increases further the density will ultimately become so great that the potential will reach a minimum (curve c in Fig. 154). Since this minimum, in the case of curve c in Fig. 154, lies higher at the point B than the cathode potential (zero level), this minimum will be reached by all the electrons shooting through the screen grid, after which they move towards the higher anode potential (V_a). As the electron current increases still further the minimum of the potential will reach the zero potential axis in the point A. The current distribution then changes in such a way that some of the electrons turn about at the potential minimum and travel back to the screen grid. As a result the electron current towards the anode diminishes, whilst the density of the charge to the left of point A increases. If the electron current passing through the screen grid becomes still greater the space charge to the left of the critical point A will increase further in density and the minimum of the potential will be displaced to the left (curve e in Fig. 154). In that case proportionately still more electrons will return at the potential minimum and a correspondingly smaller quantity of electrons will pass through it. Therefore if the density of the space charge in the space between screen grid and anode is sufficiently great it may influence the current distribution between screen grid and anode. In the points A or C the field strength and the potential are apparently equal to zero. Since this is likewise the case for a cathode one may imagine in the point A or C a cathode (parallel to the anode) which emits electrons in the space between this imaginary cathode and the anode. Such an imaginary cathode due to a potential minimum is called a **virtual cathode.** In the case of curve e, Fig. 154, we therefore have a virtual cathode in the point C and an anode at a distance d_2 from that imaginary cathode; this is equivalent to a diode with an anode voltage equal to V_a. The virtual cathode, however, also sends electrons back in the direction of the cathode, so that the virtual cathode receives electrons from the cathode and returns some of them back to it.

From the foregoing considerations it follows that with such a dense space charge in the valve section between screen grid and anode where the potential minimum reaches zero, a part of the electron current passing through the screen-grid plane returns in consequence of that minimum, so that as a result the current distribution is influenced. In the next sub-section it will be seen how this potential minimum in the space between screen grid and anode may be turned to advantage.

(d) Secondary Emission of the Anode and Screen Grid in a Tetrode

If an anode is set up behind the screen grid, as is the case in a screen-grid valve or tetrode, the electrons colliding with the anode cause secondary electrons to be released from the anode, at least if the electrons have sufficient velocity, that is to say if the anode voltage is sufficiently high (see Chapter V). Secondary electrons will be released already at a lower anode voltage than the normal screen-grid voltage (e.g. 100 V). In that case the secondary electrons released from the anode move in the direction of the higher potential of the screen-grid plane and are intercepted by the wires of that grid. Where the anode voltage is higher than the screen-grid voltage (for example, more than 20 volts higher) the relatively slow secondary electrons will return to the anode, as they are unable to overcome the potential drop from the anode to the screen grid. The released secondary electrons travelling towards the screen grid, when the anode voltage is lower than the screen-grid voltage, cause a current to flow in the direction opposed to the normal anode current. The magnitude of this current depends upon the secondary-emission factor of the material from which the anode is made, which factor in turn depends upon the anode voltage (see Chapter V, Section 22). If the secondary-emission factor in a certain range of anode voltage is greater than unity the anode current in that range will be negative.

The positive screen grid is likewise exposed to a bombardment of electrons, in consequence of which secondary electrons are released from the screen-grid wires. If the anode potential is higher than the screen-grid potential these secondary electrons will move towards the anode, thus increasing the anode current and reducing the screen-grid current. If the anode voltage is lower than the screen-grid voltage the secondary electrons from the screen grid will return to the screen grid.

Taking the plate characteristics of a tetrode (at a certain control-grid potential), it will be seen that this deviates considerably from the theoretical curve, as represented by a in Fig. 155. In most cases the shape of the curve will be according to the line b of Fig. 155. This shape is a result of the secondary electron emission of the anode and the screen grid. If the anode voltage is raised from zero the influence of the secondary emission of the anode can be very easily seen. The anode current is smaller than that indicated by curve a. Since the secondary-emission factor increases with the anode voltage the difference between the anode currents of curves a and b will increase with anode voltage. As soon as the curve a tends to become horizontal the anode current will begin to drop according to the curve b. Thus in this case the internal resistance of the valve becomes negative. This fall in the anode current with rising anode voltage continues until the anode voltage approaches the screen-grid voltage. The voltage difference between anode and screen grid is then too small to bring about any appreciable secondary electron current from screen grid to anode that is limited by the space charge (one may regard the anode as cathode and the screen grid as anode of a diode where the anode voltage is equal to V_{g2}—V_a) [1]). As a result the anode current of the valve will increase again and approach the value of curve a. In this area the influence of the secondary emission of the screen grid begins to be appreciable, and approximately when the anode voltage becomes higher than the screen-grid voltage the anode current will become greater than the value for curve a.

As the screen-grid current is usually much smaller than the anode current the increase of the latter due to the screen-grid secondary emission will also be appreciably less than the reduction of the anode current due to the anode secondary emission. The value of the anode current at the minimum Λ of curve b is related, as explained before, to the faculty of the anode surface to emit secondary electrons [2]). It should be noted that the secondary electron current may be greater

[1]) See Chapter XXXI, Section 187.
[2]) See Chapter XXXI, Section 188b.

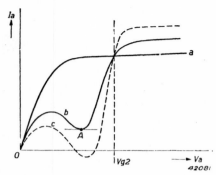

Fig. 155
Curve *a:* Plate characteristic of a tetrode at a certain screen-grid voltage in the absence of secondary emission from anode and screen grid. This curve is therefore determined by the "taking over" of current.

Curve *b:* Plate characteristic of a tetrode as measured when secondary emission takes place from anode and screen grid. Here the secondary-emission factor of the anode is less than unity.

Curve *c:* Plate characteristic of a tetrode with secondary emission from screen grid and anode where the secondary-emission factor of the anode is greater than unity. In a certain range the anode current becomes negative.

than the primary electron current, a condition observable over part of the broken line c of Fig. 155.

From the foregoing explanation it will be realized that the internal resistance of a screen-grid valve depends to a high degree upon the secondary emission of the anode and screen grid.

The phenomena connected with the secondary emission in a screen-grid valve can be remedied in two different ways, or at least their influence can be considerably reduced. One of these means, which is the technically more correct one, will be dealt with in the next section, while the other way, which is very often applied to output valves, will be briefly explained below. It has been shown that if the density of the space charge in the space between screen grid and anode is great, the potential reaches a minimum (see also, for instance, curve c of Fig. 154). This potential minimum causes a field strength at the anode which cannot be overcome by the relatively slow secondary electrons. These electrons are therefore driven back to the anode, while the primary electrons are not affected by the potential minimum, since the absolute value of the potential of this minimum in the space between the screen grid and anode is positive. The distribution of the potential, however, is very much dependent upon the electron current and upon the voltages at the electrodes, so

Fig. 156
Arrangement of the electrodes of a beam-power output valve in which the secondary emission is suppressed by the high density of the space charge in the space between screen grid and anode.

a = anode
g_1 = control grid
g_2 = screen grid
k = cathode
s = concentrating screening (beam-forming) plates at cathode potential.

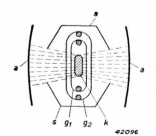

Fig. 157
a) Section through an electrode system of a screen-grid valve with suppressor grid and parallel flat electrodes.
b) *Full-line curve:* Potential distribution in section A-B for an anode voltage V_{a1} higher than the screen-grid voltage V_{g2}. *Broken-line curve:* Potential distribution in the section A-B for an anode voltage V_{a2} lower than the screen-grid voltage. These curves apply to the case of absence of space charge.

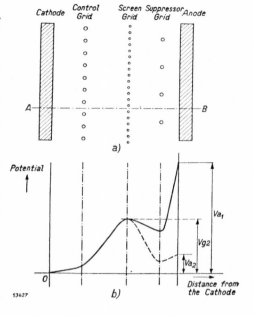

that this remedy is only effective for a certain value of current and at definite voltages. The distribution of the potential also depends upon the distance between screen grid and anode. The greater this distance the greater the influence of the space charge and the deeper the sag of the potential-distribution curve in the space between screen grid and anode. Yet this distance cannot be made too great; with the high values of the current that may occur when the valve is fully controlled the potential at the minimum would fall so sharply as to cause primary electrons to be reversed. Therefore there is a certain critical distance that is the optimum for the suppression of secondary emission for certain operating conditions of the valve. Valves in which this principle is applied are commonly called **critical-distance valves.**

In valves where the secondary emission is suppressed by a space charge of great density in the section between screen grid and anode, the anode is consequently mounted at a greater distance from the screen grid than is normally the case in screen-grid valves, whilst the electron paths in this space are concentrated by screening plates of the same potential as the cathode, so that electron beams are formed which produce the required density of space charge (see Fig. 156). The so-called **beam-power valves** are an example of such valves.

74. The Suppressor Grid

Another means of suppressing the secondary electron currents from screen grid and anode, which often have an adverse effect upon amplification, is to introduce a so-called **suppressor grid** between screen grid and anode. Usually the suppressor grid is made with very wide meshes and has the same potential as the cathode, or differing only slightly from it.

159

Due to the suppressor grid having the same potential as the cathode, a pronounced potential minimum occurs between screen grid and anode. Fig. 157b shows the potential as a function of the distance from the cathode in the section A-B of an electrode arrangement as given in Fig. 157a (parallel, flat electrodes), and the potential minimum in the plane of the suppressor grid, the space charge being left out of consideration. Owing to the presence of this minimum in front of the anode the relatively slow secondary electrons emitted from the anode are not able to overcome the field strength in the space between suppressor grid and anode, so that they will return to the anode. The secondary electrons emitted from the screen grid will likewise be incapable of overcoming the potential drop between the screen-grid plane and the suppressor-grid plane and thus will return to the screen grid.

Owing to the penetrating action of the anode and screen-grid potentials, the potential in the meshes of the suppressor grid, the potential of which is zero, will be positive. The mean potential in the plane of the suppressor grid depends upon the distances between suppressor grid, anode and screen grid and upon the mesh of the suppressor grid, as also upon the gauge of wire used. If this mean potential is too high the counteracting field between anode and suppressor grid will be inadequate to suppress the secondary electron current. With very low values of this potential, on the other hand, the primary electrons will not be able to pass the suppressor-grid plane. I.e. if the mesh is very small the mean potential in the suppressor-grid plane approaches zero value and the electrons which are deflected from the normal direction in the vicinity of the screen grid will no longer be able to reach the plane of the suppressor grid. Consequently the suppressor grid has to be so dimensioned that the primary electrons in the range of characteristics which is of practical value are allowed to pass through in sufficient quantity while the secondary electrons are held back. For this reason suppressor grids usually have a relatively wide mesh.

The suppressor grid has an adverse effect upon the primary current distribution between anode and screen grid, the screen-grid current being greater and the anode current less than would be the case in a corresponding tetrode (without secondary electron emission). On the other hand, the influence of the anode voltage upon the total cathode current I_k is reduced by the suppressor grid; the anode voltage, however, has a greater influence upon the current distribution.

75. A Second Control Grid behind a Screen Grid

In hexodes, heptodes and octodes there are two control grids, one around the cathode and the second behind a screen grid. Whereas screen-grid valves with a suppressor grid (pentodes) are so dimensioned that the electron current to the anode is as far as possible independent of the mean potential in the plane of the suppressor grid, in the case of valves having a control grid behind the screen grid it is generally desirable that the control voltage in the plane of the second control grid has the greatest possible influence upon the electron current flowing through that grid. Consequently such a grid will have a close mesh and thus a high control action.

76. A Screen Grid behind a Second Control Grid

In the types of valves with two control grids described in the previous section (hexodes, heptodes and octodes) there is a second screen grid behind the second control grid. If the plane of the first screen grid is regarded as a cathode, the second screen grid behind the second control grid has the same duty to perform as the screen grid in a screen-grid valve (tetrode or pentode).

CHAPTER XIV

Valve Capacities

77. The various Electrode Capacities in the Absence of Space Charge

If we examine a diode it will be clear that its cathode and anode may be regarded as two plates of a condenser, with the vacuum between these electrodes representing the dielectric. Thus the cathode and anode of a diode have a certain capacity with respect to each other, and this capacity will be indicated by the symbol C_{ak}. The lead-in wires from the valve-base contacts to the cathode and anode of a diode and the base contacts themselves also have a certain mutual capacity. These capacities augment the inter-electrode capacity. If the cathode of a diode is earthed then the capacity between cathode and anode is still further increased by the capacity of the anode with respect to an earthed screening contained in the bulb. Fig. 158 illustrates such a case, showing how the capacity C_{am} between metallizing and anode increases the capacity C_{ak}.

A very comprehensive valve-capacity diagram can be made by taking the electrodes as the corner points of a geo-metrical figure, say a triangle, rectangle, etc., and interposing the capacities between those points. This has been done in Fig. 159 for the diode considered above as the first example. According to this representation the diode has a capacity between anode and cathode, between anode and metallizing and between cathode and metallizing. Each of these capacities is called a **partial capacity** of the electrode concerned.

When m and k are earthed the anode capacity C_a equals the sum of the capacities C_{am} and C_{ak}. In the case of in-directly-heated valves the heater filament in the ca-thode tube also has to be regarded as an electrode and its capacities with respect to the other electrodes of the

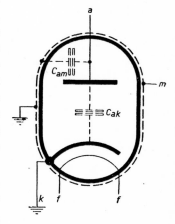

Fig. 158
Diagrammatic representation of the increase of the capacity of a diode-anode with respect to the cathode, which is earthed, due to the capacity of the anode to an earthed part, in this case the metal-lizing.

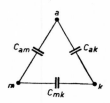

Fig. 159
Diagrammatic representation of the capacities present in in a diode with metallized bulb.

162

valve have likewise to be taken into account. If both the filament
and the cathode are earthed then the capacity between these electrodes
is of no importance; it may play a part, however, if the filament is
earthed and the cathode not.

When taking measurements it will be found that those carried out
with a cold cathode give values for the valve electrode capacities which
differ from those obtained with a hot cathode. This is due to the presence
or absence of the space charge; we will revert briefly in the next section
to the influence of the space charge upon the capacities.

Some of the electrode capacities are in parallel with H.F., I.F. or L.F.
circuits and increase the capacities existing there. Others cause couplings
between circuits connected to different valve electrodes, as a result of
which undesired effects may occur. In this way, for instance, certain
circuits may be affected by interfering voltages (see also Chapter
XXVIII) and amplified voltages may be fed back, which may result
in oscillation (voltages on one electrode may be modulated by voltages
on another which are also present on the first electrode as a result
of capacitive coupling). The capacities of a valve may influence the
tuning of connected circuits or may constitute a load on the connected
voltage source. It is therefore of importance to know exactly what these
capacities are and how they behave, and for this reason the published
data of radio valves always contain the most important of the various
electrode capacities. **Unless otherwise stated the capacities found in
publications are always those that have been measured with a cold
cathode.**

(a) The Capacities of a Triode

Disregarding, for the time being, other parts such as the metallizing
heater and the like, there are in a triode a cathode, a grid and an
anode. In this case, therefore, three capacities are to be distinguished,
viz. a capacity C_{gk} between grid and cathode, a capacity C_{ag} between anode and grid, and a capacity C_{ak}
between anode and cathode. These three capacities
form the capacity triangle drawn in Fig. 160.

(a) The Grid Capacity

The grid capacity C_g is comprised of the capacities from
the grid to all other electrodes and parts of the valve
excepting the anode, thus in a triode it is the sum of the
capacities to the cathode, the metallizing and the heater.
This capacity is sometimes termed the **input capacity.**

Fig. 160
Diagrammatic re-
presentation of the
three capacities of
a triode.

(β) The Anode Capacity

The anode capacity C_a comprises the capacities from the anode to all other electrodes and parts of the valve excepting the control grid, thus in a triode it is the sum of the capacities to the cathode, the metallizing and the filament. This is sometimes termed the **output capacity.**

(γ) The Grid-anode Capacity

Owing to the grid-anode capacity C_{ag} the anode circuit may react upon the grid circuit. If there is an impedance or resistance in the anode circuit then the anode alternating current will set up an alternating voltage across that impedance or resistance, and usually this alternating voltage is much greater than the grid alternating voltage. If in the anode circuit there is a purely ohmic resistance (see Fig. 161a) the alternating voltage of the anode with respect to the cathode is opposite in phase to the alternating voltage of the grid with respect to the cathode. If, now, the grid alternating voltage is V_{gk}, the anode alternating voltage V_{ak}, and the amplification $\left(\dfrac{-V_{ak}}{V_{gk}}\right)$ is equal to A, then $V_{ak} = -AV_{gk}$. The alternating voltage between grid and anode is equal to $V_{gk} - V_{ak} = V_{gk} + AV_{gk} = (A + 1)\,V_{gk}$. The current I_{Cag} through the grid-anode capacity is consequently equal to

$$I_{Cag} = j\omega\, C_{ag}\, (A + 1)\, V_{gk} \qquad (58)$$

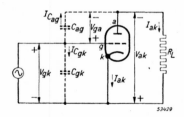

This current therefore flows from the grid via the anode to the cathode and apparently originates from a capacity $(A + 1)\, C_{ag}$ connected parallel to the grid-cathode capacity.

Fig. 161a
Diagram of a triode amplifier with anode resistance R_L, the valve capacities C_{ag} and C_{gk} being indicated by the broken lines. The positive direction of the currents through the capacities and the anode resistance are indicated by single-headed arrows, whilst the positive directions of the voltages are indicated by double-headed arrows with + and — signs.

(1) Condition for obtaining a Grid-to-cathode Capacity equal to Zero

If the anode alternating voltage were in phase with the grid alternating voltage V_{gk} and thus equal to AV_{gk}, then the voltage between grid and anode would be equal to $V_{gk} - AV_{gk} = (A - 1)\, V_{gk}$. The current through the grid-anode capacity to the grid is in that case:

$$I_{Cag} = -j\omega\, C_{ag}\, (A - 1)\, V_{gk}. \qquad (59)$$

Since there is an additional current $I_{C_{gk}} = j\omega\, C_{gk}V_{gk}$ flowing to the grid as a result of the grid-to-cathode capacity, and both currents are additive, the total capacitive current to the grid would be zero when

$$j\omega\, C_{gk} - j\omega\, C_{ag}\, (A - 1) = 0,$$

or
$$C_{ag} = \frac{C_{gk}}{A-1}. \tag{60}$$

In that case the apparent capacity between grid and cathode would also be zero. This condition can be obtained, for instance, by means of a two-stage resistance-coupled amplifier in which the anode alternating voltage of the second valve is in phase with the grid alternating voltage of the first and by coupling the anode of the last valve to the grid of the first through a capacity of value $\dfrac{C_{gk}}{A-1}$, where A is the overall gain of the amplifier.

(2) Impedance of Purely Ohmic Character in the Anode Circuit

Where there is a purely ohmic resistance in the anode circuit the alternating current from the anode to the cathode is in phase with the alternating voltage between grid and cathode V_{gk} (see Fig. 161a). The alternating voltage of the anode with respect to the cathode is then opposite in phase to the alternating voltage of the grid with respect to the cathode, and the alternating voltage of the grid with respect to the anode, which is equal to $V_{gk} - V_{ak}$, is thus in phase with the alternating voltage of the grid with respect to the cathode. Since V_{ga} and V_{gk} are in phase, the currents flowing through C_{ga} and C_{gk} will also be in phase with each other. It is, therefore, seen that with a purely ohmic resistance in the anode circuit, or with an impedance that has a purely ohmic character (oscillatory circuit at resonance) the currents passing through C_{ag} and C_{gk} augment each other.

The foregoing is shown in the vector diagram of Fig. 161b. The voltages V_{ga} and V_{gk} are in phase with each

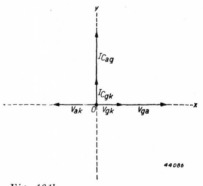

Fig. 161b
Vector diagram of the voltages between grid and cathode V_{gk}, between anode and cathode V_{ak}, and between grid and anode V_{ga}, and the current through the grid-anode capacity C_{ag} and through the grid-cathode capacity C_{gk}, when there is a purely ohmic impedance in the anode circuit.

165

other and thus their vectors lie in the same direction. The current I_{Cag} through the capacity C_{ag} which is due to the voltage V_{ag} leads by 90° on V_{ga}. The current I_{Cgk} through the grid-cathode capacity C_{gk} is similarly leading by 90° on the voltage V_{gk}, and since the two voltages are in phase with each other the currents I_{Cgk} and I_{Cag} are likewise in phase.

(3) Impedance of Inductive or Capacitive Character in the Anode Circuit

When the impedance in the anode circuit has not a purely ohmic character there will be a phase angle between the anode current I_{ak} and the voltage V_{ak} between anode and cathode. This will be the case if an oscillatory circuit forms the anode load and the frequency of the signal amplified by the valve does not correspond to the resonant frequency of the load. If that frequency is lower than the resonant frequency the circuit acts as an inductance connected in parallel with a resistance; if the frequency is higher than the resonant frequency it acts as a capacity connected in parallel with a resistance. This resistance may be, for example, the resonant resistance of the circuit, with which the internal resistance of the valve is connected in parallel. Now an inductance in the anode circuit causes a positive phase angle of the anode alternating voltage V_{ak} with respect to that voltage which would occur with a resistance in the anode circuit. Thus for $V_{ga} = V_{gk} - V_{ak}$ we get the vector drawn in Fig. 161c, which, since V_{ak} is leading, likewise has a positive phase angle (indicated here by φ) with respect to the grid-cathode voltage V_{gk}.

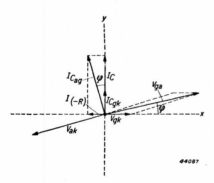

Fig. 161c
Vector diagram of the voltages between grid and cathode V_{gk}, between anode and cathode V_{ak}, and between grid and anode V_{ga}, and the currents through the grid anode capacity C_{ag} and through the grid-cathode capacity C_{gk}, when the impedance in the anode circuit is inductive. The current through the capacity C_{ag} acquires a positive phase angle φ with respect to the wattless current I_{Cgk} and then has a component $I_{(-R)}$ the phase of which corresponds to a current through a negative resistance interposed between grid and cathode in parallel with the input circuit.

This alternating voltage between grid and anode results in a current I_{Cag} passing through the anode-grid capacity, which likewise is leading on the current I_{Cgk}. This current has a wattfull component $I_{(-R)}$ the direction of which is opposite to that of the vector V_{gk}; this current might also be caused by the voltage V_{gk} across a negative resistance —R between grid and cathode. The presence of a negative resistance between grid and cathode means a reduction of the damping resistance of that circuit. This is called **regeneration.** If this negative resistance is sufficiently small it may lead to oscillation of the valve stage (thus giving rise to an oscillation of a certain constant amplitude which maintains itself without excitation by the external grid alternating voltage V_{gk}).

If the anode circuit contains an impedance which is capacitive, e.g., a capacity that is connected parallel to a resistance, the vector for V_{ga} will have a negative phase angle

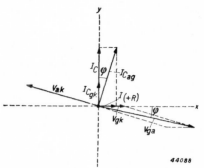

Fig. 161d
Vector diagram of the voltages between grid and cathode V_{gk}, between anode and cathode V_{ak} and between grid and anode V_{ga} and the currents through the C_{ag} and the C_{gk}, when the impedance in the anode circuit has a capacitive component. The current through C_{ag} in this case acquires a negative phase angle φ with respect to the wattless current I_{Cgk} and has a component $I_{(+R)}$, the phase of which corresponds to a current through a positive resistance interposed between grid and cathode in parallel with the input circuit.

with respect to the vector V_{gk} (see Fig. 161d). This alternating voltage will set up a current I_{Cag} through C_{ag} which lags behind I_{Cgk} and of which the wattfull component $I_{(+R)}$ has the same phase as V_{gk}. This means that this current could also have been set up by the voltage V_{gk} across a positive resistance +R between grid and cathode, which corresponds to an increase of the damping of a circuit interposed between grid and cathode. This is called **degeneration.**

The degeneration or regeneration in a stage due to the influence of the grid-anode capacity can also be derived by calculation in the following simple manner:

$$V_{ak} = —g_m Z_L V_{gk}, \tag{61}$$

in which g_m is the transconductance of the valve and Z_L the anode (load) impedance, which also contains the internal resistance R_a. The alternating voltage between grid and anode is equal to

$$V_{ga} = V_{gk} — V_{ak} = (1 + g_m Z_L) V_{gk}. \tag{62}$$

From this it follows that the current through the grid-anode capacity C_{ag} is equal to

$$I_{Cag} = j\omega C_{ag} V_{ga} = j\omega C_{ag} (1 + g_m Z_L) V_{gk}. \tag{63}$$

If now the anode impedance has a capacitive or inductive character one may write:

$$Z_L = R + j X,$$ (64)

in which X is the reactance. By replacing Z_L in Equation (63) by $R + j X$ we can write

$$I_{Cag} = j\omega\, C_{ag}\, (1 + g_m R + j g_m X)\, V_{gk} =$$
$$-\omega\, C_{ag} g_m X V_{gk} + j\omega\, C_{ag}\, (1 + g_m R)\, V_{gk}.$$ (65)

The first term of the last part of this equation apparently originates from a resistance which is equal to $- 1/\omega\, C_{ag} g_m X$ and which, through the influence of C_{ag}, is in parallel with the grid circuit. This is positive if the phase angle φ (see Fig. 161d) and consequently X are negative, thus in the case where the anode circuit has a capacitive component. In that case degeneration occurs.

For a positive value of X, when the anode circuit has an inductive component, the resistance is negative and regeneration occurs. If, therefore, for the frequency in question the anode circuit has an inductive component, then in consequence of the reaction across the C_{ag} there is regeneration in the grid-cathode circuit. Moreover, in all cases, thus also when the anode circuit is purely resistive, there is an apparent increase of the capacity between grid and cathode, because the vectors I_{Cag} in Figs 161c and d also have a component I_C in the direction of I_{Cag}. This increase of the grid capacity also follows from the term $j\omega\, C_{gk}\, (1 + g_m R)\, V_{gk}$ of Equation (65).

This increase of the grid-cathode capacity due to the reaction of the anode alternating voltage across the grid-anode capacity may be very considerable. If we consider a triode used as an audio-frequency amplifier, for instance, the triode part of the EBC 3, in which $C_{ag} = 1.3\ \mu\mu F$ and which with resistance coupling gives a 25-fold amplification, then in consequence of C_{ag} according to Equation (58) the grid-cathode capacity will increase by $1.3 \times (25 + 1) = 34\ \mu\mu F$ (by way of comparison the grid-cathode capacity of this valve is $2.0\ \mu\mu F$).

As is to be expected from the foregoing the grid-cathode capacity in R.F. amplifying circuits (see e.g. Fig. 162) may give rise to positive feedback from the anode circuit to the grid circuit (regeneration in the system). As a result the amplification in this stage

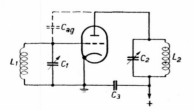

Fig. 162
High-frequency amplifying circuit with tuned circuits interposed in the grid and the anode circuits.

 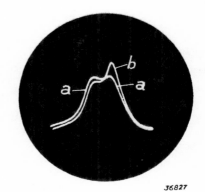

36826   *36827*

Fig. 163a   *Fig. 163b*

a) Oscillogram of the resonance curve   a) Resonance curve of an intermediate-
   of an amplifying stage with two   frequency amplifier with band-pass
   single-tuned circuits (recorded with   filters.
   a Philips frequency modulator and   b) The same curve recorded with arti-
   a Philips cathode-ray oscillograph).   ficial increase of C_{ag}.

b) Due to an increase of C_{ag} the curve
   becomes asymmetrical.

c) A further increase of C_{ag} results in the curve becoming still more asymmetrical;
   at the same time the curve becomes more peaked.

is increased and oscillation may occur, which interferes with the desired amplifying action. A further result of regeneration is an increase of selectivity. The curve representing the amplification as a function of frequency (the resonance curve) becomes more sharply peaked due to the feedback across C_{ag}; in the case of strong feedback there is, moreover, a fairly large asymmetry in this curve (see Figs 163a and b).

It can be shown that if the condition $g_m R_1 R_2 \omega C_{ag} < 2$ holds good, the stage cannot oscillate. In this term g_m is the transconductance of the valve, R_1 the impedance of the tuned grid circuit $L_1 C_1$ at resonance, R_2 the impedance of the tuned anode circuit $L_2 C_2$ at resonance with the internal resistance R_a of the valve in parallel, and ω the angular frequency to which the circuits are tuned. If $g_m R_1 R_2 \omega C_{ag} = 2$ oscillation may occur, but this is not necessarily the case and depends upon any mutual detuning of the circuits.

When, for instance, $R_1 = 400,000$ ohms, $R_1 = 10,000$ ohms, $g_m = 2$ mA/V, $\omega = 2\pi \times 1.5 \times 10^6$ (corresponding to $\lambda = 200$ m), then

$$C_{ag} < \frac{2}{2 \times 10^{-3} \times 4 \times 10^5 \times 10^4 \times 9.4 \times 10^6} \, F = 0.027 \, \mu\mu F \text{ is required}$$

to preclude oscillation.

Such a small capacity value, however, cannot be reached with a triode, a normal value for C_{ag} of a triode being 1 to 2 $\mu\mu F$, so that triodes are unsuitable for H.F. amplification. For radio- and intermediate-

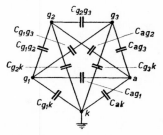

Fig. 164
Diagram of the partial capacities of a pentode

frequency amplification only capacity values of 0.003 $\mu\mu$F or less are of any use, and these can be obtained only with screen-grid valves such as pentodes for instance.

(b) The Capacities of a Pentode

Owing to the larger number of electrodes the diagram of the partial capacities of a pentode is much more complicated than that of a triode. Fig. 164 shows such a diagram. Not all these capacities are directly of importance for the application of a pentode; for instance, the capacity between g_2 and g_3 is of no importance, because these electrodes are usually connected with the cathode capacitively or directly. It is of more importance to know the input and output capacities. Of even greater importance is the capacity that causes feedback from the anode circuit, i.e. the grid-anode capacity C_{ag1}. The data published for Philips pentodes generally include the grid capacity C_{g1}, the anode capacity C_a and the maximum value of the grid-anode capacity C_{ag1}.

(α) The Grid Capacity

The grid capacity C_{g1} of a pentode is defined as the capacity of the grid with respect to all other electrodes with the exception of the anode. Thus it is assumed that except for the anode and the grid all the electrodes are connected directly or capacitively to the cathode.

(β) The Anode Capacity

The anode capacity C_a of a pentode is defined as the capacity of the anode with respect to all other electrodes with the exception of the grid.

(γ) The Grid-anode Capacity

The grid-anode capacity C_{ag1} of a pentode is defined as the capacity between anode and control grid, provided that all the other electrodes (g_2, g_3 and k) are connected up to each other.

As already explained in the case of the triodes, for H.F. amplification the value of C_{ag1} of a valve has to be extremely small. The conclusion that the C_{ag1} must be smaller than $\dfrac{2}{\omega g_m R_1 R_2}$ in order to avoid oscillation, $g_m R_2$ being the amplification of the valve, shows that if a large amplification is to be reached the grid-anode capacity must be extremely small. As mentioned when dealing with triodes, this can be attained with a screen grid such as is employed in pentodes.

The screen grid forms a screen between the anode and the control grid, hence its name. It thus causes a considerable reduction of the capacity between these electrodes, the grid-anode capacity in modern H.F. pentodes being in fact less than $0.002 \ \mu\mu F$. In manufacture particular care is taken to see that this limit is not exceeded, as otherwise there would be a risk of the valve starting to oscillate when used in a high-frequency stage. If, for instance, an I.F. amplifier consisting of a valve (pentode) with a transconductance $g_m = 2 \ mA/V$ and having $C_{ag1} = 0.002 \ \mu\mu F$, interposed between two single tuned circuits with an impedance of 400,000 ohms at the resonant frequency (e.g. 475 kc/s), then $g_m R_1 R_2 \omega C_{ag1} = 2 \times 10^{-3} \times 400,000^2 \times 3 \times 10^6 \times 0.002 \times 10^{-12} = 1.92$, which lies very close to the critical value 2. Usually, however, in I.F. amplifiers single tuned circuits are not used, but **band-pass filters** [1]), consisting of two tuned circuits mutually coupled inductively or capacitively. When these band-pass filter circuits are coupled with each other almost critically and have equal impedances, then the impedance between their input terminals at the resonant frequency is equal to one-half of the single-circuit impedance, thus in the case of circuits of 400,000 ohms it is equal to 200,000 ohms. The factor $g_m R_1 R_2 \omega \ C_{ag1}$ is therefore four times as small and there is much less risk of oscillation. These examples serve to illustrate the necessity of keeping the value of C_{ag1} small in the case of H.F. pentodes.

In power-output pentodes C_{ag1} is generally greater than that in H.F. pentodes, due to the differing dimensions of the electrodes which are necessary to make pentodes suitable for power amplification. The value of C_{ag1} of Philips output pentode EL 3N, for instance, has a maximum value of $0.8 \ \mu\mu F$. As output amplifier it normally yields a 57-fold voltage amplification, so that with a capacity value of $0.8 \ \mu\mu F$ there is an increase of the input capacity of about $46 \ \mu\mu F$. This may have a certain effect on the reproduction of high notes in low-frequency amplifiers. For this reason, as a matter of fact, the value of C_{ag1} is limited for output pentodes.

(c) The Capacities of an Octode

In a mixer valve of the octode type there are two control grids, the control grid g_1 of the oscillator part and the control grid g_4 of the mixer part. The radio-frequency alternating voltage of the signal to be received

[1]) See Appendix V.

is applied to g_4, the I.F. alternating voltage is generated at the anode, whilst g_1 and g_2 carry alternating voltages of oscillator frequency. The capacity C_{g4} of the grid is here defined as the capacity between grid and cathode that is found when all electrodes carrying no R.F. of I.F. alternating voltage or an alternating voltage of the oscillator frequency with respect to the cathode are connected to the cathode.

These are the electrodes $g_3 + g_5$, g_6, the heater and the metallizing. The anode capacity is defined as the capacity of the anode with respect to the cathode when $g_3 + g_5$, g_6, the heater and the metallizing are likewise connected to the cathode. In the data published for an octode are also included the mutual capacities C_{ag4}, C_{g1g4} and the capacities C_{g1} and C_{g2} of the oscillator control grid and of the oscillator anode. C_{ag4}, C_{g1g4}, C_{g1}, and C_{g2} are likewise the capacities found when the electrodes $g_3 + g_5$, g_6, the filament and the metallizing are connected to the cathode. The value of C_{ag4} makes it possible to determine the feedback of the intermediate-frequency anode alternating voltage to the input circuit, which may be of importance for eliminating certain whistles (see Chapter XXI). Thus it will be seen that all capacities have some significance and it is of importance to know them in order to be able to determine their influence.

(d) The Capacities of other Multiple-grid Valves and of Multiple Valves

From the foregoing the significance of the capacities quoted in publications concerning other types of multi-grid valves and multiple valves will be understood. Generally speaking the object is to indicate the electrode capacities which increase the capacities of connected circuits and those capacities which cause coupling between different circuits. The publication of these capacities is therefore usually intended to furnish the necessary data for the normal application of the valve. In the case of a double-diode-output-pentode, for instance, it is of importance to know, among others, the capacities of the diode anodes with respect to the control grid and the anode of the pentode part, but it is equally of interest to know also the capacity between the two diode anodes. According to the purpose for which a valve is designed one will therefore find a more or less extensive publication of the capacities.

78. The Influence of Space Charge upon the Capacities

In the beginning of the previous section it was pointed out that the values of the capacities of valve electrodes given in the published data are all derived from measurements taken while the cathode is cold.

Strictly speaking this is not quite complete, for these capacities apply to the cases when there is no space charge in the valve. This condition may be due to the fact that the cathode is cold, but it may also be that the cathode is heated to normal temperature but owing to a very great control-grid voltage there is no anode current flowing and consequently no space charge. It will now be explained briefly how the capacities are affected by the presence of space charge.

If we study a diode with parallel flat electrodes we find, as indicated in Fig. 11, an obliquely rising straight line from the cathode to the anode indicating the potential variation between these two electrodes when the cathode is not heated, that is, when it is not emitting any electrons. When the potential difference between cathode and anode equals V_a and the distance between these electrodes is a, the field strength F at the anode and at the cathode is equal to $\dfrac{V_a}{a}$. The charges on the anode and cathode surface per square centimetre are then given by: [1]

$$Q_k = - \varepsilon_o\, F \tag{66}$$

$$Q_a = + \varepsilon_o\, F. \tag{67}$$

Ignoring end effects, as is known, the capacity is given by

$$C = \varepsilon_o\, \frac{S}{a}, \tag{68}$$

where S is the surface area of the electrodes and a the distance between the electrodes. If now the cathode is heated to such a temperature that electrons are emitted in sufficiently large numbers that the electron current is not saturated but limited by the space charge, then a potential distribution occurs between the cathode and anode which can be represented approximately by the curve of Fig. 17. According to Equation (12) of Chapter III the current between cathode and anode is given by

$$I_a = k V_a^{3/2}.$$

Substituting for k the expression:

$$k = \left\{ \frac{\sqrt{2}}{81\,\pi \times 10^9} \sqrt{\frac{e}{m_e}} \right\} \frac{S}{d^2} = \frac{k''}{d^2}$$

and substituting for d a variable distance x from the cathode, V being the potential at a distance x from the cathode, the equation of the

---

[1] These derivations are taken from Balth. van der Pol published in Physica 3, 1923, page 253 and in Jahrb. d. drahtl. Telegr. 25, 1925.

potential-distribution curve can be obtained from the formula of the space-charge-limited current, i.e.,

$$I_a = \frac{k''}{x^2} V^{3/2}.$$

In that case we obtain

$$V = \left(\frac{I_a}{k''}\right)^{2/3} x^{4/3} = C\, x^{4/3}. \tag{69}$$

It appears that the potential in the space between cathode and anode is proportional to the $4/3$rd power of the distance x from the cathode and thus deviates from the potentials in this space indicated by the straight line in Fig. 11. At a distance $x = a$ the potential is equal to V_a and at a distance $x = 0$ it is zero, as is the case in Fig. 11.

In Equation (69) we may now express the constant C also in terms of the field strength F obtaining in the space between cathode and anode with the anode voltage V_a without space charge, multiplied by a certain factor. F being equal to $\frac{V_a}{a}$, therefore $V_a = Fa$. For $x = a$, $V_a = Ca^{4/3}$ $= Fa$. Hence $C = \frac{F}{a^{1/3}}$. If, therefore, we fill in this value of C, Equation (69) becomes

$$V = \frac{F}{a^{1/3}} x^{4/3} \tag{70}$$

By differentiation we find the field strength as a function of the distance x. This gives

$$V' = \frac{dV}{dx} = \frac{4}{3}\, \frac{F}{a^{1/3}} x^{1/3}. \tag{71}$$

For $x = 0$, i.e., at the surface of the cathode, the field strength is zero, and for $x = a$, i.e. on the surface of the anode, it is equal to $\frac{4F}{3}$. Consequently the charge density at the surface of the cathode is

$$Q_k = 0, \tag{72}$$

and at the surface of the anode

$$Q_a = \frac{4}{3}\, \varepsilon_o F. \tag{73}$$

Hence the charge density on the cathode, instead of being $- \varepsilon_o F$, is equal to zero in consequence of the space charge, and the charge density on the anode has become $4/3\, \varepsilon_o F$ instead of $\varepsilon_o F$. This means that as a result of the space charge the charge density on the anode is $4/3$ times as great and thus also the anode capacity has become $4/3$ times as great. In Fig. 165 the curve a represents the potential distribution between the two parallel flat electrodes of a diode as given by Equation (69). The tangent of the angle between the x-axis and the tangent b at the point Q of this curve, which lies on the surface of the anode, indicates the field strength at the anode. This tangent intersects the x-axis in the point P and now, according to the foregoing, OP is equal to $^1/_4$ a. The field in the vicinity of the anode will not change if we leave out the space charge and place the cathode at the point P. This figure thus illustrates how, as a result of the space charge, a capacity increase takes place, since the distance between the imaginary cathode at point P and the anode of a diode without space charge,

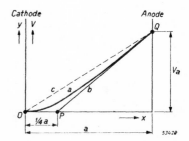

Fig. 165
Curve a: Potential distribution between two parallel flat electrodes, the potential rising in consequence of space charge according to the $^4/_3$ rd power of the distance from the cathode.
Straight line b: Tangent at the point of the potential distribution curve lying on the surface of the anode. This tangent intersects the x-axis at $x = \frac{1}{4}$ a.

which causes the same field strength at the anode, amounts to $^3/_4$ the distance between cathode and anode of a diode with space charge. Not one of the lines of force passing from the anode of a diode with space charge in the direction of the cathode actually reaches the cathode; they **all end on the electrons forming the space charge between cathode and anode.**

It might be concluded that as long as the current is limited by the space charge no change takes place in the anode capacity when the anode current is reduced, for instance by lowering the anode voltage, since the tangent on the potential distribution curve at the point **Q** will always intersect the x-axis at $^1/_4$ a from the point 0. In practice, however, one does actually find a capacity variation, for as a matter of fact the capacity drops uniformly with the decrease of anode current until a minimum is reached corresponding to the capacity value with a cold cathode. This is due to the velocity of emergence of the electrons, as a consequence of which the capacity variation takes place gradually as the anode current varies.

Similar phenomena occur with triodes. In amplifying circuits the anode has a comparatively high voltage; the anode current, however, is determined by the effective potential in the control-grid plane. The space charge between cathode and control-grid plane causes a potential minimum, whilst owing to the high velocity of the electrons between the control-grid plane and the anode there is scarcely any space charge

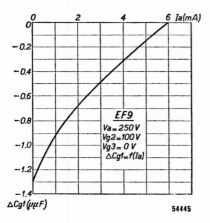

54445

Fig. 166
The control-grid capacity as a
function of the anode current,
for valve type EF 9.

at all. Consequently the capacity of the anode when the valve is cold will differ but little from the capacity while it is warm. Also in screen-grid valves the velocity of the electrons between control grid and anode is usually so high that no variation in the anode capacity takes place. The grid capacity of a triode is determined by the density of the space charge between control grid and cathode. If the anode current falls owing to an increase of the negative grid bias, then the capacity of the control grid will gradually drop with the anode current due to the velocity of emergence of the electrons, until ultimately a capacity value is reached corresponding to the capacity when the valve is cold.

Thus we find that the value of the capacity of the control grid depends on the magnitude of the anode current. The same holds, of course, also for the control grid of a pentode or any other type of multi-grid valve. This capacity variation cannot always be disregarded. In radio- and intermediate-frequency circuits it happens that the amplification of the valves is regulated by varying the negative grid bias (see Chapter XXII); as a result the anode current and the grid capacity are also changed, and this may lead to an inadmissible detuning of the resonant circuit connected to the grid. It is therefore essential to know whether a change in the grid voltage alters the grid capacity. Fig. 166 shows the capacity variation $\varDelta C_{g1}$ measured for the pentode EF 9 as a function of the anode current I_a. According to this curve the drop in the grid capacity of this valve caused by a variation of the anode current from 6 mA to 0 amounts to about 1.3 $\mu\mu$F.

CHAPTER XV

Consequences of Curvature of the Characteristic

79. Distortion

In most cases the ideal shape of the dynamic transfer characteristic of a valve is a straight line (see Fig. 167). If a sinusoidal grid alternating voltage is applied to such a valve there occurs in the anode circuit an alternating current which is likewise sinusoidal. Usually, however, the dynamic transfer characteristic is curved and the anode alternating current produced by a sinusoidal grid alternating voltage will not be sinusoidal (see Fig. 168). In that case the alternating current is said to be distorted. This distortion of an alternating curve (current plotted as a function of time) may be construed as an addition to the anode alternating current of alternating currents the frequencies of which are twice, three times, four times, ... as large as the original frequency of the sinusoidal grid alternating voltage. The original frequency is called the **fundamental wave,** whilst the frequencies that are added through distortion are termed **higher harmonics,** or simply **harmonics.** In Fig. 169 the full line represents the fundamental wave and the broken line the harmonic of a frequency which is three times that of the fundamental wave and of an amplitude a quarter of that of the fundamental wave. The dot-

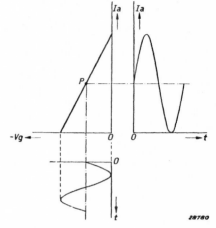

Fig. 167
A sinusoidal alternating voltage on the grid causes a sinusoidal anode alternating current if the transfer characteristic is rectilinear.

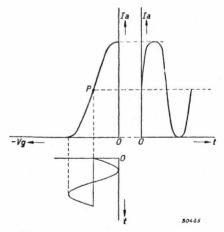

Fig. 168
Curved dynamic transfer characteristic of a pentode. Due to the curvature of the characteristic the anode alternating current is no longer sinusoidal and is therefore distorted.

177

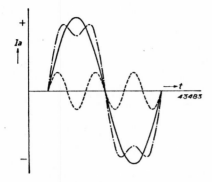

Fig. 169
Full line: Curve of a sinusoidal anode alternating current as a function of time (fundamental wave).
Broken line: Curve of a sinusoidal alternating current of three-fold frequency and an amplitude of 25% of the amplitude of the fundamental wave (25% third harmonic).
Dot-dash line: The resulting curve of the fundamental wave and 25% third harmonic.

and-dash line therefore comprises a fundamental wave with 25% third harmonic. From this figure it is seen that the upper and lower halves of the distorted curve are symmetrical. In the case of pentodes the shape of the dynamic transfer characteristic is usually similar to that shown in Fig. 168, i.e., relatively symmetrical, about the working point P, showing a curvature both at the top and at the bottom. With such a characteristic one obtains an anode-current-versus-time curve which though distorted is still fairly symmetrical. From this it may be concluded that the distortion of the anode alternating current with a pentode consists principally of odd harmonics and in particular the third harmonic, if the grid base of the valve characteristic is fully utilized.

Fig. 170 shows the dynamic characteristic of a triode. As there is no curvature at the top of the transfer characteristic, the anode-current-versus-time curve is asymmetrical and the distortion is mainly due to the second harmonic. With a triode the distortion is, in fact, due mainly to the second harmonic.

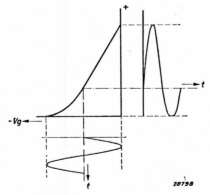

Fig. 170
Dynamic transfer characteristic of a triode. With a sinusoidal grid alternating voltage the anode current is asymmetrically distorted.

Fig. 171 shows a fundamental-frequency wave (full line) accompanied by a sinusoidal alternating current (broken line) the amplitude of which is 25% of that of the fundamental wave and the frequency twice as great (second harmonic). The dot-dash line represents the alternating current formed by the fundamental wave and the second harmonic together. This alternating current therefore comprises a fundamental wave with 25% second harmonic.

As stated earlier, in consequence of the curvature of the transfer characteristic the anode current of a valve is distorted and therefore composed of an alternating current I_1 of fundamental frequency and a number of alternating currents I_2, I_3, I_4, etc. having twice, three times, four times, etc. the frequency of I_1, i.e., a number of harmonics. The effective (r.m.s.) value of the total anode alternating current I_a is then, as is well known, equal to

$$I_a = \sqrt{I_1{}^2 + I_2{}^2 + I_3{}^2 + I_4{}^2 \ldots}$$

in which I_1, I_2, I_3, I_4 etc. represent the effective current values.

The magnitude of the total distortion will be indicated by d_{tot}, that of the second harmonic by d_2, of the third harmonic by d_3, and so on. The distortion factor d_{tot} is defined by

$$d_{tot} = \sqrt{\frac{I_2{}^2 + I_3{}^2 + I_4{}^2 + \ldots}{I_1{}^2}}$$

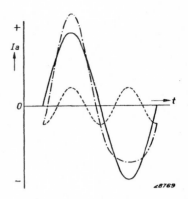

Fig. 171
Full line: Curve of a sinusoidal anode alternating current as a function of time (fundamental wave).
Broken line: Curve of a sinusoidal alternating current with twice the frequency and an amplitude of 25% of the fundamental wave (25% second harmonic).
Dot-dash line: The resulting curve of the fundamental wave and 25% second harmonic.

i.e. by the square root of the ratio of the sum of the squares of the alternating currents with harmonic frequencies and of the square of the alternating current with the fundamental frequency [1]. The distortion associated with the second harmonic is defined by

$$d_2 = I_2/I_1,$$

that associated with the third harmonic by

$$d_3 = I_3/I_1,$$

and so on, so that the total distortion may also be written as

$$d_{tot} = \sqrt{d_2{}^2 + d_3{}^2 + d_4{}^2 + \ldots}$$

The distortion arising in a valve can be determined by computation from the dynamic transfer characteristic. Generally, however, it is determined by measurements. In principle it is measured in the follow-

---

[1] Often distortion is defined as the square root of the ratio of the sum of the squares of the alternating currents with harmonic frequencies and of the square of the total alternating current I_{tot}. At distortion percentages smaller than 10 it is immaterial which of the two definitions is used.

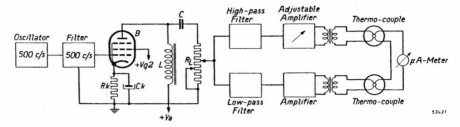

Fig. 172
Schematic diagram of a system for measuring the distortion arising in the anode current of a valve (B) owing to the curvature of the characteristic.

ing way and is diagrammatically represented in Fig. 172. The signal of an oscillator having a frequency of 500 c/s is fed to the grid of the valve to be examined via a filter that suppresses the harmonics of this signal very effectively. The anode of this valve is fed with direct current via a choke coil so dimensioned as not to cause any increase of the distortion; the alternating-current load resistance of the valve is coupled to the anode by means of a condenser. In this way the output power can be determined by measuring the alternating current in the load resistance. Since the impedance of the choke coil for 500 c/s and higher frequencies is very much greater than the load resistance R_L, the anode alternating current will flow almost entirely through R_L. The load resistance is coupled to two filters each followed by an amplifier. The channel with the low-pass filter is for the fundamental frequency, while that with the high-pass filter passes only the harmonics. The high-pass filter in the latter channel can be replaced if desired by another filter that allows only a certain harmonic of the fundamental frequency to pass. With the aid of filters adjusted to the second, third and higher harmonics the distortion can be analysed into its various components and the contribution of each of these harmonics to the total distortion can be determined.

The output side of each of the two amplifiers is coupled to the heater of a thermo-couple [1]) via a transformer. On the secondary side the two thermo-couple elements are connected in antiphase (i.e., the thermo voltages are acting in series in such a manner that the resulting voltage is the difference between the two). Their terminals are connected on

---

[1]) By thermo-couple is meant an element consisting of two different metals, between which is a potential difference that sets up a current in an external circuit and which depends upon the temperature obtaining where the metals are in contact. This point of contact in a thermo-couple is heated by a filament carrying the current to be measured. Thus the potential difference between the metals is a measure of the current passing through the heater.

one side to each other across a sensitive indicator instrument with its zero point in the middle of the scale so that the pointer may be deflected to either side.

The amplification of the high-pass channel is adjustable. This amplification is so adjusted that the pointer indicates zero. In that case the distortion can be calculated from the ratio of the gains of the two amplifiers. If, for instance, the amplifier for the harmonics has 20 times the gain of that for the fundamental frequency, then the distortion amounts to 5%.

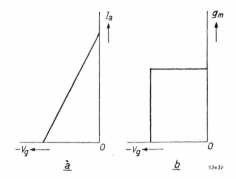

Fig. 173
a) Rectilinear transfer characteristic of a valve.
b) The corresponding transconductance characteristic.

Owing to the curvature of the characteristic there is always distortion, but with R.F., I.F. and mixer valves other phenomena also arise from it, namely hum-modulation, modulation rise, modulation distortion and cross-modulation; in the case of many mixer valves the actual mixing is in fact based on the curvature of the characteristic.

80. Application of two Signals to the same Grid

If one examines the transfer characteristic of a valve it will be clear that the transconductance would be the same throughout if this characteristic were a straight line. The transconductance of a valve can be plotted as a function of the negative grid bias. Fig. 173a represents a transfer characteristic with constant transconductance, whilst Fig. 173b gives the transconductance characteristic that can be derived therefrom.

A transfer characteristic follows a square law [1]) if the transconductance characteristic that can be derived from it is rectilinear (Figs 174a and b). If two signals, one of which has a frequency f_1 and an amplitude V_1 and the other a frequency f_2 and amplitude V_2, are applied to the grid of a valve with a square-law characteristic, in consequence of the alternating voltage V_2 the transconductance according to Fig. 174 will vary with the frequency f_2. As a result the amplification of the signal V_1 will fluctuate between a greater and a smaller value (the transconductance at point a is smaller than that at point b).

---

[1]) That is to say, the anode current is proportional to the square of the grid voltage.

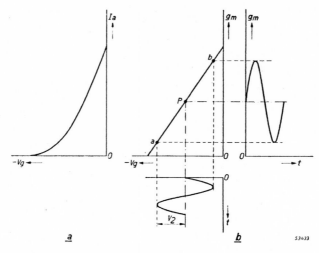

Fig. 174
a) Square-law transfer characteristic of a valve.
b) The corresponding transconductance characteristic is linear.
    With a sinusoidal alternating voltage (amplitude V_2)
    applied to the grid, the amplification of the valve for
    a small signal (ampl. V_1 not shown here) varies between
    a large and a small value with the frequency of the signal V_2.

Figs 175 a and b represent two voltages, V_1 and V_2, as functions of time, both of which are applied to the grid, whilst Fig. 175 c gives the wave form of the anode current. From this it appears that the anode current as a function of time, due to the curvature of the characteristic, is of the same form as the voltage V_1 modulated by the voltage V_2. As already shown (see Chapter IX), owing to the modulation of a signal of frequency f_1 with another signal of frequency f_2, two additional frequencies are produced, one of $f_1 + f_2$ and the other of $f_1 - f_2$. Hence, if a H.F. signal induced in the aerial and another signal having a different frequency (local-oscillator frequency) are applied to the grid of a valve with a square-law characteristic, the alternating current produced in the anode circuit will contain components having the sum frequency and the difference frequency.

In the case of a mixer valve a resonant circuit tuned to the difference frequency is connected in the anode circuit, so that across this circuit an alternating voltage of the latter frequency is formed and this can be further amplified. The sum frequency lies outside the resonance curve of the anode load and therefore is rejected.

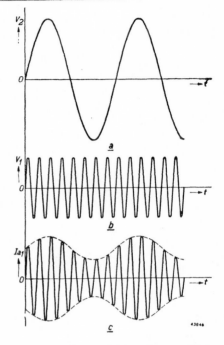

Fig. 175
a) Signal with amplitude V_2 and frequency f_2.
b) Signal with amplitude V_1 and frequency f_1.
    The amplitude of this signal is noticeably smaller than that of the signal V_2 (for instance, V_2 is a strong signal and V_1 a weak one).
c) Resulting anode alternating current with frequency f_1 and an amplitude that varies with frequency f_2. The result is therefore a modulation of the voltage V_1 by the voltage V_2.

81. Hum-modulation

If, in addition to the R.F. signal that is to be amplified, an alternating voltage originating from the a.c. mains is also applied to the grid of an amplifier valve then, if the characteristic follows a square law, in consequence of the curvature of the characteristic the R.F. signal will be modulated by the voltage of mains frequency. Since the mains frequency lies within the audible range it creates a hum in the loudspeaker. This modulation of a hum frequency on the carrier wave is called **hum-modulation.** In the manufacture of R.F. and I.F. amplifiers care has therefore to be taken that no voltage having the mains frequency can be induced in the grid circuit.

The depth of modulation of the interfering voltage in the carrier wave depends upon the curvature of the characteristic. Since in practice the shape of the transfer characteristic is usually such that the hum-modulation depth of the carrier wave produced by the mains alternating voltage varies for different values of negative grid bias and the corresponding transconductances, the hum-modulation as a function of the transconductance is published in the form of a curve. Such a curve gives the r.m.s. value of the hum voltage in volts which produces a modulation depth of 1% as a function of the transconductance (in earlier publications the alternating voltage is often given for a hum-modulation depth of 4%). Such a curve is shown in Fig. 176 for a H.F. pentode. Besides being dependent upon the curvature of the characteristic, the modulation depth of the interfering voltage in the carrier wave also varies with the intensity of the interfering voltage. Hum-modulation depth is directly proportional to the interfering hum voltage;

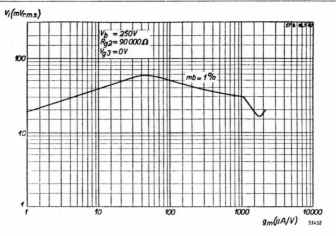

Fig. 176
Curve representing the r.m.s. value of the alternating
voltage of an interfering signal at the grid of a valve EF 9
for 1% hum-modulation as a function of the transconductance.

it is independent of the carrier-wave voltage itself on which the hum
is modulated. It can be proved that 1% hum-modulation corresponds
to $\frac{1}{4}$% distortion due to the second harmonic, so that a curve for 1%
hum modulation gives at the same time the alternating voltage on the
grid, as a function of the transconductance and consequently of the
negative grid bias, which produces an alternating current with $\frac{1}{4}$%
distortion by the second harmonic in the anode circuit. With the aid
of the grid alternating voltage for $\frac{1}{4}$% distortion by the second harmonic derived from the hum-modulation curve it is possible to calculate also the alternating voltage for a greater or smaller distortion,
because this is proportional to the grid alternating voltage. For a
distortion 10 times as great, a 10 times greater alternating voltage
can be allowed.

82. Consequences of the Curvature of the Transconductance Charac-
teristic

In practice the characteristic of a valve never truly follows a square
law. Consequently the transconductance characteristic (transconductance as a function of the negative grid bias) will not be rectilinear
but will show a certain curvature. As a result, apart from the afore-
mentioned possibilities of mixing and hum-modulation, other phenomena also occur: **modulation distortion, modulation rise** and **cross-
modulation.**

(a) **Modulation Distortion and Modulation Rise**

It is readily realised that in the case of a square-law transfer characteristic, i.e., when the transconductance characteristic $[g_m = f(V_g)]$
derived therefrom is rectilinear, the amplification is independent of
the grid alternating voltage. If the alternating voltage between grid
and cathode is sinusoidal then the average transconductance is independent of the amplitude of the alternating voltage. For a certain
negative grid bias the amplitude of the alternating current I_a in the
anode circuit that is tuned to the frequency of the grid alternating
voltage can be plotted as a function of the amplitude of the grid
alternating voltage V_1. Given a square-law characteristic a straight line
will be obtained (see Fig. 177), since $I_a = g_m V_1$. The average transconductance g_m remains the same for all grid alternating voltages,
so that the anode alternating current increases linearly with V_1. In
Fig. 177 suppose that OA is the amplitude of a carrier-wave voltage
on the grid. If this carrier wave is modulated sinusoidally this can be
represented in the figure by a sine curve, such as is shown on the right
below the horizontal axis in Fig. 177.

The amplitude of this modulation in Fig. 177 is equal to AB or AC, so
that the modulation depth of the
grid alternating voltage is equal to
$\dfrac{AB}{OA}$. This modulated carrier wave
sets up a modulated anode alternating current, and from Fig. 177
it follows that the modulation
depth of the anode alternating
current is equal to

$$\frac{A'B'}{OA'} = \frac{AB}{OA}.$$

The modulation depth of the
anode alternating current is thus,
with a square-law characteristic,
equal to the modulation depth of
the grid alternating voltage. From
Fig. 177 it also follows that the
shape of the modulation curve
of the anode current (see the left
side of Fig. 177) is the same as

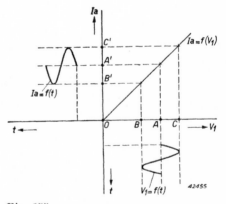

Fig. 177
The relation between the amplitude of the
anode alternating current I_a and the amplitude of the grid alternating voltage V_1
for a square-law characteristic. When, for
instance, the alternating voltage OA is
sinusoidally modulated at low frequency
with an amplitude AB = AC, a modulated
anode current is produced and its modulation depth is the same as that of the
grid alternating voltage having a modulation curve of the same shape.

185

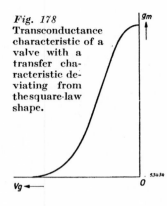

Fig. 178
Transconductance characteristic of a valve with a transfer characteristic deviating from the square-law shape.

that of the modulation curve of the alternating voltage at the grid, so that with a square-law characteristic there is no distortion of the modulation.

When the transfer characteristic does not follow a square law the transconductance as a function of the grid voltage will no longer be rectilinear but curved. The transconductance characteristic will then be as represented in Fig. 178. As a consequence of this curved characteristic the average transconductance will no longer be independent of the amplitude V_1 of the grid alternating voltage and may, for instance, rise as that voltage increases. If, in such a case, the anode alternating current I_a is plotted as a function of the grid alternating voltage V_1, then a curve is obtained as shown, for example, in Fig. 179. If, again, a high-frequency carrier wave, say OA in Fig. 179, is sinusoidally modulated so that the amplitude of the modulation is equal to AB = AC, owing to the curvature of the line giving the ratio of the grid alternating voltage to the anode alternating current the modulation of the anode current will no longer be sinusoidal (see, for instance, in Fig. 179 the wave form of the modulation current on the left of the vertical axis). In that case the modulation is distorted and one refers to **modulation distortion.** In this case the modulation of the anode alternating current therefore contains harmonics in addition to a fundamental wave of the same frequency as the modulation of the high-frequency alternating voltage applied to the grid.

Owing to the curvature of the transfer characteristic, with a sinusoidal high-frequency alternating voltage V_1 the alternating anode current will be distorted. With a square-law transfer characteristic a second harmonic (having double the frequency of V_1) will be produced in the anode circuit. Since it has been initially assumed that the resonant anode circuit is tuned to the frequency of the high-frequency alternating voltage, the harmonics in the anode alternating current will not give rise to any appreciable alternating voltages across the anode impedance and thus will not have any noticeable effect.

Accompanying the modulation distortion we have also the phenomenon of **modulation rise**, which is an increase of the modulation depth of a modulated carrier wave after this has been amplified by the valve. Owing to the curvature of the characteristic indicating the relation between the anode alternating current I_a and the grid alternating voltage V_1, the ratio of the fundamental-wave amplitude of the anode-current modulation to the amplitude of the unmodulated anode alternating current (carrier wave) becomes greater than the modulation depth of the grid alternating voltage. If the modulation depth of the grid alternating voltage is represented by m_1, and that of the anode alternating current or voltage by m_2, the percentage of modulation rise can be expressed by

$$M = \frac{m_2 - m_1}{m_1} \, 100\%. \quad (74)$$

The modulation distortion, which we shall term D, is proportional to the square of the carrier-wave voltage at the grid and proportional to the modulation depth of that voltage. For the modulation distortion we have the following simple formula

$$D = F \, V_1{}^2 m_1 \times 100\%, \quad (75)$$

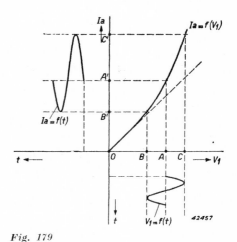

Fig. 179

The relation between the amplitude of the anode alternating current I_a and the amplitude of the grid alternating voltage V_1 when the shape of the transfer characteristic deviates from the square law. In this case when the grid alternating voltage is modulated at low frequency (e.g., in accordance with the sinusoidal curve with amplitude $AB = AC$ drawn below the horizontal axis) a modulated anode current is produced and its modulation wave deviates from the sinusoidal form, so that distortion of the modulation occurs. Moreover, the ratio of the fundamental wave of this distorted modulation to the H.F. anode-current wave (carrier) is greater than the modulation depth of the grid alternating voltage V_1 (thus modulation rise is occurring).

in which F is a factor related to the shape of the transfer characteristic (i.e., the curvature of the transconductance characteristic), V_1 the carrier-wave voltage on the grid and m_1 the modulation depth of that voltage.

The modulation rise is likewise proportional to the square of the carrier-wave voltage on the grid and depends, moreover, upon the modulation depth. For the modulation rise the formula is

$$M = {}^4/_3 \, F \, V_1{}^2 \, (1 - {}^3/_8 \, m_1{}^2), \quad (76)$$

in which F is the same factor as in Formula (75). In the case of small modulation depths the term in parentheses is practically equal to unity, so that we then have

$$M = {}^4/_3 \, F \, V_1{}^2. \quad (77)$$

(b) Cross-modulation

By cross-modulation is understood the phenomenon whereby—after amplification by the valve—the carrier wave of the desired signal contains also the modulation of a (generally) very powerful transmitter operating on an adjacent frequency. If the carrier wave of the desired transmitter is not present then, given the same tuning position and the same sensitivity of the receiver, the interfering A.F. signal will be much weaker or absent. The risk of the modulation of the undesired transmission being superimposed on the carrier wave of the desired transmission is generally greatest in the first valve of the receiver. The selectivity of the tuned circuits preceding this valve is as a rule not great enough to prevent a carrier wave of great intensity working on an adjacent frequency from reaching the grid of the input valve. Consequently the selectivity of the tuned circuits following the first valve can no longer separate the undesired modulation from the desired signal. Sometimes this phenomenon is wrongly regarded as a lack of selectivity in the receiver as a whole, whereas it is only due to the properties of the first valve and possibly inadequate selectivity before that valve. Therefore, high selectivity in the circuits preceding the first valve is desirable. For this reason two circuits coupled as a band-pass filter are often utilized before the first valve, and in the design of valves for the first stage of receivers, special attention is devoted to a favourable shape of the transfer characteristic with a view to avoiding cross-modulation.

Cross-modulation arises in the following way. Let us suppose that on the grid of a valve having a curved transfer characteristic we have a desired signal with a carrier-wave voltage V_1 and an interfering signal with a carrier-wave voltage V_2 (orginating, for example, from a local transmitter). The amplification A_1 of the desired signal V_1 in the valve can be plotted as a function of the carrier-wave voltage of the interfering signal V_2. As already stated (see Section 82a) in the case of a square-law

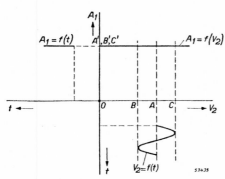

Fig. 180
Amplification A_1 of a desired signal V_1 in a valve with a square-law transfer characteristic as a function of the carrier-wave voltage V_2 at the grid. If V_2 is modulated (if consequently V_2 varies, for instance, between OB and OC) then in this case the modulation does not affect the amplification, because this is independent of V_2.

188

transfer characteristic, i.e., where the transconductance characteristic is rectilinear, the average transconductance, and thus also the amplification, is independent of the amplitude of the grid alternating voltage. Therefore in the case of a square-law characteristic A_1 will be independent of V_2 and we obtain the horizontal line drawn in Fig. 180, in which OA represents the amplitude of the carrier wave of the interfering signal V_2 at the grid and AB = AC is the amplitude of the modulation of that carrier wave.

Since the amplification is independent of the amplitude V_2, this will not be affected by the amplitude variation of V_2 due to the modulation. If, however, the transfer characteristic does not follow a square law (i.e., in the case of a curved transconductance characteristic) the amplification A_1 will depend upon the amplitude V_2 of the interfering signal. In that case the line representing the amplification A_1 as a function of the amplitude V_2 of the interfering signal will be curved, as shown in Fig. 181; here OA is again the amplitude of the interfering carrier wave V_2 and AB = BC is the amplitude of the modulation of that carrier wave. The amplitude variation of V_2 caused by this modulation results in a variation of the amplification A_1 of the desired signal V_1, as may be seen in Fig. 181, this amplification fluctuating with the modulation frequency of V_2. This leads to a modulation of the signal V_1 the frequency of which corresponds to the modulation frequency of V_2.

Thus we find the modulation of V_2 superimposed on the signal V_1. This modulation transferred from V_1 to V_2 will as a rule be distorted owing to the curvature of the A_1/V_2 characteristic. The modulation depth produced by the interfering carrier wave upon the desired carrier wave is proportional to the square of the amplitude V_2 of the interfering signal and to its modulation depth. The modulation depth of the desired carrier wave caused by the interfering transmitter will now be indicated by m_k. According to the foregoing this modulation depth is equal to

$$m_k = CV_2{}^2 m_2, \qquad (78)$$

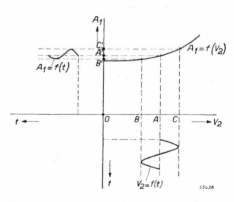

Fig. 181
Amplification A_1 of a desired signal V_1 in a valve with a transfer characteristic deviating from the square-law form as a function of the carrier-wave voltage V_2 at the grid. If V_2 is modulated, for instance, as indicated by the full-line curve below the horizontal axis, the amplification A_1 of the signal V_1 is likewise modulated. This is equivalent to a modulation of the signal V_1.

189

where C is a constant depending on the shape of the valve characteristic. This formula holds only for small amplitudes of the desired
carrier wave (see Chapter XVI).

The degree of interference caused by cross-modulation is determined
by the ratio of the modulation depth m_k on the desired carrier wave due
to cross-modulation and the modulation depth m_1 of the desired carrier
wave, thus by

$$\frac{m_k}{m_1} = \frac{CV_2{}^2 m_2}{m_1}. \tag{79}$$

If the modulation depth of the interfering carrier wave is equal to
that of the desired carrier wave, as approximately occurs in practice
in many cases, then Equation (79) is simplified and becomes

$$\frac{m_k}{m_1} = CV_2{}^2. \tag{80}$$

This ratio, where $m_2 = m_1$, is termed the **cross-modulation factor** and
is represented by the letter K. Thus the cross-modulation factor is
defined as the ratio of the modulation depth brought about by the
interfering transmitter on the desired carrier wave to the modulation
depth of the desired carrier wave itself, assuming that the modulation
depth of both carrier waves is equal. According to Equation (80) the
cross-modulation factor is independent of the signal strength of the
desired station, but is proportional to the square of the signal strength
of the interfering station. If the modulation depth of both stations
is 30% ($m_1 = m_2 = 0.30$) a cross-modulation factor of 1% for a given
voltage of the carrier wave from the indesired station means, therefore,
that the modulation depth of the undesired station on the desired
carrier wave amounts to $0.01 \times 0.3 \times 100 = 0.3\%$.

Just as with modulation distortion and modulation rise, cross-modulation is determined by the curvature of the transconductance characteristic. Consequently these factors are related to each other, i e. 1%
cross-modulation corresponds to $\frac{1}{2}\%$ modulation rise and $^3/_8\%$ modulation distortion; hence the factor C of the Equations (78), (79) and
(80) is equal to $^8/_3$ times the factor F of Equation (75) and for (78)
can also be written

$$m_k = {}^8/_3 FV_2{}^2 m_2. \tag{81}$$

For (80) we may write

$$K = \frac{m_k}{m_1} = {}^8/_3 FV_2{}^2. \tag{82}$$

Generally we shall require to know what voltage from an interfering
station produces a certain percentage of cross-modulation on the grid

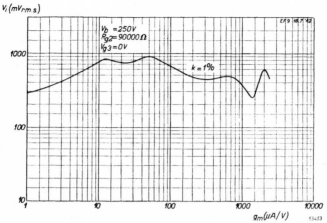

Fig. 182
Curve indicating the r.m.s. value of the voltage of a
modulated interfering carrier wave at the grid of valve
EF 9 producing 1% cross-modulation as a function of the
transconductance of that valve, the modulation depth of
the two carrier waves being 30%.

of a valve. According to Equation (82) this voltage equals $\sqrt{\dfrac{3K}{8F}}$. For
a given ratio of the interfering modulation level to the desired modulation level the value of the voltage of the interfering carrier wave
producing that ratio can be calculated. Since the curvature of the
transconductance characteristic varies for different values of negative
grid bias and consequently for different transconductances, the factor
F in Equation (82) will likewise vary with the grid bias or the transconductance. For a fixed value of the cross-modulation factor K this
means that the voltage of the interfering carrier wave producing this
cross-modulation factor depends upon the value of the negative grid
bias and upon the corresponding transconductance. The r.m.s. value
of the interfering carrier-wave voltage V_2 resulting in a certain cross-
modulation factor K can be plotted as a function of the negative grid
bias or of the transconductance (in Philips publications on valves V_2
is nearly always given as a function of the transconductance). Such a
curve is called a **cross-modulation curve.**
Cross-modulation curves are published on a double logarithmic scale
for 1% cross-modulation (see Fig. 182), because the ratio of 1 : 100
between the interfering and the desired output signals complies best
with the requirements met in practice (formerly the r.m.s. value of
the interfering carrier-wave voltage at the grid of the valve as a function
of the transconductance was given for 6% cross-modulation).

191

The cross-modulation curve applies also to the modulation rise and to the modulation distortion; the r.m.s. voltages of the desired carrier wave for a certain value of modulation rise and of modulation distortion can also be read from the cross-modulation curve. It can be shown that the distortion by the third harmonic of the anode alternating current of a valve with sinusoidal input voltage is equal to

$$d_3 = \frac{2}{9} F V_i^2, \tag{83}$$

where F is again the same factor as in Equations (81) and (82). Therefore the distortion of an unmodulated sinusoidal input voltage V_i due to the third harmonic is $^1/_{12}$ of the cross-modulation factor. Consequently, 1% cross-modulation corresponds to 0.083% distortion due to the third harmonic (6% cross-modulation corresponding to 0.5% distortion). Accordingly the cross-modulation curve can also be used to determine the third-harmonic distortion occurring with an unmodulated sinusoidal voltage on the grid. If a R.F. amplifying valve with variable transconductance, the cross-modulation and hum-modulation curves of which are published, is to be used in an A.F. stage with controlled amplification, then with the aid of these curves it is possible to determine the distortion values due to the second and third harmonics for any negative grid bias or amplification and for any value of the grid alternating voltage.

For those valves which are used as variable-gain amplifiers it is stated in the data how far the transconductance can be reduced without serious distortion occurring. The range of transconductance variation in which no serious distortion takes place will be called the **useful gain-control range.** For the H.F. pentode EF 9, for instance, with a supply voltage of 250 volts the limit of the useful gain-control range is given as $^1/_{500}$ of the initial transconductance in the non-regulated condition. At the same time the corresponding negative grid bias is given. In the preliminary design of a receiving set the indication of the limit of the useful gain-control range gives an idea of the scope of gain control for each valve.

In the final design of receiving sets one must, of course, take account of the cross-modulation curves of the various valves. If the signal voltages to be handled are very high it is possible that in certain sets the gain-control range indicated cannot be used in its entirety. The limit of the useful gain-control range likewise indicates where the cross-modulation curves and the curves for the related magnitudes (modulation distortion etc.) of various valves of the same type begin to show a fairly large dispersion.

CHAPTER XVI

Representation of the Transfer Characteristic by an Exponential Series, and its Application

83. The Exponential Series Used

It is well known that a function $y = f(x)$ can be approximated as a series by a suitable choice of exponents. The function $y = \cos x$, for instance may be represented by

$$y = 1 - \frac{x^2}{1 \times 2} + \frac{x^4}{1 \times 2 \times 3 \times 4} - \frac{x^6}{1 \times 2 \times 3 \times 4 \times 5 \times 6} +$$

$$\frac{x^8}{1 \times 2 \times 3 \times 4 \times 5 \times 6 \times 7 \times 8}, \text{etc.,}$$

the accuracy of the approximation depending upon the number of terms chosen; two or three terms generally suffice.

The static or dynamic transfer characteristic of a valve is likewise a function $y = f(x)$ or $I_a = f(V_g)$, and this can also be represented by an exponential series, the degree of accuracy being all the greater as more terms are used. The advantage of expressing a transfer characteristic as a series is that various effects can then be deduced both as to existence and magnitude, and conclusions drawn as to what should be done to get any desired effect or avoid any undesired ones. If, for instance, there is on the grid a sinusoidal alternating voltage $V_m \cos \omega t$, where V_m = amplitude, then with the aid of a transfer characteristic written as a formula we can determine the nature of the alternating current at the anode. If the transfer characteristic proves then to be non-linear this current will no longer be sinusoidal.

Various exponential series could be considered for representing a characteristic. In this book we will use the series

$$y = y_0 + \alpha x + \beta x^2 + \gamma x^3 + \delta x^4 + \varepsilon x^5 \ldots \text{etc.}$$

In most cases this yields simple results that are found by standard mathematical means.

Thus we may write a (static or dynamic) transfer characteristic as

$$I_a = I_0 + \alpha V_g + \beta V_g^2 + \gamma V_g^3 + \delta V_g^4 + \varepsilon V_g^5 \ldots \quad (84)$$

193

This equation applies for the working point P (see, e.g., Fig. 183) where the anode current is I_o, for if in (84) $V_g = 0$ then $I_a = I_o$. Given a valve with a characteristic expressed by (84), where the coefficients a, β, γ, δ, ε, etc. are assumed known, and applying to the grid a sinusoidal alternating voltage $V_g = V_m \cos \omega t$, then by substituting $V_m \cos \omega t$ for V_g we get

$$I_a = I_o + a V_m \cos \omega t + \beta V_m{}^2 \cos^2 \omega t + \gamma V_m{}^3 \cos^3 \omega t + \dots \quad (85)$$

Considering the first four terms only,

$$\text{since } \cos^2 \omega t = \tfrac{1}{2} + \tfrac{1}{2} \cos 2\,\omega t$$
$$\text{and } \cos^3 \omega t = \tfrac{3}{4} \cos \omega t + \tfrac{1}{4} \cos 3\,\omega t,$$

and substituting these in (85) we obtain :

$$I_a = I_o + \frac{\beta}{2} V_m{}^2 + \left(a V_m + \frac{3}{4} \gamma V_m{}^3\right) \cos \omega t + \frac{\beta}{2} V_m{}^2 \cos 2\omega t +$$

$$+ \frac{\gamma}{4} V_m{}^3 \cos 3\omega t \quad (86)$$

From this equation several conclusions are to be drawn, viz:

a. The d.c. component of anode current varies with the square of the amplitude of the grid alternating voltage; rectification consequently takes place. With very small alternating voltages, where $\frac{\beta}{2} V_m{}^2$ is extremely small, the anode d.c. component will be practically equal to I_o.

b. The transconductance of the valve, which is equal to the amplitude of the fundamental component of anode alternating current of angular frequency ω divided by the amplitude of the grid alternating voltage, is

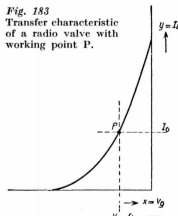

Fig. 183
Transfer characteristic of a radio valve with working point P.

$$g_m = \frac{a V_m + \frac{3}{4} \gamma V_m{}^3}{V_m} = a + \frac{3}{4} \gamma V_m{}^2, \quad (87)$$

and thus varies with the square of the amplitude of the grid alternating voltage V_m. If the value of V_m is very small then in most cases also $\frac{3}{4} \gamma V_m{}^2$ is likewise extremely small and the transconductance is practically equal to a. In the case of a square-law characteristic, where only the first three terms of Equation (85) occur, the term $\frac{3}{4} \gamma V_m{}^2$ of Equation

(87) does not exist and consequently $g_m = a$. The transconductance is then independent of the amplitude of the grid alternating voltage.

c. In addition to a direct-current component $I_o + \frac{\beta}{2} V_m^2$ and a fundamental wave of amplitude $a V_m + \frac{3}{4} \gamma V_m^3$, the anode current contains a second harmonic (frequency 2ω, amplitude $\frac{\beta}{2} V_m^2$) and a third harmonic (frequency 3ω, amplitude $\frac{\gamma}{4} V_m^3$). Still higher harmonics would be found in the anode current if more terms were used in Equation (85). Equation (86) thus shows that, as explained in Section 79, Chapter XV, the curvature of the characteristic results in distortion of the anode alternating current.

84. Distortion due to Curvature of the Characteristic

As follows from Section 79, distortion is defined by:

$$d_{tot} = \sqrt{\frac{I_2^2 + I_3^2 + I_4^2 + \cdots}{I_1^2}} = \sqrt{d_2^2 + d_3^2 + d_4^2 + \cdots}$$

where $d_2 = \frac{I_2}{I_1}$ and $d_3 = \frac{I_3}{I_1}$ (I_1 = fundamental, I_2, I_3 = amplitudes of second and third harmonics, respectively).
Hence, as follows also from Equation (86),

$$d_2 = \frac{\beta/2 \, V_m^2}{a V_m \left(1 + \frac{3}{4} \gamma / a V_m^2\right)} = \frac{\beta}{2a} \frac{V_m}{1 + \frac{3}{4} \gamma / a V_m^2} \tag{88}$$

and

$$d_3 = \frac{\gamma/4 \, V_m^3}{a V_m \left(1 + \frac{3}{4} \gamma / a V_m^2\right)} = \frac{\gamma}{4a} \frac{V_m^2}{1 + \frac{3}{4} \gamma / a V_m^2} \tag{89}$$

For small values of V_m Equations (88) and (89) may be simplified to

$$d_2 = \frac{\beta}{2a} V_m \tag{90}$$

and

$$d_3 = \frac{\gamma}{4a} V_m^2 \tag{91}$$

85. Mixing two Grid Alternating Voltages of Different Frequency

When there are two sinusoidal alternating voltages, $V_1 \cos \omega_1 t$, and $V_2 \cos \omega_2 t$, on the grid of a valve, then by substituting $V_1 \cos \omega_1 t +$

$V_2 \cos \omega_2 t$ for V_g in Equation (85) and for simplicity disregarding the terms with V_g^3 and higher, we obtain

$$I_a = I_0 + aV_1 \cos \omega_1 t + aV_2 \cos \omega_2 t + \beta V_1^2 \cos^2 \omega_1 t +$$

$$\beta V_2^2 \cos^2 \omega_2 t + 2 \beta V_1 V_2 \cos \omega_1 t \cos \omega_2 t \qquad (92)$$

Substituting in this equation

$$\cos^2 \omega_1 t = \tfrac{1}{2} + \tfrac{1}{2} \cos 2\omega_1 t,$$
$$\cos^2 \omega_2 t = \tfrac{1}{2} + \tfrac{1}{2} \cos 2\omega_2 t, \text{ and}$$
$$\cos \omega_1 t \cos \omega_2 t = \tfrac{1}{2} \cos (\omega_1 - \omega_2) t + \tfrac{1}{2} \cos (\omega_1 + \omega_2) t,$$

we obtain

$$I_a = I_0 + \frac{\beta}{2} (V_1^2 + V_2^2) + aV_1 \cos \omega_1 t + aV_2 \cos \omega_2 t +$$

$$\frac{\beta}{2} V_1^2 \cos 2 \omega_1 t + \frac{\beta}{2} V_2^2 \cos 2 \omega_2 t + \beta V_1 V_2 \cos (\omega_1 - \omega_2) t +$$

$$\beta V_1 V_2 \cos (\omega_1 + \omega_2) t \qquad (93)$$

In the anode circuit there are therefore alternating currents with six different frequencies, viz, the fundamental frequencies, and the second harmonics of the two original grid alternating voltages; and, in addition, the frequencies equal to the difference and the sum of those at the grid.

As already indicated in Chapter XV, Section 80, in the case of a mixing valve the practice is to insert in the anode lead a resonant circuit tuned to the difference frequency in order to eliminate all others. In such a case, therefore, the only term of importance in Equation (93) is $\cos (\omega_1 - \omega_2)t$. If the amplitude of the R.F. signal on the grid of a mixing valve is V_1, then the conductance of the valve circuit operating with an anode circuit tuned to $(\omega_1 - \omega_2)$ [i.e., the anode-current component of frequency $(\omega_1 - \omega_2)$ divided by the grid-voltage amplitude of frequency ω_1] is equal to

$$g_c = \frac{\beta V_1 V_2}{V_1} = \beta V_2. \qquad (94)$$

The quantity βV_2 is called the conversion conductance (see Chapter XXI).

86. Amplitude Modulation by Grid Injection of two Alternating Voltages of Different Frequency

When there are at the grid two alternating voltages $V_1 \cos \omega_1 t$ and $V_2 \cos \omega_2 t$, ω_2 being for instance small with respect to ω_1, and $V_2 \cos \omega_2 t$ being in that case the modulating signal, and when we have in the anode lead a resonant circuit which has an impedance only at the frequencies ω_1, $(\omega_1 - \omega_2)$ and $(\omega_1 + \omega_2)$ (only currents of these frequencies being consequently of interest), Equation (93) is simplified into

$$I_a = a\,V_1 \cos \omega_1 t + \beta\,V_1 V_2 \cos(\omega_1 - \omega_2)t + \beta\,V_1 V_2 \cos(\omega_1 + \omega_2)t \quad (95)$$

According to Equation (28) of Chapter IX, a modulated R.F. signal with modulation depth m is represented by

$$i = I_o \cos\omega_o t + \tfrac{1}{2} m\,I_o \cos(\omega_o - p)t + \tfrac{1}{2} m\,I_o \cos(\omega_o + p)t, \quad (96)$$

when current flows are substituted for voltages. Putting $I_o = aV_1$, $i = I_a$, $\omega_o = \omega_1$ and $p = \omega_2$ then Equation (96) becomes

$$I_a = aV_1 \cos \omega_1 t + \tfrac{1}{2} m\,aV_1 \cos(\omega_1 - \omega_2)t + \tfrac{1}{2} m\,aV_1 \cos(\omega_1 + \omega_2)t. \quad (97)$$

From (95) and (97) it follows that

$$\tfrac{1}{2} m\,a = \beta\,V_2 \text{ or } m = \frac{2\,\beta}{a}V_2. \quad (98)$$

In that case, if $\beta\,V_2$ is substituted for $\tfrac{1}{2} m\,a$ in the second and third terms of Equation (97), we obtain Equation (95).

Where $V_2 \cos \omega_2 t$ is a hum voltage on the grid of the valve, then the hum-modulation depth is equal to $\dfrac{2\,\beta}{a}$ times the amplitude of the hum voltage.

87. Modulation Distortion and Modulation Rise

If there is on the grid a modulated alternating voltage $V_m(1 + m \cos pt) \cos \omega t$ and in the anode lead a resonant circuit tuned to ω, then, by substituting $V_m(1 + m \cos pt)$ for V_m, the third term of Equation (86) which is alone important becomes

$$I_a = a\,V_m \left\{ (1 + m \cos pt) + \tfrac{3}{4} \frac{\gamma}{a} V_m^2 (1 + m \cos pt)^3 \right\} \cos \omega t. \quad (99)$$

By expansion we obtain

$$I_a = aV_m \cos \omega t \left[\left\{ 1 + \tfrac{3}{4} \frac{\gamma}{a} V_m^2 (1 + \tfrac{3}{2} m^2) \right\} + \right.$$
$$m \cos pt \left\{ 1 + \tfrac{9}{4} \frac{\gamma}{a} V_m^2 (1 + \tfrac{1}{4} m^2) \right\} + \tfrac{9}{8} \frac{\gamma}{a} m^2 V_m^2 \cos 2pt +$$
$$\left. \tfrac{3}{16} \frac{\gamma}{a} m^2 V_m^2 \cos 3pt \right] \quad (100)$$

Within the square brackets there are four terms, the first two of which have the form $A + B\,m \cos pt$ or $A\left(1 + \dfrac{B}{A} m \cos pt\right)$, which means that the depth of modulation is equal to $\dfrac{B}{A}$ m. Consequently the modulation depth of the anode current is equal to

$$m_a = \frac{1 + \frac{9}{4}\frac{\gamma}{a} V_m{}^2 \left(1 + \frac{1}{4} m^2\right)}{1 + \frac{3}{4}\frac{\gamma}{a} V_m{}^2 \left(1 + \frac{3}{2} m^2\right)}\, m \qquad (101)$$

According to Equation (74) in Chapter XV the modulation rise M equals $\dfrac{m_2 - m_1}{m_1}$ or, in the terms used above, $\dfrac{m_a - m}{m}$, which means that

$$M = \frac{\dfrac{1 + \frac{9}{4}\frac{\gamma}{a} V_m{}^2 \left(1 + \frac{1}{4} m^2\right)}{1 + \frac{3}{4}\frac{\gamma}{a} V_m{}^2 \left(1 + \frac{3}{2} m^2\right)}\, m - m}{m} = \frac{\frac{3}{2}\frac{\gamma}{a} V_m{}^2 \left(1 - \frac{3}{8} m^2\right)}{1 + \frac{3}{4}\frac{\gamma}{a} V_m{}^2 \left(1 + \frac{3}{2} m^2\right)} \qquad (102)$$

For small values of V_m the denominator is about equal to unity, so that in this case

$$M = \frac{3}{2}\frac{\gamma}{a} V_m{}^2 \left(1 - \frac{3}{8} m^2\right) \qquad (103)$$

If the depth of modulation is also small then the modulation rise is equal to

$$M = \frac{3}{2}\frac{\gamma}{a} V_m{}^2 \qquad (104)$$

(For m = 0.3, $\frac{3}{8} m^2$ equals $0.375 \times 0.09 = 0.03375$ and is almost negligible compared with unity.)

The third and fourth terms within the square brackets in Equation (100) contain respectively the double and triple values of the modulation frequency p. Since they relate to the amplitudes of the second and third harmonics of the modulation of the anode alternating current, they show that the modulation will be distorted. The second-harmonic modulation distortion equals

$$D_2 = \frac{\frac{9}{8}\frac{\gamma}{a} m^2 V_m{}^2}{m\left\{1 + \frac{3}{4}\frac{\gamma}{a} V_m{}^2 \left(1 + \frac{1}{4} m^2\right)\right\}} \qquad (105)$$

For small values of V_m the denominator can again be taken as approximately unity. In that case (105) becomes

$$D_2 = \frac{9}{8}\frac{\gamma}{a} m V_m{}^2, \qquad (106)$$

in which $\frac{9}{8}\frac{\gamma}{a}$ is the factor F in Equation (75) of the previous chapter. In the same way we find that for small values of V_m, the third-harmonic modulation distortion equals

$$D_3 = \frac{3}{16}\frac{\gamma}{a}mV_m{}^2 \text{ or } D_3 = \frac{1}{6}F\,V_m{}^2\,m. \qquad (107)$$

Generally, distortion due to the third harmonic is negligible compared with that caused by the second harmonic, which may therefore be regarded as the determining factor here (Equation 106).

88. Cross-modulation

As indicated in Section 82(b), Chapter XV, cross-modulation arises from the presence of two R.F. signals on the grid of the valve, namely $V_{m1}\cos\omega_1 t$ and $V_{m2}\cos\omega_2 t$, of which the latter may be, for instance, the interfering voltage. If this interfering voltage is modulated we may write it in the form $V_{m2}\cos\omega_2 t\,(1 + m_2\cos pt)$. Departing again from Equation (84) and substituting $V_{m1}\cos\omega_1 t + V_{m2}\cos\omega_2 t$ for V_g— while considering only the terms with $\cos\omega_1 t$ and ignoring those of a higher order than the third—we obtain

$$I_a = aV_{m1}\cos\omega_1 t\,(1 + \frac{3}{4}\frac{\gamma}{a}V_{m1}{}^2 + \frac{3}{2}\frac{\gamma}{a}V_{m2}{}^2) \qquad (108)$$

Since cross-modulation is only noticed when a weak signal is being received, V_{m1} is small and $\frac{3}{4}\frac{\gamma}{a}V_{m1}{}^2$ can be disregarded, so that (108) is simplified and becomes:

$$I_a = a\,V_{m1}\cos\omega_1 t\,(1 + \frac{3}{2}\frac{\gamma}{a}V_{m2}{}^2) \qquad (109)$$

Hence it follows, as explained in Chapter XV, Section 82b, and in Fig. 180, that the amplification A_1 for the desired signal depends upon the amplitude of an interfering signal V_2, provided that a term with γ occurs in the equation for the anode current, which means that the characteristic does not follow a square law, or is not a straight line. If the interfering signal is modulated then in Equation (109) V_{m2} has to be replaced by $V_{m2}\,(1 + m_2\cos pt)$, and we obtain

$$I_a = a\,V_{m1}\cos\omega_1 t\,[1 + \frac{3}{2}\frac{\gamma}{a}V_{m2}{}^2(1 + \frac{1}{2}m_2) + 3\frac{\gamma}{a}V_{m2}{}^2\,m_2\cos pt] \qquad (110)$$

The expression between the square brackets is again of the form

$$A + Bm_2\cos pt = A\,(1 + \frac{B}{A}m_2\cos pt)$$

199

so that the modulation depth of the anode alternating current of frequency ω_1 is equal to $\dfrac{B}{A} m_2$, or

$$m_k = \frac{3 \dfrac{\gamma}{\alpha} V_{m2}^2}{1 + \frac{3}{2} \dfrac{\gamma}{\alpha} V_{m2}^2 \left(1 + \frac{1}{2} m_2\right)} m_2.$$ (111)

For relatively small values of V_{m2} Equation (111) can be simplified to

$$m_k = 3 \frac{\gamma}{\alpha} V_{m2}^2 m_2.$$ (112)

According to Equation (82) of Chapter XV the cross-modulation factor is $K = \dfrac{m_k}{m_1}$, where m_1 represents the modulation depth of the desired signal $V_{m1} \cos \omega_1 t$, so that if $m_1 = m_2$ the cross-modulation factor is:

$$K = 3 \frac{\gamma}{\alpha} V_{m2}^2 = \frac{8}{3} F V_{m2}^2.$$ (113)

89. Determining Coefficients of the Exponential Series

In the foregoing sections the various factors derived from curvature of the characteristic have been expressed in terms of coefficients α, β and γ of an exponential series representing anode current as a function of grid voltage, but to get quantitative as well as qualitative results the actual values have to be known. These can be determined, for instance, by measuring the distortion caused by second and third harmonics, d_2 and d_3, in the anode alternating current for a given grid bias with a small superimposed sinusoidal voltage $V_m \cos \omega t$ at the grid, as indicated in Chapter XV, Section 79 (see Fig. 172). By this means it is possible to determine the ratios $\dfrac{\beta}{\alpha} = \dfrac{2d_2}{V_m}$ and $\dfrac{\gamma}{\alpha} = \dfrac{4d_3}{V_m^2}$, where V_m is again the amplitude of the grid alternating voltage. The coefficient α is the transconductance of the valve, arrived at by dividing the fundamental-frequency component of anode alternating current by the grid alternating voltage. The quotients $\dfrac{\beta}{\alpha}$ and $\dfrac{\gamma}{\alpha}$ can also be determined from curves for hum-modulation and cross-modulation published for the valve concerned (see e.g. Figs 176 and 182).

According to Equation (98) $m_b = \dfrac{2 \beta}{\alpha} V_2$ or $\dfrac{\beta}{\alpha} = \dfrac{m_b}{2V_2}$. From Fig. 176 it follows that with $g_m = \alpha = 1000 \, \mu\text{A/V}$ the r.m.s. value of the grid alternating voltage for 1% hum-modulation amounts to 30 mV. The peak value

$V_2 = V_i \sqrt{2}$ is therefore 0.0425 V, so that $\dfrac{\beta}{a} = \dfrac{0.01}{2 \times 0.0425} = 0.118$.

Since $a = 0.001$ A/V, therefore $\beta = 0.118 \times 10^{-3} = 118 \times 10^{-6}$ A/V².

According to Equation (113) $K = 3 \dfrac{\gamma}{a} V_{m2}{}^2$ or $\dfrac{\gamma}{a} = \dfrac{K}{3 V_{m2}{}^2}$. From Fig. 182 it follows that with $g_m = 1000$ μA/V $(a = 10^{-3})$ the r.m.s. value of the interfering carrier-wave voltage on the grid of the valve EF 9, for 1% cross-modulation, amounts to 370 mV, corresponding to a peak value $V_{m2} = 0.524$ V. Consequently $\dfrac{\gamma}{a} = \dfrac{0.01}{3 \times (0.524)^2} = \dfrac{0.01}{0.825} = 0.0121$. Therefore, since $a = 0.001$, $\gamma = 0.0121 \times 10^{-3} = 1.21 \times 10^{-5}$ A/V³.

The curves for hum-modulation and cross-modulation show that the coefficients a, β and γ vary with the working point on the valve characteristic. From Figs 176 and 182 we find that for $g_m = 100$ μA/V $(a = 10^{-4})$ $\beta = 7 \times 10^{-6}$ and $\gamma = 3.9 \times 10^{-7}$. Measuring the distortion produced by second and third harmonics with apparatus such as that shown in Fig. 172, we can determine the coefficients a, β and γ for any setting of the valve operating point.

Such apparatus are not, however, always available, and we are then obliged to work with the hum-modulation and cross-modulation characteristics. These, however, hold good only for certain electrode voltages, and for static valve characteristics, so that we cannot derive the necessary data for any appreciably different cases, for instance with resistance coupling of a valve, or for different screen-grid voltage. If the static or dynamic characteristic is known for the adjustment with which we desire to work, then the amplitudes of the fundamental wave and the harmonics in the anode alternating current can be determined graphically, and from those the distortions d_2, d_3, etc., and thus, if required, one can calculate the coefficients a, β and γ.

From Equation (86) it follows that the anode alternating current can be represented by the equation:

$$I_a = I_g + I_1 \cos \omega t + I_2 \cos 2\omega t + I_3 \cos 3\omega t + \dots, \quad (114)$$

where

$$I_g = I_o + \frac{\beta}{2} V_m{}^2, \quad (115)$$

$$I_1 = a V_m \left(1 + \frac{3}{4} \frac{\gamma}{a} V_m{}^2\right), \quad (116)$$

$$I_2 = \frac{\beta}{2} V_m{}^2, \quad (117)$$

$$I_3 = \frac{\gamma}{4} V_m{}^3, \text{ etc.} \quad (118)$$

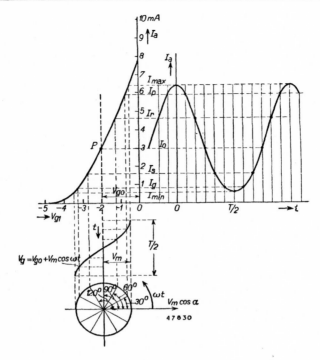

Fig. 184
Transfer characteristic of pentode EF 6 for $V_{g2} = 100$ V. The working point P chosen is $I_o = 3$ mA, $V_{go} = -2$ V. On the grid there is a sinusoidal alternating voltage with an r.m.s. value of 1 V. Below the V_{g1}-axis the grid voltage is drawn as a function of time and to the right of the I_a-axis is the anode current also as a function of time. By dividing the cycle T of the alternating grid voltage into 12 equal parts and determining the anode-current value for each period of time $t = n \dfrac{T}{12}$ (n = 0, 1, 2 ....), with the aid of the formulae given in the text it is possible to determine the amplitudes of the fundamental wave and the harmonics, as well as the direct-current component.

Fig. 184 shows the transfer characteristic of a pentode (EF 6). The working point has been chosen at $I_a = 3$ mA, $V_g = -2$ V. Below the V_g-axis a half cycle has been drawn of a sinusoidal alternating voltage $V_m \cos \omega t$ with amplitude $V_m = 1.42$ V, which is projected on the transfer characteristic. To the right of the I_a-axis we have the resultant anode alternating current drawn as a function of time for the interval T/2. Its wave form is noticeably distorted. If the half-cycle T/2 is divided into 6 equal parts, each corresponds to $\omega t = 30°$.

Now

$\cos 0° = 1, \quad \cos 30° = \tfrac{1}{2}\sqrt{3}, \quad \cos 60° = \tfrac{1}{2}, \quad \cos 90° = 0, \quad \cos 120° = -\tfrac{1}{2},$

$\cos 150° = -\frac{1}{2}\sqrt{3}$ and $\cos 180° = -1$, so that the grid voltage has successively the values:

$$V_m, \quad \tfrac{1}{2}\sqrt{3}\,V_m, \quad \tfrac{1}{2}V_m, \quad 0, \quad -\tfrac{1}{2}V_m, \quad -\tfrac{1}{2}\sqrt{3}\,V_m, \quad -V_m.$$

Now we call:

I_{max} anode current for $V_g = V_{go} + V_m$,

I_{min} ,, ,, ,, $V_g = V_{go} - V_m$,

I_p ,, ,, ,, $V_g = V_{go} + \tfrac{1}{2}\sqrt{3}\,V_m$,

I_q ,, ,, ,, $V_g = V_{go} - \tfrac{1}{2}\sqrt{3}\,V_m$,

I_r ,, ,, ,, $V_g = V_{go} + \tfrac{1}{2}V_m$,

I_s ,, ,, ,, $V_g = V_{go} - \tfrac{1}{2}V_m$ and

I_o ,, ,, ,, $V_g = V_{go}$.

These anode current values can be read from the characteristic (see Fig. 184).

By derivation:

$$\begin{aligned}
I_g &= 1/12 \left\{(I_{max} + I_{min}) + 2(I_p + I_q) + 2(I_r + I_s) + 2I_o\right\} \\
I_1 &= 1/6 \left\{(I_{max} - I_{min}) + \sqrt{3}(I_p - I_q) + (I_r - I_s)\right\} \\
I_2 &= 1/6 \left\{(I_{max} + I_{min}) + (I_p + I_q) - (I_r + I_s) - 2I_o\right\} \\
I_3 &= 1/6 \left\{(I_{max} - I_{min}) - 2(I_r - I_s)\right\} \\
I_4 &= 1/6 \left\{(I_{max} + I_{min}) - (I_p + I_q) - (I_r + I_s) + 2I_o\right\} \\
I_5 &= 1/6 \left\{(I_{max} - I_{min}) - \sqrt{3}(I_p - I_q) + (I_r - I_s)\right\} \\
I_6 &= 1/12 \left\{(I_{max} + I_{min}) - 2(I_p + I_q) + 2(I_r + I_s) - 2I_o\right\}
\end{aligned} \qquad (119)$$

Once these values have been determined by means of the transfer characteristic we find the distortion from the second harmonic $d_2 = \dfrac{I_2}{I_1}$ and that from the third harmonic $d_3 = \dfrac{I_3}{I_1}$, and so on.

With the aid of the values found for I_1, I_2 and I_3 and Equations (116), (117) and (118) it is possible to calculate the coefficients α, β and γ of the exponential series:

$$\alpha = \frac{I_1 - 3I_3}{V_m}, \qquad (120)$$

$$\beta = \frac{2I_2}{V_m^2} \text{ and} \qquad (121)$$

$$\gamma = \frac{4I_3}{V_m^3}. \qquad (122)$$

In this way we can determine the coefficients a, β and γ for a given static or dynamic characteristic. For the characteristic shown in Fig. 184 we find:

I_{max} = 6.45 mA, I_p = 5.95 mA, I_r = 4.65 mA, I_o = 3 mA,
I_{min} = 0.6 mA, I_q = 0.8 mA, I_s = 1.55 mA.

From this we calculate:

I_g = 3.24 mA; I_1 = 2.98 mA; I_2 = 0.267 mA; I_3 = 0.058 mA, so that:

$$a = \frac{2.98 - 0.174}{1.414} = 1.985 \text{ mA/V}; \quad \beta = \frac{2 \times 0.267}{(1.414)^2} = 0.27 \text{ mA/V}^2$$

$$\text{and } \gamma = \frac{4 \times 0.058}{(1.414)^3} = 0.0825 \text{ mA/V}^3.$$

The distortion of the anode current due to the second harmonic d_2 is equal to

$$\frac{I_2}{I_1} \times 100\% = \frac{0.267}{2.98} \times 100\% = 9\%,$$

and that due to the third harmonic d_3 equals

$$\frac{I_3}{I_1} \times 100\% = \frac{0.058}{2.98} \times 100\% = 1.95\%.$$

CHAPTER XVII

Final or Power Amplification

90. The Purpose of the Final Stage and the Valves Used

The final stage or power amplifier in a receiver must be capable of supplying sufficient power to the loudspeaker. The power output required depends upon the loudspeaker system used and the sound volume desired. It will be clear that the more sensitive the loudspeaker, the less power output will be needed for a given volume of sound. Generally speaking, in normal receivers a power output to the loudspeaker of about 4 watts is required for reproducing the loudest passages in music or speech. With a loudspeaker of normal sensitivity this produces the right volume of sound in living rooms of normal dimensions. In larger rooms a higher power output may often be required, whereas in many cases a lower power output suffices. (For receivers fed from dry batteries, e.g., portable radios, the output power is limited to much less than 4 watts so as to avoid using very heavy batteries; to get a reasonable volume of sound an extremely sensitive loudspeaker is often used.)

The final stage is generally controlled (modulated) by the pre-amplifying stage without any appreciable supply of energy, but the incoming alternating voltages have to be converted into alternating-current power of sufficient magnitude to operate the loudspeaker. In H.F. or L.F. pre-amplifying stages on the other hand the power supplied to the coupling resistance or coupling impedance is much less. It is therefore necessary to use for the final stage one or more special valves constructed to deliver a large power output.

In the final stage of receiving sets usually only one power-amplifying valve is used, while in the highest-class receivers two are often used and in public-address amplifiers nearly always two. Where two power valves are employed they are connected in **push-pull**, as this offers very many advantages. In some cases two or more valves are connected in parallel (see also Section 96). In the following pages a distinction will be made between **single-ended stages,** i.e., those where only one valve is used in the final stage, and **push-pull stages.**

Nowadays triodes and pentodes, or tetrodes with pentode properties, are used almost exclusively as power-amplifying valves. Triodes have almost entirely disappeared from receiving sets because the pentode offers such great advantages over the triode. For the same reason triodes are being used less and less in public-address amplifiers, since the advantage of the lower internal resistance of a triode, which for some applications of amplifiers is essential (greatly fluctuating load, e.g. in radio relay systems), can also be obtained by employing pentodes with feedback (see Chapter XXVII).

The advantages that a pentode has over a triode are:
1) higher efficiency (about twice as great) and
2) greater sensitivity, i.e., for a given power output a much smaller grid alternating voltage is needed for a pentode than for a triode.

91. Matching the Loudspeaker to the Output Valve

Nowadays electrodynamic loudspeakers are almost exclusively used. Such a loudspeaker comprises a small coil attached to a conical diaphragm and placed in a constant magnetic field. The alternating current from the amplifier flows through this coil, which for constructional reasons is generally so dimensioned that a large current at a relatively low voltage flows through it. The coil is consequently made of a small number of windings of heavy copper wire, so that its impedance (a.c. resistance) is relatively low and generally of the order of 2—10 ohms. Now this impedance is much lower than that necessary to obtain the maximum power output from the anode circuit of the final-stage valve, so that the loudspeaker has to be matched to the final stage of the receiver. For this purpose an iron-cored transformer is used, which transforms the high alternating voltage in the anode circuit into a lower one for the loudspeaker circuit (see Fig. 140). This matching transformer must produce on the primary side a load impedance of such a value that the valve can supply its maximum power with the least possible distortion. If the **optimum load resistance** of the output valve is termed R_L and the impedance of the loudspeaker coil for the frequency considered is R_l, the loudspeaker transformer must have a ratio of

$$n = \sqrt{\frac{R_L}{R_l}} . \tag{123}$$

This holds only for the case of an ideal transformer, and no account is taken here of the various losses occurring in the transformer. Owing to these losses the impedance on the primary side with this calculated transformer ratio becomes somewhat greater, so that the ratio chosen must be slightly lower. The impedance of the loudspeaker coil is not the same at all frequencies and consequently the impedance on the primary side of the loudspeaker transformer also depends upon the frequency.

Furthermore it is only within a limited range that this impedance is a purely ohmic resistance. Fig. 185 shows the variation of the impedance Z_L on the primary side of the combination of a loudspeaker and a matching transformer the primary winding of which is shunted by a condenser of 2000 $\mu\mu$F. This diagram also gives the phase angle of this impedance at various frequencies, i.e., the angle between current and voltage. From these curves it is seen that in the most important frequency range the impedance of the loudspeaker combination varies by a factor of 2 to 3. Cheaply constructed loudspeakers often show a larger impedance variation. Generally the impedance of a loudspeaker is indicated at a certain frequency, 800 or 1000 c/s.

92. The Optimum Value of Load Impedance with Triodes in a Single-ended Stage

As indicated above, a certain impedance is required in the anode circuit for obtaining the maximum power output. This impedance, assuming that it is purely resistive, is called the **optimum load impedance.** The value of the optimum load impedance can be determined experimentally for any valve with given operating voltages. It is also possible to derive the approximately correct value of the optimum impedance from the valve data. For **power amplifying triodes** the optimum load impedance value is usually equal to twice the internal resistance of the valve.

$$R_L = 2\,R_a.\,^1)\qquad (124)$$

This holds, theoretically, for triodes with straight plate characteristics (see **Fig. 186**) and with a straight transfer characteristic. In order to prove that the optimum load is equal to twice the internal resistance we have to determine for this family of plate

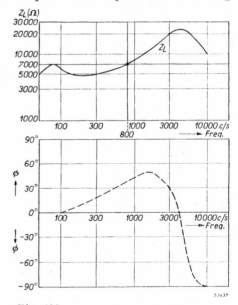

Fig. 185
The impedance Z_L and the phase angle φ of that impedance as functions of the frequency, for the combination of a loudspeaker with a matching transformer the primary winding of which is shunted by a condenser of 2000 $\mu\mu$F.

---

[1]) See B. D. H. Tellegen, Final Amplifier Problems, T. Ned. Radio-Genootschap 3, 1928, pages 141—160, or W. J. Brown, Proc. Phys. Soc., Vol. 36, 1924.

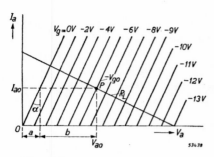

Fig. 186
Idealized plate characteristics of a triode, with the straight line representing the load resistance R_L. The working point P is fixed by the voltages V_{ao} and V_{go} and by the corresponding anode current I_{ao}.

characteristics the required negative grid bias V_{go} and the anode load R_L for maximum power output at the anode supply voltage V_{ao} (assuming that R_L is formed by an ideal transformer loaded on the secondary side with a resistance; R_L therefore opposes only the a.c. and not the d.c.).

It is easy to realize that for a given direct anode voltage V_{ao} and direct grid voltage V_{go} the anode direct current must be equal to:

$$I_{ao} = \frac{V_{ao} - \mu V_{go}}{R_a}, \qquad (125)$$

in which μ is the amplification factor and R_a the internal anode resistance of the valve.

According to Fig. 186 $V_{ao} = a + b$. Now $a = I_{ao} \tan \alpha = I_{ao}R_a$ and $b = \mu V_{go}$, so that $V_{ao} = I_{ao}R_a + V_{go}$, or $I_{ao}R_a = V_{ao} - \mu V_{go}$, or

$$I_{ao} = \frac{V_{ao} - \mu V_{go}}{R_a}.$$

At maximum power output the following conditions have to be satisfied:

a) The a.c. amplitude I_{max} may not be greater than the anode d.c., as otherwise during the negative halves of the anode-current cycles the peaks of the sine curve of the anode a.c. would be flattened; the resulting anode current cannot be smaller than zero (see Fig. 187).

b) The grid-alternating-voltage amplitude V_{max} may not be greater than the grid direct voltage V_{go}, as otherwise during the positive halves of the grid-voltage cycles the peaks of the grid alternating voltage would penetrate the positive-grid-voltage region and cause the flow of grid current (see Fig. 188).

When regarding the triode as an a.c. generator with an internal resistance R_a and an external load R_L, according to Equation (44) in Chapter XII the alternating current is equal to

$$I = g_m V_g \frac{R_a}{R_a + R_L} = \frac{\mu V_g}{R_a + R_L}, \qquad (126)$$

$$I_{max} = \frac{\mu V_{max}}{R_a + R_L}. \qquad (127)$$

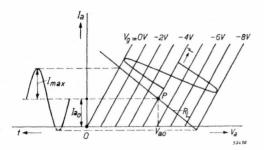

Fig. 187
Right: Idealized plate characteristics of a triode, with a load resistance R_L through the working point P. The grid alternating voltage is also given as a function of time.
Left: The anode current as a function of time t. Owing to the unfavourable slope of the resistance line R_L and the wrong location of the working point P the peak of the sine curve of the anode alternating current is cut off during the negative cycle; thus distortion takes place.

According to the conditions prescribed under a) and b) in Equation (127) I_{max} has to be replaced by I_{ao} and V_{max} by V_{go}. We then obtain:

$$I_{ao} = \frac{\mu V_{go}}{R_a + R_L} \quad (128)$$

From (125) and (128) it follows that

$$\frac{\mu V_{go}}{R_a + R_L} = \frac{V_{ao} - \mu V_{go}}{R_a}$$

or that

$$V_{go} = \frac{V_{ao}}{\mu} \times \frac{R_a + R_L}{2R_a + R_L} \quad (129)$$

If we substitute this value of V_{go} in (128) we obtain for the alternating-current amplitude the formula:

$$I_{max} = I_{ao} = \frac{V_{ao}}{2R_a + R_L} . \quad (130)$$

Now the power developed in the load resistance R_L is equal to

$$P_L = \frac{I_{max}^2}{2} R_L,$$

or

$$P_L = \frac{V_{ao}^2}{2} \times \frac{R_L}{(2 R_a + R_L)^2} \quad (131)$$

P_L is maximum for $R_L = 2R_a$ [1]) and this maximum is then equal to

$$P_{Lmax} = \frac{V_{ao}^2}{16 R_a} . \quad (132)$$

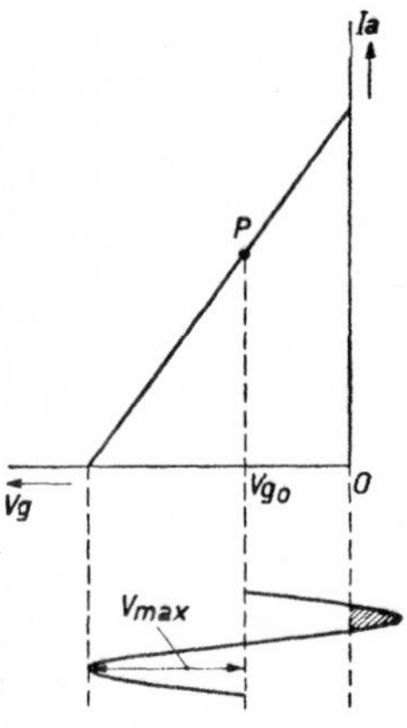

Fig. 188
Dynamic transfer characteristic of a triode with working point P. If the grid-alternating-voltage amplitude V_{max} is greater than V_{go} grid current occurs.

---

[1]) As is well known, this maximum can be found by differentiating P_L with respect to R_L and equating the differential quotient to zero.

From this equation it follows that the maximum output power is inversely proportional to the internal resistance of the valve and directly proportional to the square of the anode direct voltage. It is therefore economical to use a triode with low internal resistance operated with a high anode direct voltage.

From Equation (130) it follows that with the most favourable load impedance $(R_L = 2R_a)$ $I_{ao} = \dfrac{V_{ao}}{4\,R_a}$. Now $\dfrac{V_{ao}}{R_a}$ is equal to the anode current flowing when $V_g = 0$. Therefore the working point of a triode must be so chosen that the anode d.c. amounts to one-quarter of the anode current flowing when $V_g = 0$. From (132) it then follows that the maximum power output is equal to:

$$P_{Lmax} = \tfrac{1}{4}\,I_{ao} \times V_{ao}. \tag{133}$$

Since the power supplied, drawn from the H.T. source, is equal to $I_{ao}V_{ao}$, the efficiency η of the valve at maximum power output is 25%.

In practice values for the maximum power output are to be found somewhat smaller than $\dfrac{V_{ao}^2}{16\,R_a}$, due in the first place to the fact that the characteristics of triodes have not the ideal shape upon which the above calculation, has been based, the curvature of the transfer characteristic causing distortion. Moreover the amplitude of the grid alternating voltage may never reach the value $V_g = 0$, because at a low value of negative grid voltage a considerable grid current begins to flow. This grid current causes a very marked distortion, which should be avoided.

In the above calculation it was assumed that the current $I_{ao} = \dfrac{V_{ao}}{4\,R_a}$ is permissible so far as the maximum anode dissipation is concerned. If that is not the case then I_{ao} must be kept lower. By means of a similar calculation to that employed above it can be shown that the most favourable value for R_L must then be greater than $2R_a$. The power output is in that case less than $\dfrac{V_{ao}^2}{16\,R_a}$ and the efficiency greater than 25%.

If an output triode is completely controlled (or in other words fully modulated) the distortion at the optimum load impedance is usually 5%. By complete control or full modulation is understood that the grid alternating voltage is of such a value that it swings between the grid-current starting point and the zero-anode-current point.

93. The Optimum Load with Pentodes in a Single-ended Stage

For the determination of the theoretical optimum load impedance of pentodes we likewise depart from the idealized conditions. We will assume that the pentode has an infinite internal resistance and that grid current occurs only when the grid voltage is positive. Further we assume that the transfer characteristic is straight (see Fig. 189), and that the screen-grid voltage V_{g2} has a fixed value.

Since the internal resistance is assumed to be infinitely large, the transfer characteristic of Fig. 189 also represents the dynamic characteristic of the valve. Again we have the condition that the amplitude of the anode alternating current must not be greater than the anode direct current I_{ao}. Further, the grid-alternating-voltage amplitude must not be greater than V_{go}, and finally the anode-alternating-voltage amplitude must not exceed the anode direct voltage V_{ao}, as otherwise during the negative half of the anode-alternating-voltage cycle the peak of the sine curve would be flattened.

If we imagine that an alternating voltage is applied to the grid of such a value that the valve yields its full power output, then the anode current must vary between zero and the maximum value at $V_g = 0$. The grid alternating voltage must therefore vary between $V_g = 0$ and the value at which the anode current becomes equal to zero.

Assuming further that in the anode circuit a load resistance is introduced which offers an ohmic resistance R_L to alternating current but none to direct current, the maximum power dissipated in that resistance then equals:

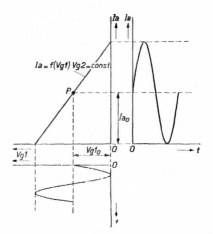

$$P_{Lmax} = \frac{(I_{max})^2}{2} R_L = \frac{I_{ao}{}^2}{2} R_L, \quad (134)$$

I_{max} being the amplitude of the anode alternating current. The amplitude of the anode alternating voltage V_{max} equals $I_{max}R_L = I_{ao}R_L$. According to the foregoing conditions this amplitude must not exceed the anode direct voltage. Thus we have:

$$I_{ao}R_L = V_{ao} \quad (135)$$

Fig. 189
Idealized transfer characteristic of a pentode. To the left below, the grid alternating voltage as a function of time is represented at full excitation of the grid base. To the right above, the resultant anode alternating current as a function of time is shown.

or:
$$R_L = \frac{V_{ao}}{I_{ao}}.$$
(136)

According to Equation (134), when $\dfrac{V_{ao}}{I_{ao}}$ is substituted for R_L, the maximum power output is equal to:

$$P_{Lmax} = \tfrac{1}{2} I_{ao} V_{ao}.$$
(137)

Since the power supplied by the d.c. source to the valve is equal to $I_{ao} V_{ao}$ the maximum efficiency equals 50%.

This efficiency figure applies only for a pentode possessing the ideal properties given above. Actually, however, pentodes have a more or less curved transfer characteristic and consequently their efficiency is always below the theoretical value of 50%.

When considering the efficiency of pentodes in comparison with triodes account has also to be taken of the power dissipated by the screen grid, as this too has to be supplied by the d.c. source feeding the anode. With many Philips pentodes, such as the EL 3N, EL 5 and EL 6, the screen-grid current equals $^1/_9$th to $^1/_{10}$th of the anode current and the screen-grid voltage equals the anode direct voltage. The total power supplied by the d.c. source to the pentode is therefore about 10% greater than $V_{ao} I_{ao}$ and consequently the theoretical efficiency is also 10% less than 50%, or about 45%. Further, it has to be borne in mind that the efficiency figure of 45% holds only for the maximum permissible grid swing. With small grid alternating voltages the efficiency decreases proportionately to the square of the grid alternating voltage (this applies both to triodes and to pentodes in a single-ended stage).

In practice grid current does not generally occur in pentodes until there is a distortion of more than 10% (see Section 94b).

94. Distortion in Output Valves in a Single-ended Stage

(a) Concerning the Admissible Distortion [1])

What is understood by distortion has already been explained in Chapters XV and XVI, where the **distortion factor** d_{tot} is defined. The distortion factor d_{tot} is not in every case a correct criterion for the quality of the reproduction. Distortion due to discontinuous phenomena, such as the occurrence of grid current during part of the alternating-

---

[1]) Cf. A. J. Heins van der Ven: Output Stage Distortion, Wireless Engineer 16, 1939, pp. 383—390 and 444—452.

voltage cycle at the grid or a flattening of the peaks of the sine curve
of the anode voltage is much more disturbing than might be deduced
from the magnitude of the distortion factor alone. This may be ascribed
to the fact that with such discontinuous phenomena a great many
harmonics of a high order occur.

Apart from these discontinuous phenomena however, the distortion
factor is valuable as a measure for judging the quality of the power
supplied by an output valve to the loudspeaker, which is then dependent
upon a more or less pronounced curvature of the characteristic in its
operating range.

Now, according to what has been said in Chapter XV, with triodes
mainly the second harmonic predominates and with pentodes the
third harmonic. The question then arises which of these two harmonics
is the most detrimental to the quality of reproduction. Several authors
have made a study of this subject [1]). According to their conclusions
the greatest distortion is admissible in speech. When non-distorted
and distorted reproductions are compared about 5% of second harmonic or 3% of third harmonic are just perceptible. In the case of the
reproduction of music 4% second or third harmonics can just be distinguished from the non-distorted reproduction.

The interference due to distortion does not lie so much in the formation
of harmonics, because the tones brought out by musical instruments
or by speech also contain harmonics. There arise, however, in addition
to the harmonics themselves, new tones—the so-called **combination
tones**—as a consequence of the sum and difference frequencies caused
by the presence of voltages of different frequencies at the grid of a
valve with a curved characteristic (see Chapter XVI, Section 85). This
results in a change of timbre. If the distortion is not strong the
accompanying change of timbre is not marked. Distortion becomes really
troublesome when the formation of combination tones leads to huskiness in the reproduction. Such is the case if one of the original tones
has a low pitch. Then the sum and difference frequencies will differ
little from the original frequency. If, for example, the grid of an output
valve carries simultaneously a frequency of 100 c/s and another of
1,000 c/s then for a square-law transfer characteristic new tones are
formed having frequencies of 900 and 1,100 c/s in addition to the
harmonics with frequencies of 200 and 2,000 c/s. This means that the
frequency of 1,000 c/s appears to have been modulated by 100 c/s.
When this modulation frequency lies between 40 and 100 c/s a certain

[1]) W. Janovsky, E. N. T. 1929, 6, page 421; F. Massa, Proceedings Inst. Radio Eng.
1933, 21, page 682; H. v. Braunmühl and W. Weber, Akust. Zeitschr. 1937, 2, page 135.

huskiness arises, which is very troublesome. If the transfer characteristic is not purely square-law, additional tones of 800 and 1,200,
700 and 1,300 c/s, etc. may arise.

Experience shows that for normal radio reception a distortion factor
of 4 to 5% at the maximum grid-voltage amplitude occurring is
sufficiently low. In cases where very high demands are to be met as
regards reproduction quality the distortion factor should not be more
than 1 to 1.5% at the maximum grid-voltage amplitude. It is always
important, however, that there should be no flattening of the anode-
alternating-voltage curve due to grid current or to the anode alternating-
voltage amplitude exceeding the anode direct voltage.

(b) The Relation between Power Output and Distortion

It will be evident that the distortion taking place in the anode alternating current of an output valve will be greater with high grid alternating voltages than with low ones; when the grid alternating voltage
gradually rises from zero to a maximum value the distortion increases
similarly. (It sometimes occurs, however, that after having reached a
certain value the distortion decreases again over a certain range of
values of the grid alternating voltage, after which it begins to increase
anew.) It is evident that the reason for distortion being less with low
alternating voltages than with high ones is that the former traverse
an almost straight part of the characteristic whilst the latter pass also
through those parts of the characteristic having a marked curvature.
In the case of very high alternating voltages distortion also arises
through the amplitude of the anode alternating voltage becoming too
great with a consequent marked flattening of the anode voltage wave
form and/or due to the occurrence of grid current.

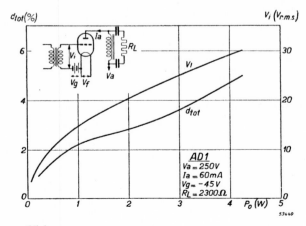

In the data published for
an output pentode it is
stated that the power
output of this valve a-
mounts to a certain watt-
age at 10% distortion.
It sometimes happens,

Fig. 190
The total distortion d_{tot} and
the requisite grid alternating
voltage V_i as a function of the
power output of triode AD 1
used as a single-ended final-
stage amplifier.

however, that grid current occurs before that distortion figure is reached, and in that case the power output is given for the grid-modulation voltage at which grid current begins. In that case the distortion that then takes place is also quoted. Sometimes, too, the power output is indicated for the grid-modulation voltage at which grid current starts when the distortion is greater than 10%. For triodes it is usual to quote the power output at 5% distortion, as already mentioned in Section 92.

For the designer of receiving sets or power amplifiers it is, of course, of great importance also to know how much the distortion is for smaller power outputs. Therefore it has become customary to indicate the distortion as a function of the power output in the form of a curve. At the same time the grid alternating voltage required for each power output is shown in the form of a curve.

Fig. 190 shows such curves for the Philips output triode AD 1 and Fig. 191 those for the Philips output pentode EL 3N. As it may, moreover, be of importance to know how this distortion is split up into second and third harmonics, curves are frequently given also for distortion due to the separate harmonics; such curves are included in Fig. 191.

(c) **Power Output and Distortion for Values of the Matching Impedance Differing from the Optimum Value**

If one regards the loudspeaker as a constant purely resistive impedance then the choice of the optimum matching is extremely simple. In reality, as will be seen from Fig. 185, this impedance is not of the same value for all frequencies and, moreover, at different frequencies it causes different phase angles between current and voltage. As already indicated in Chapter XII, when the anode impedance has a reactive component the dynamic plate characteristics will be ellipsoidal (see **Fig. 138**) and a

Fig. 191
The total distortion d_{tot}, the distortion due to the second harmonic d_2, that due to the third harmonic d_3, and the requisite grid alternating voltage V_i as functions of the power output of pentode EL3N used as a single-ended final-stage amplifier.

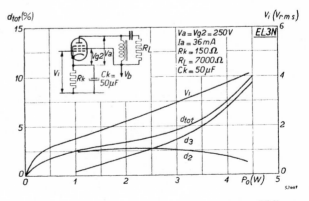

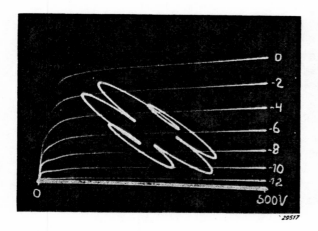

Fig. 192
Oscillogram of the dynamic
plate characteristic of a
power pentode loaded by a
loudspeaker when two alternating voltages of frequencies having a ratio of 6 to
1 are simultaneously applied
to the grid.

certain area of the
plate-characteristic di-
agram of the output
valve will be covered.
If voltages with differ-
ent frequencies are
applied to the grid
simultaneously the area of the family of plate characteristics occupied
will be still greater. Fig. 192 gives an idea of the form of the dynamic
plate characteristic when two alternating voltages with frequencies
having a ratio of 6 to 1 are applied to the grid.

The influence of the value of the impedance in the anode circuit of
an output pentode can best be shown by a family of curves, the most
suitable being those which give the power output P_o as a function of
the load resistance for a given constant distortion d_{tot}. For values of
load resistance smaller than the optimum value V_{ao}/I_{ao} the power output
P_o of a pentode is limited to $\frac{1}{2} I_{ao}^2 R_L$, because the anode current cannot
become negative and consequently the anode a.c. amplitude can be
at the most equal to the anode direct current I_{ao}. With load resistances
greater than V_{ao}/I_{ao} the power output P_o is limited to $\frac{1}{2} \dfrac{V_{ao}^2}{R_L}$, because
the anode voltage cannot become negative and the anode-alternating-
voltage amplitude can at the most be equal to the anode direct voltage
V_{ao}. A curve indicating distortion as a function of load resistance for
a certain power output P_o has therefore little practical significance,
since the distortion can only be a useful measure of the quality of
reproduction when it remains small and consists of harmonics of a
low order (e.g., up to the fifth). If a curve is constructed showing the
distortion as a function of the load resistance for a certain constant
power output P_o, then at small values of the load resistance the limiting
value of the power output $\frac{1}{2} I_{ao}^2 R_L$ is smaller than the output power
P_o. At high values of the load resistance the limiting value $\frac{1}{2} \dfrac{V_{ao}^2}{R_L}$
is also smaller than the given constant output power P_o. In these

cases there would be a large number of harmonics of a very high order and the distortion factor would no longer be any guide to the quality of reproduction. If one examines the plate characteristics of an output pentode it will be seen that the anode voltage cannot drop below a certain positive value, which for most modern pentodes is about 20 volts. Therefore, instead of the above limit of $\frac{1}{2}\dfrac{V_{ao}^2}{R_L}$ for such valves it is more correct to take the limit as $\frac{1}{2}\dfrac{(V_{ao}-20)^2}{R_L}$.

Fig. 193 shows a number of curves giving the power output of the Philips output pentode EL 3N as a function of the load resistance R_L for distortions of 5, 2.5 and 1%. The same diagram also shows the limits $\frac{1}{2} I_a^2 R_L$ and $\frac{1}{2}\dfrac{(V_a-20)^2}{R_L}$ (here the symbols I_a and V_a have the meaning of I_{ao} and V_{ao} respectively). It will be evident that the higher a curve lies for a certain distortion the more favourable the valve will be as regards its power output for that distortion factor. If the curve showing the impedance as a function of the frequency for a certain loudspeaker is available, then with the aid of a family of curves as shown in Fig. 193 one can determine for every frequency the maximum power output for any given distortion, e.g., 5% or 2.5%. In this way the related intensities of high and low notes in the reproduction of music or speech, as compared with the intensity at a frequency where the loudspeaker has the optimum impedance value, can be ascertained.

By varying the ratio of the matching transformer the frequency at which the impedance value of the loudspeaker combination corresponds to the optimum load for the valve can be selected. By this means it is possible to vary the ratio between the maximum output powers at different frequencies. In Fig. 185, for instance, at a

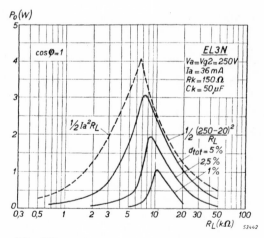

Fig. 193
The power output P_o as a function of the load resistance R_L for harmonic distortions of 1, $2\frac{1}{2}$ and 5% and limit lines (broken) $\frac{1}{2} I_a^2 R_L$ and $\frac{1}{2}\dfrac{(V_a-20)^2}{R_L}$ for the power pentode EL 3N. These curves apply to the case when the loudspeaker impedance does not produce any phase angle (cos $\varphi = 1$).

217

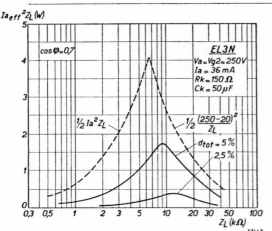

Fig. 194
The apparent power output $I_{a\,eff}{}^2 Z_L$ as a function of the load impedance Z_L for $2\frac{1}{2}$ and 5% distortion in the plate current and limit lines (broken) $\frac{1}{2} I_a{}^2 Z_L$ and $\frac{1}{2}\dfrac{(V_a - 20)^2}{R_L}$ for the power pentode EL 3N when the loudspeaker combination has $\cos\varphi$ equal to 0.7.

frequency of 800 c/s the loudspeaker has an impedance Z_L of 7,000 ohms, which corresponds to the optimum load impedance for the EL 3N. At 3,000 c/s, however, the impedance has risen to 20,000 ohms and, according to the curve in Fig. 193, if a maximum distortion of 5% is permissible, with this load resistance an output power of only 1 watt can be obtained. For the same distortion a higher power output would be obtainable for high frequencies if the matching impedance for 800 c/s were chosen lower, so as to lower the whole curve of Fig.185.From the curve for 5% distortion in Fig. 193, however, it follows that having regard to the intensity of the low notes this would not be very desirable; generally speaking, the high notes in music or speech to be reproduced are of less intensity than the low notes. The influence on the power output of the phase angle between the current and voltage across the primary winding of the loudspeaker transformer produced by the loudspeaker impedance, is best expressed when the power output is measured for different distortion factors and different phase angles as a function of the load impedance Z_L. Since the acoustic power is proportional to the current through the loudspeaker coil it is possible to measure the distortion of the alternating current. In order, however, to draw a comparison between different cases, wherever $\cos\varphi$ is less than unity we have to work with the apparent power $I_{a\,eff}{}^2 Z_L$ instead of the effective output power as a function of the load resistance. This simplifies the comparison with the theoretical maximum power. Fig. 194 shows the result of such measurements taken with the output pentode EL 3N for the case where $\cos\varphi$ of the loudspeaker combination is 0.7 (corresponding to a phase angle φ of about 45°). From these curves it immediately follows that in consequence of the phase angle between current and voltage the maximum apparent output power is considerably lower than in the absence of a phase angle. This will always have to be taken into account in practical cases.

218

From what has been said above it should be clear that a simple statement of the output power of a valve for a certain distortion at the optimum matching resistance, where there is no phase angle between current and voltage, can hardly be a guide to the performance of the valve, and account must certainly be taken of the behaviour of the valve at other values of the load impedance and of cos φ. This is of special importance in view of the fact that valves giving apparently identical output-power values may, when examined from the aspects discussed above, show considerable differences.

95. The Requirements to be met by Output Pentodes used in Single-ended Stages

It has already been indicated that, particularly in the absence of ohmic resistances, we have to consider the whole area of the plate characteristics. It is to be assumed, however, that the part of the family of characteristics for low anode current and low anode voltage will hardly ever be used. Fig. 138 shows that the dynamic characteristic for small amplitudes is an ellipse, whilst for larger amplitudes there is a noticeable deviation from the ellipsoidal shape. If the plate characteristics in the area used were parallel straight lines, lying at equal distances for certain equal increments of the grid-bias potential, there would be no distortion of the ellipse so long as this did not tend to go beyond the horizontal or the vertical axes. It is not possible, however, to obtain such families of characteristics in practice, because the anode current increases by the $^3/_2$ power of the effective potential in the control-grid plane. Consequently the plate characteristics can never be equally spaced.

Nevertheless, from the point of view of freedom from distortion it is desirable that the plate characteristics should be as equally spaced as possible, that is to say, the transfer characteristic should be as straight as possible and have the sharpest possible bend at the lower extremity.

Moreover, it is desirable that the plate characteristics should be as straight as possible. **In order to achieve this the influence of the secondary emission of the anode in the working range of the anode current and anode voltage must be entirely eliminated,** as secondary emission phenomena may cause considerable deviations. This point is highly important, as the secondary emission may sometimes cause a deviation in the centre of the plate characteristics with a resulting distortion of the dynamic curve and consequently the occurrence of harmonics, even with small amplitudes of the grid alternating voltage. This will be the case particularly if the load line represents a high resistance.

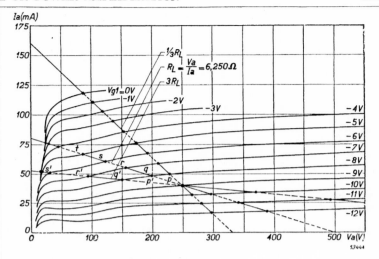

Fig. 195
Plate characteristics of a power tetrode with suppression of the
secondary emission by the space charge between screen grid and
anode. At the optimum load $R_L = V_a/I_a$ we find that p = q = r =
s = t and consequently there is little distortion. For a load resistance equal to 3 R_L we find that p' and r' are greater than q', which
entails distortion.

Fig. 195 shows a family of plate characteristics of a tetrode output
valve in which the secondary emission has to be suppressed by the
space charge between screen grid and anode (see Chapter XIII, Section
73d, Fig. 156). On this family of characteristics load lines are drawn
corresponding to the optimum resistance $R_L = V_a/I_a$, to 3 R_L and to
$^1/_3$ R_L. The plate characteristics for equal increments of V_{g1} intersect
the load lines, and when the intercepts are equal there will be no
distortion.

From Fig. 195 it is to be seen that this is practically the case for the
load resistance R_L (p = q = r = s = t). For R = 3 R_L, however,
p' and r' are larger than q'. This means that distortion occurs due to
the influence of secondary emission. With too small a value of the load
resistance, for instance $^1/_3$ R_L, the distortion is less pronounced. In
a tetrode with limitation of secondary emission by space charge it is
never possible to suppress the secondary emission over the whole
area of the plate characteristics used, as is evident from Chapter XIII,
Section 73d. This is possible, however, with a pentode and therefore
this type of valve is more suitable for power amplification than the
tetrode with limitation of secondary emission by space charge. This
will not be so manifest with the optimum load impedance as with

220

the higher values of that impedance and also when a phase angle between current and voltage occurs. Fig. 196 gives a comparison of the efficiency $\left(\dfrac{\text{output power}}{\text{input power}} \times 100\% = \dfrac{P_o}{P_a} \times 100\%\right)$ with 5% distortion and a load impedance having $\cos \varphi$ equal to 0.7, a tetrode (6L6) and for a pentode (Philips EL 3N) as a function of the load impedance Z_L, expressed as a percentage of the optimum load impedance. Curve I in this diagram represents the efficiency for the EL 3N and curve II that for the 6L6. This comparison shows that the pentode is to be preferred.

As is clear from Section 94a, the sudden occurrence of grid current results in a very troublesome form of distortion, as harmonics of a very high order then occur. For this reason the available grid swing of the dynamic transfer characteristic must be sufficiently large so that the maximum output power is not restricted by the occurrence of grid current, but only by the less troublesome occurrence of harmonics due to the curvature of the characteristic. Too wide a control-grid-voltage sweep then has a less adverse effect.

With large amplitudes of the grid alternating voltage the output power will then be limited by the crowding together of the plate characteristics for large and for small anode currents. This produces a top and a bottom bend of the dynamic transfer characteristic (see Fig. 135). It is therefore of importance that the region where the load line crosses the crowded parts of the plate characteristics at low anode voltage, and where as a result the top bend is marked, should lie as far as possible to the left in the plate characteristic diagram, so that the plate characteristics for small grid potentials are sloping down at the lowest anode voltage possible. This requirement has been given serious consideration in the designing of the modern Philips pentodes, whilst at the same time suppressing the secondary emission in the

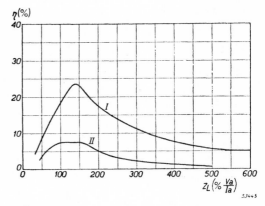

Fig. 196

The efficiency $\left(\eta = \dfrac{P_o}{P_a} \times 100\%\right)$, for 5% distortion, of a pentode (curve I) and of a tetrode (curve II) as a function of the load impedance (expressed as a percentage of the optimum impedance V_a/I_a) for a load impedance having $\cos \varphi$ equal to 0.7.

221

whole working area of plate characteristics, thereby raising the maximum efficiency.

Suppression of the secondary emission has been attained by coating the anode with a substance having a low secondary-emission factor (by blackening of the anode with a layer of carbon) and by forming a potential minimum between screen grid and anode with the help of the combined effect of a suppressor grid and space charge.

96. Output Stages in Push-pull Connection

In the foregoing we have been dealing with single-ended output stages consisting of one valve, a pentode or a triode. Of course it is possible to connect two or more valves in parallel in order to get a higher power output, and sometimes this is done where the anode voltage is low.

Such output stages are in essence single-ended final stages. Further it is possible to connect two valves in push-pull. Push-pull connection offers the advantage of a higher output from the stage than is attainable with a single output valve, whilst moreover the distortion may be appreciably less.

Fig. 197 represents the principle of a push-pull output stage employing two pentodes. The grid alternating voltage is applied to the grids of the two output valves via a transformer, the secondary of which has a centre tap. This centre tap is earthed for low-frequency (at cathode potential); when the voltage at one grid changes in the positive direction that at the other grid changes symmetrically in the negative direction. In the anode circuit we again have an output or matching transformer, the primary of which is tapped exactly in the centre. The anode direct voltage is applied to the anodes via this tap. Across the upper half of the primary of the output transformer an alternating voltage occurs which increases in the positive direction if the grid voltage of the upper valve changes negatively. At the same time the grid voltage of the lower valve changes in the positive direction, as a consequency of which the anode voltage of the lower half of the primary of the output transformer changes negatively. Consequently the voltage at point a increases symmetrically with respect

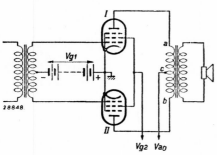

Fig. 197
Basic circuit of a push-pull final stage using two pentodes.

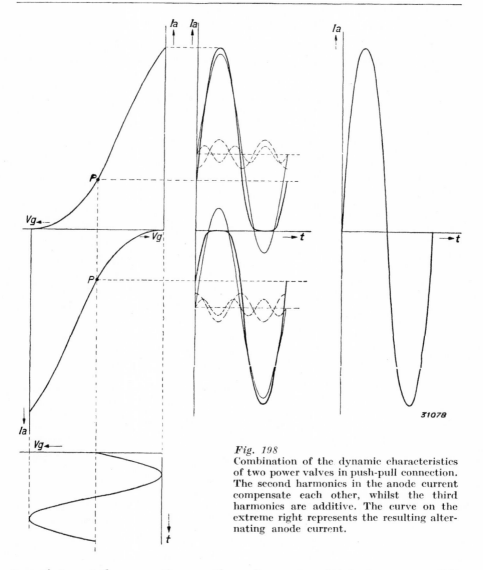

Fig. 198
Combination of the dynamic characteristics
of two power valves in push-pull connection.
The second harmonics in the anode current
compensate each other, whilst the third
harmonics are additive. The curve on the
extreme right represents the resulting alternating anode current.

to point c at the same time as the voltage at point b decreases with
respect to point c. The alternating currents of the two primary transformer windings each induce alternating currents in the secondary, and
in such a way that these are added together. With push-pull stages both
halves of the primary of the output transformer carry equal direct
currents. Their directions are however opposed to each other so that
no d.c. pre-magnetization of the transformer core occurs, which is
favourable as regards distortion in the transformer.

223

If, now, the dynamic transfer characteristics of the two valves are combined in such a way [1]) that the characteristic of the lower one is turned through 180° and the two working points lie on the same vertical line (see Fig. 198), and an alternating voltage is applied to the grids, distorted alternating anode currents are generated in the anode circuits of each of the two valves, as indicated in Fig. 198. In this diagram it has been assumed that the distortion of the anode current consists exclusively of the second and third harmonics. If the two alternating anode currents are added together the two second-harmonic components neutralize each other, whilst the two third-harmonic components are added together (Fig. 198).

In the case of push-pull output stages the even-harmonic components neutralize each other and the odd-harmonic components, and thus also the fundamental-frequency components, are added together. It is therefore advantageous to use in push-pull output stages valves whose distortion consists principally of the second harmonic. With such valves the resultant distortion will consist exclusively of the third and higher odd harmonics and thus be extremely small. With triodes it is mostly the second harmonic that is predominant, so that in so far as distortion is concerned these valves are eminently suitable for push-pull stages. With most pentodes the third harmonic will be greater than the second if the load resistance is equal to V_a/I_a.

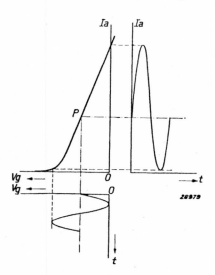

For push-pull final stages using pentodes the load resistance per valve is therefore chosen lower than V_a/I_a; the pentode has then the advantage of an appreciably greater efficiency.

97. Class-A Operation

In Class-A operation the negative grid bias of the valve is chosen in such a way that anode current flows during the whole cycle of the alternating grid voltage. Consequently the amplitude of the alternating

Fig. 199
Location of the working point P for Class-A operation.

---

[1]) In this connection it is to be noted that with push-pull stages the dynamic characteristic cannot be determined in such a simple manner as with output pentodes in a single-ended stage, since in those cases the load line on the plate diagram is no longer straight.

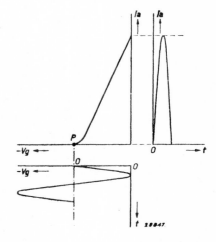

Fig. 200
Location of the working point P for Class-B operation.

grid voltage does not exceed the grid bias, at which the anode current is zero (Fig. 199).

Usually with Class-A operation the instantaneous grid voltage does not exceed the grid-current starting point. Therefore the grid signal voltage can swing only between the grid-current starting point and the point where the anode current is zero.

In the ideal case, according to Sections 92 and 93, when operation of the valve is limited to the negative grid-bias range the efficiency (the relation between the power delivered to the valve and the power output) at maximum grid excitation is 50% for pentodes, whilst for triodes the efficiency is lower.

98. Class-B Operation

With an amplifying valve the working point can be selected in such a way that the negative grid bias has approximately the value at which the anode current becomes zero (see **Fig. 200**). This mode of operation is described as Class-B. If the alternating grid voltage is sinusoidal the alternating anode current is found to consist exclusively of half waves. These half waves are equivalent to a sinusoidal anode alternating current with a frequency equal to that of the grid signal and containing a large number of higher harmonics, almost entirely even ones. Therefore if one single valve were used in a Class-B output amplifier, there would be considerable distortion. When, however, two valves are used in push-pull connection the two half-waves combine and a complete sinusoidal curve is formed with practically no distortion. In this manner an output amplifier is obtained having an efficiency much higher than that resulting from the use of two valves in a Class-A push-pull amplifier. Fig. 201 shows how the second valve supplies the other half-wave of the anode alternating current as soon as the anode current of the first valve ceases. The two valves supplement each other on the anode side, so that there is practically no distortion. When there are small signals at the grid the distortion is sometimes slightly greater, owing to the relatively greater effect of the curvature in the lower part of the characteristics.

225

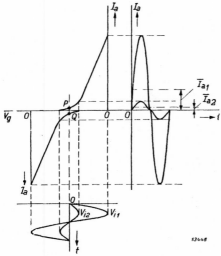

Fig. 201
Combination of the dynamic characteristics of two Class-B power valves connected in push-pull With a strong signal
V_{i1} there is comparatively little distortion of the anode current; with a weak
signal (e.g., V_{i2}) a relatively larger distortion results from the curvature in
the lower bend of the characteristic.

The advantage of a Class-B final
stage is that with two valves a
much greater power can be supplied than with two identical
valves in a Class-A amplifier,
without overloading the valves.
This is connected with the fact
that in a Class-B final stage the
power input increases as the signal
voltage is increased, whereas in a
Class-A final stage that power
remains constant. The anode dissipation (the difference between the
power input and power output)
is a measure of the heating of the
anode, and the output power of a
valve is limited by the maximum
value of the anode dissipation.
The power output is dissipated in
the loudspeaker and therefore does
not contribute towards the heating
of the valve itself. In the case of a
Class-A amplifier the power input is
constant, because the anode direct current remains constant. If, therefore, there is no signal at the grid the valve does not supply power to
the loudspeaker and consequently the anode dissipation equals the
power input ($I_{ao} \times V_{ao}$). With the maximum efficiency of 50% of a
Class-A amplifier the maximum power output is thus equal to 50%
of the power input and also 50% of the maximum anode dissipation
of the valve.

In a Class-B amplifier (see also Fig. 201) semi-sinusoidal anode current
pulses flow through each valve. In this case the mean value of the
anode current of a valve determines the heating of its anode. This
mean value is much lower than the peak value of the anode current
pulses, in the case of a half sinusoidal wave per period being equal to
$1/\pi$ times the peak value. When the mean value of the anode current

per valve amounts to $\overline{I}_a = \dfrac{I_a}{\pi}$ (I_a = peak value of the anode current)

the input power per valve is

$$P_i = \frac{I_a V_{ao}}{\pi},\qquad (138)$$

(V_{ao} = anode direct voltage) and the power input supplied to the whole stage is twice as large.

If the maximum anode current amplitude is $I_{a\,max}$, which is limited by grid current, distortion or anode dissipation, then the maximum power that can be supplied is equal to

$$P_{i\,max} = \frac{I_{a\,max}\,V_{ao}}{\pi} \,. \tag{139}$$

The output power naturally depends upon the load resistance R_L. We will assume that the optimum value is chosen. The amplitude of the alternating voltage across this load resistance may never be greater than the anode direct voltage, in order to avoid flattening of the peaks, so that the maximum output power per valve will amount to

$$P_{o\,max} = \tfrac{1}{2} \times \frac{I_{a\,max}\,V_{ao}}{2} = \frac{I_{a\,max}\,V_{ao}}{4} \,. \tag{140}$$

With small amplitudes the output power is equal to

$$P_o = \frac{I_a V_a}{4} \,, \tag{141}$$

where V_a is the amplitude of the anode alternating voltage.

The efficiency η with the maximum output power is thus found to be

$$\eta_{(Po\,=\,max)} = \frac{P_{o\,max}}{P_{i\,max}} 100\% = \frac{\pi}{4} \cdot 100\% = 78.5\% . \tag{142}$$

In order to attain the maximum a.c. power indicated by Formula (140) the load resistance (i.e. the load resistance per anode, from anode to centre tap on the output transformer and not from anode to anode) must be such that with the amplitude of the anode a.c. $I_{a\,max}$ obtaining for this power output the alternating-voltage amplitude is equal to V_{ao}. Thus it is found that the load resistance per anode should be equal to

$$R_L = \frac{V_{ao}}{I_{a\,max}} \,. \tag{143}$$

From (141) it follows that

$$P_o = \frac{I_a^2}{4} R_L . \tag{144}$$

Now the input d.c. power is equal to the a.c. power output plus the

227

power that is converted into heat in the valve itself, i.e., the anode dissipation. From this it follows that

$$P_a = P_i - P_o = \frac{I_a \cdot V_{ao}}{\pi} - \frac{I_a^2}{4} R_a. \tag{145}$$

The anode dissipation P_a has a maximum when

$$I_{a\,(P_a\,=\,max)} = \frac{2}{\pi} I_{a\,max} \tag{146}$$

(this can be found by differentiating P_a from Formula (145) with respect to I_a and putting the differential coefficient $\frac{dP_a}{dI_a}$ equal to zero). By taking the value found for I_a in (146) and substituting it in (145) one finds for the maximum anode dissipation

$$P_{a\,max} = \frac{1}{\pi} P_{i\,max}. \tag{147}$$

From Equations (146) and (147) it follows that the maximum anode dissipation does not occur at the maximum anode a.c. amplitude $I_{a\,max}$ but at $\frac{2}{\pi} I_{a\,max} = 64\%$ of $I_{a\,max}$ and that it amounts to $\frac{100}{\pi} = 32\%$ of the max. power input $P_{i\,max}$. By plotting the magnitudes of P_i, P_o and P_a given by Equations (138), (144) and (145) as functions of the amplitude I_a of the anode a.c. the curves I, II and III of Fig. 202 are obtained. From this diagram it is clear that the maximum power input occurs at the maximum amplitude of the anode current; the maximum anode dissipation, however, appears to lie at about 64% of the max. ampli-

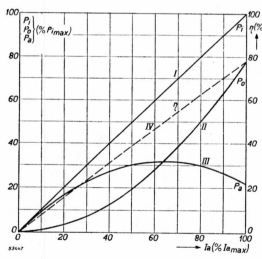

Fig. 202
The power input P_i (curve I), the power output P_o (curve II) and the power dissipated in the valve P_a (curve III) expressed as a percentage of the maximum power input $P_{i\,max}$ as a function of the amplitude of the output alternating current expressed as a percentage of the amplitude at the maximum output power for Class-B amplification. The efficiency η is likewise shown (curve IV).

tude of the anode a.c. At full grid excitation the efficiency of the Class-B final stage according to Equation (142) is theoretically equal to 78.5% (ratio of output power to input power), so that in such a final stage with two valves a power of $2 \times \dfrac{78.5}{32} = 4.9$ times the max. anode dissipation of one valve should theoretically be attainable. With two Class-A-operated amplifier valves the theoretical maximum power to be supplied is equal to the max. anode dissipation of one valve.

99. Control in the Grid-current Region

Usually amplifying valves are driven only as far as the point where grid current begins to flow. The quoted theoretical maximum efficiency of 78.5% then applies exclusively when two pentodes are used as Class-B amplifiers. When two triodes are used the maximum efficiency is lower, because then only a small part of the anode-voltage range available can be used. If, however, the grids of triodes are driven right into the region of positive grid current the entire available anode-voltage range can be used and the efficiency may then in fact approximate 78.5%. The operation of valves in this manner right into the region of positive grid voltage involves the flow of grid current and the source of grid-signal voltage has to supply a certain amount of power. To meet this, therefore, the Class-B output stage is preceded by a so-called **driver valve** capable of supplying a certain power. In order to get an approximately correct matching, this valve is coupled to the Class-B output stage via a step-down transformer provided with a centre tap on the secondary winding. Owing to the fact that as soon as the starting point is passed the grid current always begins to increase fairly suddenly, voltage pulses which are liable to cause very undesirable distortion occur across the impedance of the transformer. In order to limit these voltage pulses, the impedance between the secondary terminals of the transformer must be kept as low as possible. Therefore a driver valve is chosen with the lowest possible internal resistance and a transformer that steps down and has a very small stray field. Further, the two secondary transformer windings are shunted with damping resistances and condensers, which constitute a low-impedance path for the very high harmonics.

When applying the principle of Class-B amplification with grid current, triodes can be used which have a very high amplification factor. At zero grid voltage such a triode has a very low anode current, so that this point of the characteristic can be regarded as the cut-off point

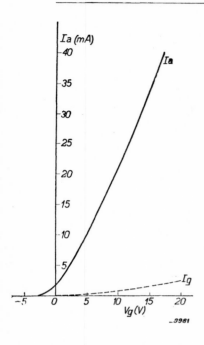

Ia (mA)

Fig. 203
Transfer characteristic of a triode part of the double power triode KDD 1 for Class-B power amplification with grid current. The grid-current characteristic is represented by a broken line.

of the anode current for Class-B operation. Thus the grid-bias battery can be dispensed with. An example of such an output valve is the double triode KDD 1 for battery sets. The characteristic of one triode part of this valve is shown in Fig. 203. Grid-current-operated Class-B final stages are almost exclusively applied in battery sets, because for these sets high efficiency of the output stage is of paramount importance, whilst the property of a Class-B final stage that the anode current increases and decreases approximately proportionally to the grid signal is also of importance. This saves the anode battery, as the valve is only occasionally fully driven and the average anode battery drain is consequently much lower.

In the case of the Class-A amplifier the anode battery drain is practically independent of the grid signal, so that for battery sets this final stage is less economical. It is to be noted that in consequence of the current supply required by the driver stage a Class-B final stage with grid current is not so economical as might be expected. Two pentodes in Class-B without grid current would be just as economical, with the same efficiency, whilst distortion due to grid current is avoided. Furthermore, the sensitivity is greater, particularly because it is possible to use step-up transformers. For these reasons this solution has been given preference in some modern developments of output valves for battery sets (cf. Philips double output pentode DLL 21).

For a.c. sets the Class-B final stage has the drawback that a fixed negative grid bias is essential; owing to the highly fluctuating anode current this can be obtained only at the cost of additional measures. Owing to the fluctuating anode current it is also necessary to keep the internal resistance of the anode-supply rectifier low in order to avoid too great a variation of the anode voltage. In high-power amplifiers, therefore, gas-filled rectifying valves are often used, the internal resistance of which is very low.

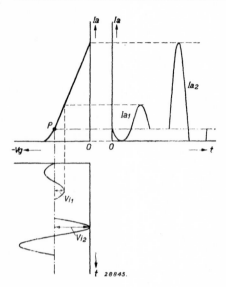

Fig. 204
Location of the working point P for Class-AB amplification. With a weak signal (e.g., V_{l1}) current will flow during the whole grid-alternating-voltage cycle, whilst if the signal is strong, current will flow during only part of the alternating-voltage cycle; this part of the cycle, however, is greater than one-half.

100. Class-AB Operation

If the working point of an amplifying valve is so chosen that with small signals the anode current flows during the whole cycle of the grid alternating voltage but with large signals it becomes zero during part of the cycle, the mode of operation is described as Class-AB. From this definition it follows that with large signals the anode current must be equal to zero during less than half the grid alternating voltage cycle (see Fig. 204). Once again, in output amplifiers two push-pull connected valves must be used in order to eliminate the distortion which would otherwise occur due to the absence of anode current crests.

Class-B amplification has the disadvantage that a considerable percentage of distortion occurs when the signals are small. This distortion can be avoided by applying Class-AB operation, for then the valves operate in Class-A with small signals. Class-AB operation has the additional advantage that the anode-current variation due to the grid-signal-voltage changes is smaller than that with Class-B operation, so that less stringent requirements have to be met by the rectifier circuit. Furthermore this system offers the possibility of applying automatic negative grid bias (see Fig. 205).

With automatic negative grid bias this voltage increases as the anode current increases, so that a common cathode resistance has to be found of such a value that at full grid excitation the maximum power output is reached.

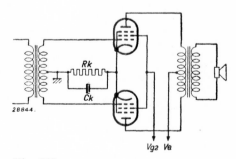

Fig. 205
Fundamental circuit of the push-pull stage with self bias.

On the other hand, care has to be taken that the maximum anode dissipation is not exceeded and that the distortion at full grid excitation is not too great. (With the maximum signal the working point must still lie on the characteristic, that is to say, the anode current corresponding to the working point may only just be equal to zero.)

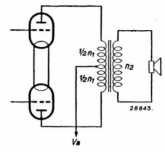

Fig. 206
Two power triodes connected in push-pull with an output transformer.

The maximum power output is usually not so great with Class-AB as with Class-B operation without grid current, as in the former case one has to take account of the fact that in the absence of a signal the maximum anode dissipation must not be exceeded. Usually, however, the maximum power output is greater than that with two Class-A valves connected in push-pull.

101. Matching of the Loudspeaker to Class-B and Class-AB Output Stages

For two triodes in a Class-B power amplifier the following considerations apply for matching the loudspeaker. Let us put the total number of primary turns of the transformer as n_1 and the number of secondary turns as n_2 (see Fig. 206). Since in principle the working point is chosen at $I_{ao} = 0$ (quiescent current), the valves work for half a cycle alternately. If no grid current is allowed the amplitude which can be applied at each grid may be at most equal to V_{go} (see Fig. 207).

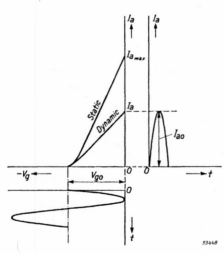

Fig. 207
Static and dynamic characteristics of a power triode for Class-B operation.

The greatest power output is obtained if $R_L = R_a$ (see Appendix II/7). In that case the slope of the dynamic characteristic amounts to half that of the static characteristic. The amplitude of the anode current is therefore

$$I_a = \tfrac{1}{2} I_{a\,max} = \tfrac{1}{2} \frac{V_{ao}}{R_a}. \qquad (148)$$

The alternately acting halves of

the loaded push-pull output transformer may also be regarded as a single load resistance R_L with the total alternating current flowing through this resistance having an amplitude equal to I_a. The maximum power output is then

$$P_{o\,max} = \tfrac{1}{2}\,I_a{}^2 R_L = \tfrac{1}{8}\,\frac{V_{ao}{}^2}{R_a} = \tfrac{1}{8}\,V_{ao}\,I_{a\,max}. \qquad (149)$$

Owing to the fact that with Class-B operation the valves are not working for alternate half cycles, one half of the primary of the output transformer can be left out of consideration during each half cycle. Therefore the load formed by one half of the primary winding must have the indicated value of $R_L = R_a$ where the other half is not connected. Thus we obtain

$$\left(\frac{\tfrac{1}{2}n_1}{n_2}\right)^2 = \frac{R_a}{R_1} = \frac{V_{ao}}{I_{a\,max}\,R_1}, \qquad (150)$$

hence

$$\frac{n_1}{n_2} = \sqrt{\frac{4\,V_{ao}}{I_{a\,max}\,R_1}}, \qquad (151)$$

in which R_1 represents the resistance of the speech coil of the loudspeaker. From this it also follows that with Class-B operation the anode impedance on the total primary side of the transformer (calculated) between the anodes is equal to four times the optimum matching impedance for each single valve.

For the application of Philips valves in push-pull stages (Class-A, Class-B or Class-AB) the figure quoted as optimum matching impedance is always the impedance between the two anodes.

With pentodes in Class-B the anode alternating current is equal to $I_{a\,max}$, regardless of the load resistance R_L, if the grid is fully excited. The voltage amplitude across R_L may not, however, become greater than V_{ao} (the anode voltage may not become negative). The requirement for optimum matching is therefore that with an a.c. amplitude $I_{a\,max}$ there must be an alternating-voltage amplitude equal to V_{ao}, thus

$$R_L = \frac{V_{ao}}{I_{a\,max}} \quad \text{[see Equation (143)].}$$

From this the ratio of the output transformer can be calculated as follows:

$$\left(\frac{\frac{1}{2}\,n_1}{n_2}\right)^2 = \frac{R_L}{R_l} = \frac{V_{ao}}{I_{a\ max}\,R_l} \tag{152}$$

and

$$\frac{n_1}{n_2} = \sqrt{\frac{4\,V_{ao}}{I_{a\ max}\,R_l}}. \tag{153}$$

The same formulae apply for triodes in Class-B push-pull connection with grid current.

CHAPTER XVIII

Rectification of the R.F. or I.F. Signal

102. Rectifying Action of the Diode

As already explained in Chapter IX, the modulation signal is separated from the carrier wave in the detector valve. In order to effect this the modulated H.F. or I.F. signal has to be rectified. The direct voltage thereby obtained, however, has not a constant value. Due to rectification the lower half-cycles of the alternating voltage of the signal, for instance, are suppressed, in which case only the upper half-cycles remain, and their amplitudes increase and decrease according to the superimposed modulation frequency (see Fig. 208b). If the rectifying circuit consists of a rectifying valve with a buffer or reservoir condenser and a leak resistance and these elements are proportioned so that only the H.F. fluctuations of the half-waves are smoothed and not the L.F. fluctuations, only a unidirectional voltage fluctuating in accordance with the modulation frequency remains across the resistance (see Fig. 208c). If this fluctuating unidirectional voltage is applied to a coupling element consisting of a resistance and condenser (CR element) (Fig. 209) the steady component of voltage will be blocked by the condenser C and across

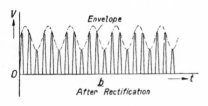

a
Before Rectification

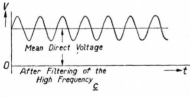

After Rectification

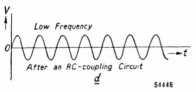

Mean Direct Voltage

After Filtering of the High Frequency
c

Low Frequency

After an RC-coupling Circuit
d

54446

Fig. 208

a) Modulated carrier wave.

b) Owing to the rectifying action of the detector valve the negative half-waves are not passed through, so that only the signal remain.

c) Due to the time constant of the leak resistance and the condenser used for detection there remains across the leak resistance a negative direct voltage with a low-frequency modulation superimposed upon it.

d) After the blocking condenser there is only the L.F. modulation remaining.

the resistance R there will be a L.F. alternating voltage of the same frequency as that with which the carrier wave was modulated (see Fig. 208d). The process whereby the H.F. component of the rectified H.F. signal is smoothed by means of a condenser with a resistance in parallel—which results in a direct voltage fluctuating in accordance with the modulation frequency—may be explained in the following manner (see Fig. 210).

Fig. 209
CR circuit.

In Fig. 210a a sinusoidal alternating voltage is produced by the generator G, connected in series with a valve V (the rectifying or detector valve) which allows the generator current to pass only in the direction of the arrow. This generator G in series with the valve V produces between the points 1 and 2 a pulsating direct voltage, the wave form of which is represented by the full-line curve in Fig. 210b. The internal resistance of G and V in series is here assumed very small compared with R in the conducting direction of V and infinitely large in the opposite or blocking direction. If now the condenser C is connected in parallel with R the former will be charged when the voltage pulse at terminal 1 is positive and as soon as the voltage of this pulse drops again it will partly discharge itself through the resistance R until it is re-charged by the next voltage pulse. Thus the voltage at the condenser does not follow the full-line curve of Fig. 210b but the broken line and consequently has a more or less constant value. Obviously the discharge current through the condenser becomes smaller when the resistance is higher, so that the condenser discharges more slowly and after a certain discharge period the voltage across the condenser will have dropped less. It will also be evident that a certain discharge current in a given interval of time will cause less voltage drop with a large condenser than with a small one. In order to obtain a direct voltage as constant as possible one must therefore have a large condenser and a high resistance value.

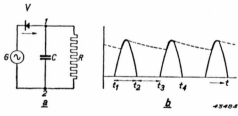

Fig. 210

a) Generator G of a sinusoidal alternating voltage having a low internal resistance. This generator is connected in series with a valve V (e.g., a rectifying valve) which allows the current from the generator to pass only in the direction of the arrow. In parallel with the series combination of G and V is a condenser C and a resistance R.

b) *Full line:* Pulsating direct voltage, which in the absence of the condenser C is produced by the generator G in series with the valve V between the terminals 1 and 2, plotted as a function of time.
Broken line: Voltage variation across the resistance R (between the terminals 1 and 2) when a condenser C is connected in parallel with R.

For the process of detection, however, these components may not be so large as to influence the much slower modulation fluctuations superimposed upon the rectified voltage.

The voltage across a condenser that discharges through a resistance is a function of time and, as already explained above, depends upon the capacity value C and the resistance value R. The formula for the voltage across the condenser is

$$V_C = V_{Co}\, \varepsilon^{-\frac{t}{RC}} \tag{154}$$

in which V_{Co} is the voltage between the condenser plates at the beginning of the discharge, ε the base of natural logarithms (ε = 2.7182 . . . .), t the time of discharge in seconds, R the resistance in ohms and C the capacity in farads.

The constant RC of a smoothing filter consisting of a resistance and a condenser is equivalent to the time interval in seconds required for the voltage across the condenser to diminish to 1/2.72 of its original value, for when t = RC Formula (154) becomes $V_C = V_{Co}\, \varepsilon^{-1} = V_{Co}/2.72$. This time is very easily calculated from the product of the capacity in farads and the resistance in ohms. For example, a condenser of 100 $\mu\mu$F and a resistance of 1 megohm will have an RC constant of $100 \times 10^{-12} \times 10^6 = 10^{-4}$ second, that is to say, the voltage across the condenser will have dropped to 1/2.72 of its original value after 1/10,000th of a second. A wavelength of 2,000 metres corresponds to a frequency of 150,000 c/s so that the duration of a cycle is 1/150,000th of a second. In this short time with the given CR smoothing circuit the condenser voltage will remain practically constant. At shorter wavelengths the conditions for smoothing will be still more favourable.

The highest modulation frequency occurring in practice in radio broadcasting is usually about 5,000 c/s. The condenser with leak resistance must therefore have such an RC constant that the condenser discharge is not slower than the drop of the sinusoidal wave for 5,000 c/s if the leak-resistance voltage is to follow precisely the modulation envelope for 5,000 c/s. If the RC constant is less than $\frac{1}{2\pi}$ times the duration of the modulation cycle the leak-resistance voltage follows the modulation envelope exactly when the modulation depth is 70%. With smaller modulation depths the RC constant may be chosen proportionately greater. From this it follows that for the CR circuit described above having an RC constant of 10^{-4} second and for a modulation depth of 70% the upper frequency limit amounts to 1,600 c/s; in practice it is common to use a condenser of 100 $\mu\mu$F and a leak resistance of

0.5 megohm; consequently the time constant is half the value indicated above and since a modulation depth of 70% at the high frequencies hardly ever occurs, modulation frequencies of 5,000 c/s and higher are reproduced without distortion.

In the rectification of a signal use is made of the property of the radio valve to pass electrons exclusively from the emitting cathode to the anode and not in the reverse direction. For this purpose a diode (valve with two electrodes) is therefore adequate. If a diode is connected in series with a resistance across a resonant circuit, as represented in Fig. 211, and if in that circuit an alternating voltage of the resonant frequency is applied, then the diode will allow the current to pass only in the direction of the arrow. As a result a pulsating direct voltage is developed across the series resistance R. If now the resistance R is shunted by a condenser C (see Fig. 212) a smoothing circuit having a certain RC constant is obtained, and if this constant is chosen correctly the direct voltage across R will be practically steady. Thus we have at the anode of the diode a negative direct voltage and a H.F. alternating voltage (see Fig. 214).

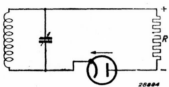

Fig. 211
Connection of a diode, in series with a resistance, to an oscillatory circuit.

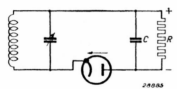

Fig. 212
Connection of a diode, in series with a resistance with a condenser in parallel with it, to an oscillatory circuit.

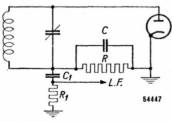

Fig. 213
Circuit of a diode with leak resistance and condenser. The L.F. modulation is taken off via a filter consisting of C_1 and R_1.

Assuming that the direct voltage across the condenser remains constant, only the shaded voltage peaks will cause a passage of current through the diode; consequently, it is due to them that the direct voltage across the leak resistance is developed. If the diode alternating voltage drops the voltage peaks disappear and the current through the leak resistance falls until a new equilibrium is reached. When the alternating voltage across the resonant circuit is modulated it increases and decreases in sympathy with the modulation frequency; with the right choice of R and C the modulation frequency will remain across R as a direct voltage fluctuation. The low-frequency voltage superimposed on the direct voltage

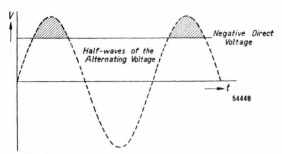

Fig. 214
H.F. alternating voltage and negative direct voltage at the anode of the diode as a function of time. Only the shaded voltage crests cause a current to flow through the diode.

can be separated by means of an RC circuit. Nowadays it is usual to connect the leak resistance R of the rectifying circuit between the cathode of the diode and the resonant circuit, so that the complete rectifying circuit is as represented in Fig. 213. The negative direct voltage across R is then blocked by the condenser C_1 and can be used in the set for special purposes (see Chapter XXIII).

103. Effect of the Current-Voltage Curve of the Diode

The current which causes the voltage drop in the leak resistance flows through the diode and is governed by the alternating voltage applied and the properties of the diode. The variation of H.F. voltage caused by the modulation should in principle result in an absolutely identical variation of the direct voltage across the leak resistance. In order to achieve this, however, account has to be taken of the properties of the diode, and these will now be examined.

The direct current passing through the diode can be plotted as a function of the direct voltage applied. Such a curve, where I_a is plotted on a logarithmic scale, is shown in Fig. 215. Owing to detection, a negative direct voltage is produced across the leak resistance and this therefore exists between the anode and the cathode of the diode. At the same time there is the applied H.F. alternating voltage between anode and cathode. As a consequence of the superposition of an alternating voltage upon a negative direct voltage periodic current pulses flow through the diode from anode to cathode; the mean value of the current pulses flowing through the diode can be measured. When the diode receives large signals its anode will acquire such a strongly

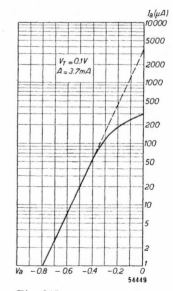

Fig. 215
Current through a diode as a function of the direct voltage applied.

239

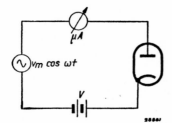

Fig. 216
Circuit for measuring the
direct current through a
diode as a function of the
direct voltage for a number
of values of the simultaneously applied alternating
voltage.

negative potential that only the peaks of the
alternating voltage will cause a current to
flow through it.

If, as shown in the diagram of Fig. 216, the
direct current through the diode is measured
and plotted as a function of the applied
negative direct voltage for a series of alternating voltage values (V_m = peak value of
the alternating voltage applied), a family of
curves is derived as shown in Fig. 217. These
curves therefore show the relation between
the mean diode current I, the direct voltage
V across the diode and the amplitude of the
H.F. signal V_m.

As already explained, owing to rectification the direct voltage across
the diode is produced by the voltage drop across the leak resistance
R, so that when the leak resistance is connected to the cathode the
average current I through the diode should satisfy the equation

$$-V = IR. \qquad (155)$$

In Fig. 217 this equation is represented by the straight line OP for
R = 0.5 megohm. The point P gives, therefore, the value for I where
the leak resistance is 0.5 megohm and the H.F. amplitude is 0.5 volt.

If now there is a modulated
H.F. alternating voltage
across the diode the voltage
V_m will increase and decrease
in accordance with the
modulation frequency, and
the direct voltage across the
leak resistance will fluctuate
correspondingly. Thus a direct voltage upon which
the alternating voltage of
modulation frequency is
superimposed is set up
across the leak resistance.
If there is a single modulation frequency, for instance
500 c/s, and the modulation
depth is 40%, with an un-

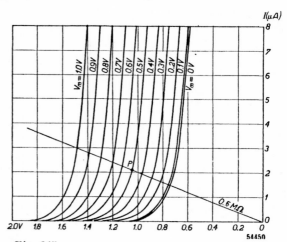

Fig. 217
Family of curves representing the direct current
through the diode as a function of the negative
direct voltage applied, for different values of the
simultaneously applied alternating voltage.

modulated carrier-wave amplitude of 0.5 volt the voltage will fluctuate between 0.7 and 0.3 volt. Consequently the direct voltage across the leak resistor fluctuates 500 times per second between the point of intersection of the straight line (OP) with the curve for $V_m = 0.7$ V and that with the curve for $V_m = 0.3$ V. For undistorted reproduction of the modulation of the carrier wave across the leak resistor the resistance line should be uniformly divided by the curves $I = f(V)$. The distances from $V_m = 0.5$ to 0.6 and from 0.5 to 0.4 should therefore be equal, as also the distances from $V_m = 0.5$ to 0.9 and from 0.5 to 0.1. From the curves of Fig. 217 it appears that in the case of large signals this requirement is satisfied but not in the case of small values of V_m, because the curves $I = f(V)$ corresponding to these small values lie closer together and there is therefore no linear relation between the increase of V_m and that of the direct voltage V. With signals smaller than 0.03 volt the direct voltage increases approximately in proportion to the square of the amplitude of the H.F. voltage; with signals greater than 1 volt the direct voltage increases directly proportionally to the H.F. alternating voltage. With values lying between these figures the direct-voltage increase is neither directly proportional nor proportional to the square of the H.F. voltage amplitude. Only when the relation between the H.F. signal amplitude and the direct voltage is linear will the carrier-wave modulation be reproduced undistorted.

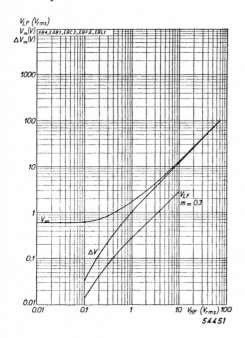

With weak signals (< 0.03 V), square-law detection will cause serious distortion; with a modulation depth of 80% there is 20% second-harmonic distortion. For good quality of reproduction it is therefore necessary to ensure that even with weak signals the amplitude is greater than 1 volt.

For all Philips diodes characteristics are published from which the **increase** of the direct voltage across the leak

Fig. 218
Direct voltage $V_=$ and increase $\varDelta$ V of the direct voltage at the terminals of the leak resistance of a diode, as a function of the unmodulated H.F. alternating voltage. L.F. alternating voltage V_{LF} at the leak-resistor terminals as a function of the 30% modulated H.F. voltage. These curves apply to a leak resistance of 0.5 megohm.

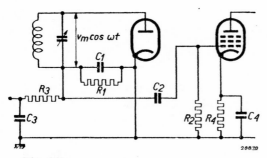

Fig. 219
Connections of a diode rectifier with coupling elements to the succeeding L.F. amplifier and to the automatic-volume-control circuit.

resistance with respect to the direct voltage without a signal can be read off as a function of the effective (r.m.s.) value of the H.F. alternating voltage at the diode for a leak resistance of 0.5 megohm (see Fig. 218). These curves are obtained by measurement but can also be derived from the curves of Fig. 217 by reading off the increase of the direct-voltage values corresponding to the points of intersection of the 0.5-megohm-resistance line with the curves for V_m. For diodes, curves are also published indicating the direct voltage at the leak resistance as a function of H.F. voltage. Furthermore, a curve is given for each diode from which the L.F. alternating voltage at the leak resistance as a function of the r.m.s. value of the H.F. alternating voltage, 30% modulated, can be read off; this curve can also be derived from that giving the direct-voltage increase across the leak resistance as a function of H.F. voltage applied to the diode.

104. Influence of the Coupling Element upon Detection

In the foregoing it was assumed that the leak resistance of the diode is the only resistance in the diode circuit. For the L.F. range we have usually in parallel with that resistance the leak resistance of the amplifying valve that follows the diode (see Fig. 219) and sometimes also the smoothing resistance of the automatic-volume-control circuit (see Chapter XXIII). These resistances, it is true, are separated so far as d.c. is concerned by condensers, but as these condensers have to pass the low frequencies, (or, in other words, for good smoothing have to be chosen so large that their impedance to low frequencies is small) the resistance for the L.F. fluctuations across the leak resistance is changed. The L.F. resistance is calculated from the parallel connection of the diode leak resistance, the grid leak resistance of the following valve and any smoothing resistance used for the automatic-volume-control circuit. Where the modulation depth of the carrier wave is large this factor may result in considerable distortion, as will become evident from a study of Fig. 220, in which the curves $I = f(V)$ are plotted once more with V_m as parameter.

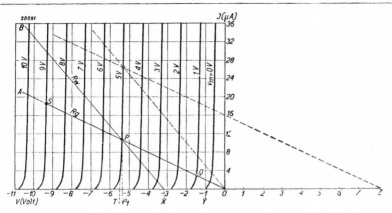

Fig. 220
Family of curves I = f(V) of a diode with V_m as parameter. The line OA represents the diode leak resistance which constitutes also the d.c. resistance of the circuit. The line XB corresponds to the a.c. resistance. As soon as the L.F. amplitude exceeds PX serious distortion arises, owing to the modulation crests not being reproduced. This figure applies to linear rectification.

Here the line OA represents the current-voltage characteristic of the d.c. resistance R_g of the diode circuit, i.e., of the 0.5-megohm leak resistance. When the voltage of the H.F. signal carrier wave amounts to, say, 5 volts the negative direct voltage across the leak resistance adjusts itself to the point P. If, now, the L.F. load is provided by the leak resistance R_g, the modulation fluctuations follow that line; thus with a modulation depth of 80% the point indicating the relation between the negative direct potential V, the H.F. alternating voltage V_m and the direct current I will fluctuate between the points S and Q on the line R_g. As the lines for equal increments of V_m in this region cut the load line into equal sections, when the modulation is sinusoidal the L.F. alternating voltage will likewise be sinusoidal. If, however, the a.c. resistance deviates from R_g and equals, for instance, R_w, the point indicating the relation between the direct current I, the negative direct potential V and the H.F. alternating voltage V_m will move along the line R_w. Given a certain critical modulation depth the "operating point" fluctuates as far as the point X. It will be evident that as the modulation depth exceeds this critical value there will be serious distortion, because the peaks of the sinusoidal modulation curve are then no longer reproduced across the leak resistance (see Fig. 221). The

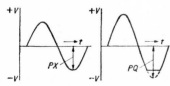

Fig. 221
Distortion due to flattening of the modulation crests.

243

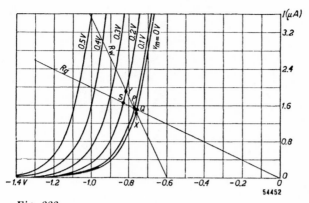

Fig. 222
D.C. resistance and a.c. resistance with squarelaw detection. When the modulation depth is great the modulation crests are not cut off.

modulation depth that can be applied without distortion arising from the flattening of the sinusoidal modulation curve crests is approximately equal to

$$m_{max} = \frac{R_w}{R_g}. \qquad (156)$$

If R_w deviates from R_g the flattening of the modulation crests can be avoided, as shown in Fig. 220, by applying to the leak resistance a bias potential which is positive with respect to the cathode (OZ in Fig. 220).

The permissible modulation depth is greater for small signals and may be equal to unity. From Fig. 222, which applies to small signals, it is seen that there is no distortion as a consequence of absence of the modulation crests. With small signals, however, there is distortion, already explained, due to non-linear rectification. Consequently diode detection is free from distortion only when the amplitude of the carrier wave is sufficiently large and the L.F. a.c. load of the diode deviates little from the leak resistance R.

105. Damping caused by the Diode Detector

(a) Origin of the Damping

The damping of an oscillatory circuit increases when a resistance is connected in parallel (see Chapter XXV, Section 148). If there is a diode with leak resistance and smoothing condenser (see Fig. 223) in parallel with the oscillatory circuit, in consequence of the continuous discharge of the condenser C, a direct current I_g will flow through leak resistance R. The power dissipated by the leak resistance is then $P = I_g^2 R$, which has to be supplied by the oscillatory circuit. This entails an increase of the damping of the circuit.

(b) Equivalent Resistance for the Rectifying Circuit

The damping caused by the rectifying circuit may be represented by a resistance R_d, called the **equivalent resistance of the diode circuit.** It is apparent that the value of the equivalent resistance R_d varies for different values of alternating voltage.

(a) Damping with Large Signals

According to Section 103, with H.F. amplitudes greater than 1 volt the detection is linear. In that case the negative potential between the anode and cathode of the diode has such a large value that current flows through the diode during only a very small part of the a.c. voltage cycle. As may be proved both mathematically and by measurement, for such signals there is a damping resistance

$$R_d = \tfrac{1}{2} R \, \frac{V_m}{V_{c=}}, \tag{157}$$

in which R is the leak resistance, V_m the diode-alternating-voltage amplitude, and $V_{c=}$ the direct voltage on the smoothing condenser or leak resistance. With very large signals ($V_m > 10V$) the quotient $V_m/V_{c=}$ approximates unity, so that R_d then equals $\tfrac{1}{2} R$.

From this it follows that a leak resistance of high value has to be used in order to keep the damping as low as possible. A satisfactory value in practice is 0.5 megohm. If this resistance is chosen too large there is the drawback that in consequence of the limited value of the grid leak resistance of the following amplifying valve distortion arises from the flattening of modulation crests when the modulation depth is great (see Section 104).

(β) Damping with Small Signals

With small signals (amplitude < 0.1 V) the negative direct voltage produced across the leak resistance is so low that the diode allows current to pass during the whole alternating-voltage cycle. [As explained in Chapter IV, Section 17, where there is a low negative voltage at the anode, current (the potential-barrier current) flows owing to the emergence velocity of the electrons from the cathode.] From the anode-current-anode-voltage curve

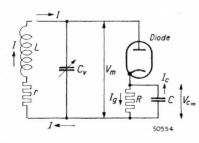

Fig. 223
Connection of a diode detector in parallel with the preceding oscillatory circuit.

245

(plate characteristic) of Fig. 215 it appears that with a negative voltage of 0.7 V this current has a value of 2.5 μA. In Fig. 224 the same curve is plotted afresh on a linear scale, with the resistance line for $R = 0.5$ megohm added. The point where the resistance line intersects the plate characteristic lies at $V_d = -0.74$ V, so that according to this figure the diode adjusts itself to this voltage when there is no signal. With small signals, such as are represented by V_{i2}, current flows during the whole cycle, even when taking into account the fact that the point P is shifted somewhat farther to the left owing to the direct current generated by the rectifying action. Since current is flowing during the whole of the alternating voltage cycle, with small signals the damping of the rectifying circuit on the oscillatory circuit is determined by the alternating current resistance R_a of the diode (for

an alternating voltage the leak resistance is shunted by the condenser C).

The alternating-current or internal resistance R_a can be determined approximately by drawing the tangent to the curve through the point P. From this figure it is then seen that the a.c. resistance amounts to 55,000 ohms, whereas measurement on a normal diode gives a value of 70,000 ohms (see also Fig. 225).

Fig. 225 gives the damping resistance as a function of the H.F. voltage at the diode with a leak resistance of 0.5 megohm. The dot-dash line gives the damping resistance as a function of H.F. voltage for the case where a positive bias of 1.3 V is applied to the anode of the diode. [A positive bias may be applied in order to avoid the modulation crests not being reproduced when the a.c. and d.c. resistances are unequal (see Section 104)].

From Fig. 225 it appears that

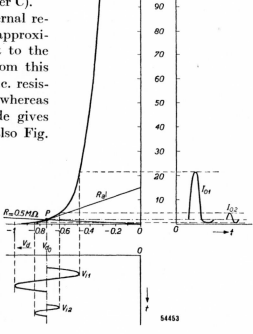

Fig. 224
Current I_d through the diode as a function of applied direct voltage V_d. The line for the leak resistance $R = 0.5$ megohm is also drawn. With small alternating voltages current flows through the diode during the whole cycle (see I_{01} and I_{02}). The a.c. resistance of the diode is then approximately equal to the internal resistance R_a.

246

with small signals the damping resistance without bias is 70,000 ohms and with large signals $\frac{1}{2} \times 0.5 = 0.25$ megohm. In order to keep the damping as small as possible it is therefore advisable to work with large signals on the diode (this is also advisable from the point of view of minimizing distortion). This figure further shows that with small signals a diode to

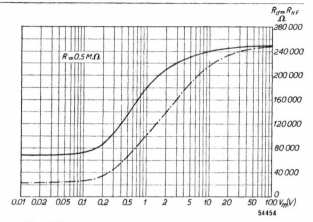

Fig. 225
Full line: H.F. resistance or damping of a diode detector as a function of the signal voltage.
Dot-dash line: H.F. resistance as a function of the a.c. voltage, if the leak resistance is connected to + 1.3 V.

which a positive bias is applied gives much greater damping.

(c) Damping when the Leak Resistor is Parallel to the Circuit

A rectifying circuit as represented in Fig. 226 is often employed. It is used, for example, when one side of the oscillatory circuit is earthed or with rectifiers for delayed automatic volume control. In that case the damping resistance of the rectifier alone amounts to $R/2$ with large signals. Moreover, the leak resistor R is then in parallel (see Fig. 227), so that the resultant damping is equal to

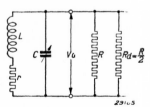

Fig. 226
Principle of the circuit of a diode detector connected to an earthed oscillatory circuit.

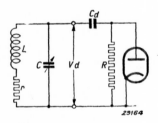

Fig. 227
Equivalent circuit of a diode detector which is connected to an earthed oscillatory circuit. The total damping resistance is formed by the combination of R and R_d.

$$R_{res} = \frac{\dfrac{R}{2} \cdot R}{\dfrac{R}{2} + R} = \tfrac{1}{3} R \qquad (158)$$

With small signals R is usually much greater than the a.c. resistance of the diode, so that in practice this latter figure can be used for calculations.

247

106. Anode Detection

An alternating voltage can also be rectified by utilising the curvature of the transfer characteristic of a valve. In Fig. 228 for instance, by means of the negative grid voltage (adjusted to $V_g = V_{go}$) the working point P has been so chosen as to give zero anode current when no grid alternating voltage is present.

If now an alternating voltage is applied to the grid, anode current can flow only during the positive half-cycles. Consequently this gives rise to a unidirectional pulsating anode current which is equivalent to a mean steady anode current I. If a coupling element with sufficiently large time constant is provided in the anode circuit, for example a resistance parallel to a condenser (see Fig. 229), the anode-current pulses will give rise to a mean direct voltage $\overline{V}$ across that coupling element. Obviously, the anode-current peaks increase in size according to the rise in amplitude of the grid alternating voltage and this is accompanied by an increase in the mean steady anode current.

If, in Fig. 229, V_i is a L.F. modulated H.F. alternating voltage and the time constant R_1C_1 of the anode coupling element is so chosen that the H.F. ripple across the condenser is smoothed out, but not the modulation ripple, then there arises across the coupling element R_1C_1 a L.F. alternating voltage superimposed on the direct voltage $\overline{V}$. This L.F. alternating voltage can be separated from the direct voltage $\overline{V}$ by means of a condenser-resistance coupling (C_2 and R_2 in Fig. 229), as is done in the case of diode detection.

The anode detector has the advantage that there is no damping of the grid oscillatory circuit, provided that care is taken to hold the amplitude of the alternating grid voltage below the absolute value of V_{go}, that is to say, provided that no grid current flows.

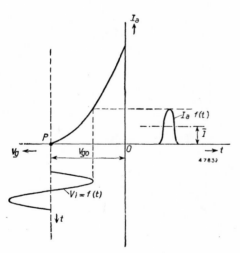

Fig. 228
Transfer characteristic of an amplifying valve. The working point P is adjusted to the grid voltage V_{go}, at which the anode direct current is zero. Consequently only the positive half-cycles of an alternating grid voltage V_i can cause anode current to flow. Thus there arises a unidirectional pulsating anode current equivalent to a continuously flowing direct current $\overline{I}$.

In order to avoid grid current at a certain ampli-
tude of grid alternating
voltage, the working point
P has to be chosen suffi-
ciently far to the left.
Should the working point
happen to be at the same
time the cut-off point, the
direct anode voltage must
be μ times as great as the
grid bias chosen, where

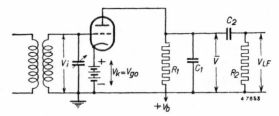

Fig. 229
Circuit diagram of an anode detector. In the anode
circuit a coupling element R_1C_1 has been introduced
with a time constant large enough to obtain a
steady direct voltage V.

μ is the amplification factor. To get the desired result with the lowest
possible anode voltage, we must choose a valve with small amplification
factor, but then this means a low gain, so that it is necessary to seek
for each individual case the best compromise between amplification and
anode voltage.

For rectifying a signal with an r.m.s. value of 5 V, or a peak value
of 7 V, when using a triode with an amplification factor of 20 and with
grid current beginning to flow at —1.3 V, an anode voltage of at least
$20 \times (7 + 1.3) = 166$ V is needed to avoid grid-current damping.

Another advantage of the anode detector according to Fig. 229 is that
amplification takes place in the valve simultaneously with the detection.
With a pentode used for anode detection (see Fig. 230) a detector-
amplification factor of about 8 can be obtained. By detector amplification is to be understood the ratio of the L.F. output alternating
voltage to the 30% modulated H.F. input alternating voltage. When
using the pentode AF 7 in a circuit as illustrated in Fig. 230, at a H.T.
supply voltage $V_b = 250$ V,
a screen-grid voltage $V_{g2} =$
100 V, a cathode resistance
$R_k = 10,000$ ohms and an
anode coupling resistance
$R_L = 0.32$ megohm, the
detector amplification a-
mounts to 8.4. With an
alternating output voltage
of 14 V the distortion is
5.6%.

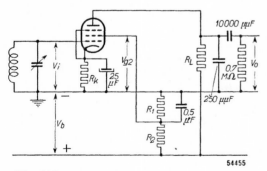

Fig. 230
Circuit diagram of a H.F. pentode used for anode
detection.

Generally speaking, the L.F.
distortion with anode detec-

tion is greater than that with diode detection, because there is no question of any linear detection, even with comparatively large alternating grid voltages.

When a pentode is used as anode detector, the screen grid has to be fed via a voltage divider of low internal resistance, because when feeding through a series resistance the screen-grid voltage drops as a result of the increase in screen-grid current accompanying a rising alternating grid voltage, and with that the grid range is also reduced. The application of automatic grid bias (self bias) as indicated in Fig. 230 has the advantage that with increasing signal the negative bias rises, consequently larger signals can be rectified; without involving grid current, than is possible with a fixed negative grid bias and the same anode voltage. This is mainly of importance when it is desired to rectify very large alternating voltages, say of the order of 80 to 100 volts, while retaining a high input impedance and thus avoiding grid current. With a fixed grid bias and a valve having an amplification of 20, for an alternating grid voltage of 100 volts we would require an anode voltage of $100 \sqrt{2} \times 20 = 2800$ V, which is prohibitive for valves of normal dimensions.

The same alternating grid voltage however can be rectified with a direct anode voltage of 350 V and a valve having an amplification factor of 20 by applying self bias. The only conditions are that the transconductance of the valve and the cathode resistance must be large enough. It is to be deduced that the product $g_m R_k$ (transconductance × cathode resistance) must satisfy the requirement:

$$g_m R_k > \frac{\pi^2}{2\sqrt{2}} \left(\frac{\mu V_i \sqrt{2}}{V_a} \right)^{3/2}, \tag{159}$$

where V_i is the r.m.s. value of the alternating input voltage, μ the amplification factor and V_a the direct anode voltage.

$$(V_a = V_b - V_k \approx V_b - V_i \sqrt{2}).$$

Further, for the circuit of Fig. 231 it can be deduced that if the grid is earthed, thus $V_i = 0$,

$$I_a = \frac{V_b}{R_a + (\mu + 1) R_k} \tag{160}$$

and

$$V_k = \frac{V_b}{R_a/R_k + \mu + 1}. \tag{161}$$

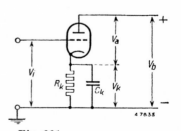

Fig. 231
Fundamental circuit diagram of an anode detector with self bias.

If $R_a \ll R_k$ then (160) resolves into:

$$I_a = \frac{V_b}{(\mu + 1)\, R_k} \qquad (162)$$

and (161) into:

$$V_k = \frac{V_b}{\mu + 1}. \qquad (163)$$

From this last equation it appears that the cathode voltage V_k is then independent of the value of the cathode resistance.

A family of rectification curves can be measured for anode detection, just as has been done for diode detection in Fig. 220.

Fig. 232 shows the principle of the measuring circuit and Fig. 233 gives a family of curves for the experimental triode EC 51. This indicates the relation between the mean direct anode current $\bar{I}_a$, the negative bias voltage V_g applied to the grid and the r.m.s. value $V_{o\,rms}$ of the alternating voltage $V_m \cos \omega t$ likewise applied to the grid.

The direct anode current and hence the detection characteristic can be determined from this family of curves as a function of the alternating grid voltage for any value of cathode resistance. From the characteristic we can also find the detection curve for a fixed bias, by dropping a vertical line through the point V_{go} on the x axis corresponding to the assumed value of grid bias (see the dotted line in Fig. 233).

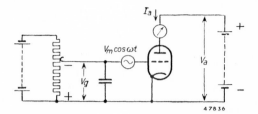

Fig. 232
Circuit for measuring the direct current I_a through a triode as a function of the direct grid voltage V_g for a number of values of the grid alternating voltage $V_m \cos \omega t$ applied simultaneously.

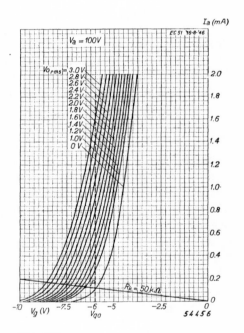

Fig. 233
Family of curves representing the direct anode current I_a through a triode as a function of the negative direct voltage applied to the grid, for various values of the alternating voltage applied simultaneously to this grid. These curves apply for the experimental triode EC 51 with $V_a = 100$ V.

107. Grid Detection

Grid detection is in essence the same as diode detection. The grid and cathode of an amplifying valve, e.g. a pentode, are used as two electrodes of a diode, the anode current of the valve being thereby controlled by the direct voltage on the grid resulting from rectification. Fig. 234 shows the principle of the circuit. A direct voltage is produced across the leak resistance, which may be, for instance, 1 megohm, the grid being negative with respect to the cathode. If the time constant of $C_{g1}R_{g1}$ is so chosen as to be large for the carrier-wave frequency of the input alternating voltage but small for the modulation frequency, then there arises across R_{g1} an alternating voltage that fluctuates with modulation frequency. At the same time across R_{g1} there is the applied alternating voltage with the carrier-wave frequency, and this has to be smoothed out in the anode circuit by means of the capacity C_a ($= 250\ \mu\mu F$). The modulation-frequency voltage across R_{g1} is amplified in the anode circuit.

With such a circuit used for pentodes showing a transconductance of about 2 mA/V and having an anode coupling resistance of about 0.2 megohm a detector amplification of 20 to 30 can be reached. The screen-grid voltage has to be chosen fairly low, about 37.5 V, so as not to obtain too large a steady anode current in the absence of an alternating input voltage. In that case, owing to grid current, the grid adjusts itself to a negative voltage of about —0.7 V. Moreover, under these conditions, with a higher screen-grid voltage the working point would come to lie higher in the upper bend of the transfer characteristic, so that the transconductance would be very small and the L.F. distortion very large for small signals (see Fig. 235).

For rectifying signals that are not so small it is desirable to have a larger grid range of the transfer characteristic. In the case of a fixed screen-grid voltage, as results from Fig. 235, the working point shifts towards the lower bend of the transfer characteristic, due to rectification. This results in a simultaneous occurrence of anode detection. The L.F. alternating voltage obtained in the anode circuit as a result of this anode detection is in anti-phase with respect to

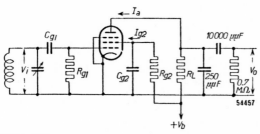

Fig. 234
Circuit diagram of a pentode used as grid detector.

that obtained by grid detection, so that the resulting alternating voltage is much smaller than that obtained with grid detection alone. In order to avoid this undesirable effect the supply to the screen-grid has to be effected by means of a series resistance. Since with increasing input alternating voltage the negative grid voltage also increases, the direct anode current and the screen-grid current will drop. When the screen grid is fed via a series resistance the screen grid voltage is able to rise, and consequently the grid range expands automatically. This avoids shifting of the working point to the lower bend of the characteristic. Fig. 235 gives a family of transfer characteristics for various screen-grid voltages of valve EF 6 used with an anode series resistance of 0.2 megohm. The obliquely ascending curve represents the anode current as a function of

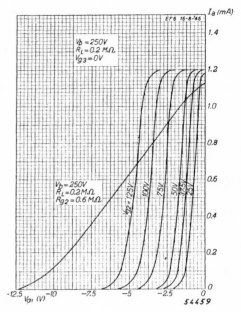

Fig. 235

Transfer characteristics of pentode EF 6 for various screen-grid voltages and for an anode series resistance of 0.2 megohm connected to the H.T. supply of 250 volts.

the negative grid voltage when the screen grid is fed via a series resistance of 0.6 megohm from a H.T. supply source of 250 V. It is to be seen that for each negative bias the screen-grid voltage adjusts itself in such a way that the working point lies roughly in the straight part of the transfer characteristic corresponding to that particular voltage.

In this way one gets the least distortion and anode detection is avoided. The maximum L.F. alternating anode voltage obtained with a certain admissible percentage of distortion (say 5%) depends upon the length of the straight part of the transfer characteristic between the upper and lower bends. When using an EF 6 as grid detector, for $V_b = 250$ V, $R_L = 0.2$ megohm, $R_{g2} = 0.6$ megohm, a maximum alternating anode voltage of 16 V_{rms} can be obtained. The detector amplification is then about 32, but it becomes smaller as the signal is weaker, because the detection is not linear. Since the alternating grid voltages are small, the grid rectification follows a square law and the distortion is much greater than in the case of diode detection, where a linear detection can be obtained by applying larger signal voltages.

With grid detection it is not possible to apply large signals, because these would cause too large a grid swing by the L.F. alternating voltage, accompanied by excessive distortion. With grid detection the sensitivity of the circuit can be considerably increased by employing back-coupling, to which we shall revert in Chapter XX when dealing with oscillators.

CHAPTER XIX

Mains-voltage Rectifiers

108. Introduction

Generally the only source of supply for energizing the valves of a receiving set is the alternating voltage from the lighting mains. An a.c. supply has the advantage that with the aid of transformers the voltage can easily be stepped up or down to the desired value. The electrodes of radio valves, however, have to be fed with direct current. In order to transform the a.c. voltage into d.c. voltage use is made of the rectifying action of diodes discussed in the preceding chapter.

Since the power required for feeding radio valves is much greater than the power supplied to the leak resistor by rectification of the R.F. or I.F. signal, for rectifying the mains voltage valves which are able to pass a much higher current have to be employed. These valves are called rectifying valves.

Although in principle it is feasible to use rectifying valves which are filled with gas, thereby reducing the power dissipation in the valve, high-vacuum valves are usually employed in receiving sets, because gas-filled rectifying valves lead to high-frequency disturbances.

109. Single-phase Rectifying Valves

In essence the circuit of a mains-voltage rectifier is the same as that of a rectifier for R.F. or I.F. signals. Fig. 236 shows a diagram of the circuit of a single-phase rectifier valve. The a.c. voltage from the mains is changed by the transformer to such a value that after rectification the desired direct voltage is obtained. The condenser C serves as a reservoir. If such a condenser were not provided the power supplied to the resistance R would be pulsating. The positive half waves of the alternating voltage between anode and cathode produce a flow of current through the valve in the direction of the arrow, from anode to cathode; the negative half waves do not produce any current. Without a condenser current would flow through the load resistor during alternate half-cycles only. The condenser, which acts as a reservoir, is charged during the

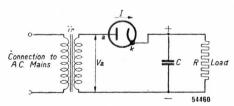

Fig. 236
Diagram of a single-phase rectifying valve circuit.

positive half-cycles and continuously discharged via the load resistor, so that the current supply to this resistor is uninterrupted. It has already been explained in the preceding chapter that the condenser voltage does not remain constant. This voltage drops during the discharge until the condenser is recharged by the next successive positive voltage pulse and the condenser voltage rises again.

Thus the condenser voltage fluctuates between two limits; across the condenser is a d.c. voltage with a superimposed a.c. voltage the wave form of which (voltage as a function of time) is not sinusoidal. This superimposed alternating voltage, which has the shape of a series of saw-teeth, is called the **ripple** of the rectified voltage. In the previous chapter it was also mentioned that the fluctuations of the d.c. voltage become smaller as the resistance R is made greater. A large resistance R corresponds to a low current, so that conversely the ripple with a higher current will be large. The larger the condenser the smaller the ripple. In a circuit as illustrated in Fig. 236 the frequency of the ripple voltage equals that of the mains voltage. During the positive half-cycles the voltage rises, and during the negative half-cycles it falls.

From what has been stated above it follows that the ripple voltage depends entirely upon the capacity of the condenser and upon the current supplied, and not upon the condenser voltage. Between the anode and cathode of the rectifying valve there are in fact three voltages, viz.:

1. the a.c. voltage V_{tr}, which is supplied by the transformer and reaches the cathode via the condenser C;

2. the negative d.c. voltage, which reaches the anode via the transformer winding;

3. the ripple voltage.

During the negative half-cycles of the transformer voltage the d.c. voltage and the negative voltage are added, so that between anode and cathode there is a high negative voltage at the peaks of the negative half-cycles. This so-called **peak-inverse voltage** is very little affected by the ripple voltage. With this peak-inverse voltage there must be no breakdown and the insulation in the rectifier and the distance between cathode and anode must be calculated to safeguard against it.

During the positive half-cycles the negative d.c. voltage and the positive peak voltage counteract each other more and less and the voltage between anode and cathode is very low. This is graphically

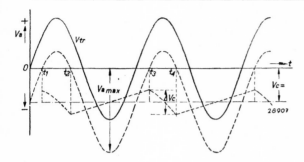

Fig. 237
Full line: Transformer voltage applied to the rectifying valve anode as a function of time.
Dot-dash line: The mean negative voltage of anode with respect to cathode.
Broken lines: The resulting a.c. voltage at the anode of the rectifying valve and the ripple of the direct voltage across the reservoir condenser.

illustrated in Fig. 237. During the intervals $t_2 - t_1$, $t_4 - t_3$, etc. the condenser is charged and consequently the voltage of the negative condenser plate with respect to the cathode rises. During the intervals $t_3 - t_2$, $t_5 - t_4$, etc. the condenser voltage drops. From this illustration it is evident that between the anode and cathode only low positive voltages and high negative voltages occur. From the same figure it is seen that the intervals $t_2 - t_1$ during which the condenser is being charged are much shorter than the intervals $t_3 - t_1$ during which the reservoir condenser is being discharged via the load resistance. During the charging period a charge has to be supplied to the condenser which is equal to that lost during discharge. Since the current intensity means the displacement of charge through a conductor per unit time, it follows that during the charging of the condenser considerably higher current intensities occur than during the discharge.

The mean value of the charging current is equal to the ratio of the intervals

$$\frac{t_3 - t_1}{t_2 - t_1}$$

times the direct current supplied, and as this ratio is very high the rectifying valve must be capable of supplying the high value of the current throughout, particularly since the peak value of the current is still higher than the mean value during the period $t_2 - t_1$. The variation of the ripple voltage is likewise indicated in Fig. 237 and the variation of the condenser voltage resulting therefrom is indicated by ΔV_C. Hence the peak value of the ripple voltage is equal to $\frac{1}{2} \Delta V_C$. The plotted wave form of the ripple voltage applies only approximately.

As the ripple voltage is conducted to the various electrodes of the valves together with the d.c. voltage, if it assumes high values the loudspeaker will reproduce it in the form of hum. In order to avoid this interference condensers of high capacity are usually employed, for instance electrolytic condensers of 16 or 32 μF. Usually even these high capacities are inadequate for smoothing the ripple sufficiently and a **smoothing circuit**, consisting of a choke coil or a resistance with a condenser, is intro-duced after the reservoir condenser. Generally this second condenser is also electrolytic.

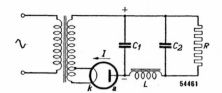

Fig. 238
Rectifying circuit with smoothing circuit consisting of a choke coil (L) and a condenser (C_2).

Fig. 238 shows the diagram of a recti-fying system with smoothing circuit. The choke coil has a high im-pedance to the alternating ripple voltage, whilst the impedance of the condenser C_2 is very low to that voltage; these two together form a voltage divider for the ripple voltage to the condenser C_1 and only a small part of this voltage will reach the load R.

The degree of smoothing provided by a filter consisting of a choke coil of inductance L (corresponding to the d.c. flowing) and capacity C_2 is given by

$$\frac{V_2}{V_1} = \frac{1}{\omega^2 L C_2 - 1}, \qquad (164)$$

in which V_1 is the alternating voltage input to the filter and V_2 the alternating voltage across the output side of the filter, i.e., across the condenser C_2.

Now according to Fig. 237 the ripple voltage across the reservoir condenser C_1 is not sinusoidal; it is approximately sawtooth-shaped. This voltage, however, can be analysed into a sinusoidal fundamental wave with higher harmonics. The harmonics have at least double the frequency of the fundamental wave and according to Equation (164) the second harmonic is therefore smoothed about four times more efficiently than the fundamental wave. Moreover, the amplitudes of the harmonics will be smaller than that of the fundamental wave. In practice, therefore, one has to take into account mainly the smoothing of the fundamental wave.

In the diagram of Fig. 238 the cathode of the rectifying valve is connected to the transformer winding and the anode to the first smoothing condenser.

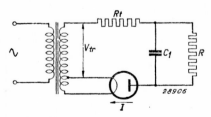

Fig. 239
Rectifying circuit in series with a resistance R_t, representing the in-ternal resistance of the transformer and that of the valve.

Fig. 240
Direct voltage across the reservoir condenser
as a function of the current supplied from
the rectifier for different no-load voltages
and for different internal resistances of the
transformer ($R_t = R_s + n^2 R_p$).

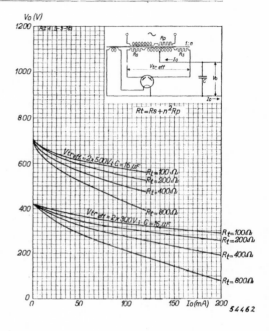

The advantage of this is that the valve-heater winding can be connected to the transformer secondary winding. This is the circuit usually employed with single-phase rectifying valves.

Instead of a choke coil a resistance can be used in smoothing filters. The advantage of a resistance is its low cost, but the voltage drop across the resistance is greater than across the choke, so that in order to get a certain direct voltage across the second smoothing condenser C_2 provision has to be made for a higher voltage across the first condenser. The degree of smoothing of a filter consisting of a resistance R and capacity C is given by

$$\frac{V_2}{V_1} = \frac{1}{\sqrt{(\omega R C)^2 + 1}}. \tag{165}$$

In the foregoing no account has been taken of the internal resistance of the rectifying valve and the transformer. In the diagram of Fig. 239 the resistance R_t represents the total resistance formed by the **internal resistance** of the rectifying valve and that of the transformer. The latter is determined not only by the ohmic resistance of the secondary winding, but also by that of the primary winding. It is true that the internal resistance should also include the stray reactance of the transformer, but this is relatively low and is generally ignored. The internal resistance R_t of the transformer can therefore be written as

$$R_t = R_{(sec.)} + n^2 R_{(prim.)}, \tag{166}$$

in which n represents the ratio of the number of turns on the secondary to that on the primary.

259

Owing to the voltage drop in this internal resistance the direct voltage at the condenser will be lower than if there were no internal resistance. Further the d.c. voltage is lower if the size of the reservoir condenser is reduced. When normal electrolytic condensers are used this effect is not noticeable (if there were no reservoir condenser the average d.c. voltage would amount to $1/\pi$ times the peak value of the alternating voltage).

For a rectifying valve curves can be plotted which give the d.c. voltage across the first smoothing condenser as a function of a certain alternating voltage across the transformer secondary winding, for any desired current and for different values of the transformer internal resistance $R_s + n^2\,R_p$, in which R_s represents the resistance of the secondary winding and R_p that of the primary. Fig. 240 shows such a family of curves with the transformer internal resistance and the voltage induced in the secondary winding as parameters. From these curves it can be seen that if no current is supplied the d.c. voltage across the reservoir condenser is equal to the peak value of the alternating voltage across the secondary winding.

110. Double-phase or Full-wave Rectifying Valves

So far we have been dealing with single-phase rectifying valves, and it has been seen that only the positive half-cycles of the alternating voltage cause a flow of current between anode and cathode, the current being blocked during the negative half-cycles by the valve and the condenser discharging itself via the load resistance. Now two single-phase rectifiers can be connected in such a way that the reservoir condenser is charged by both half-waves, in which case the circuit is referred to as a double-phase or full-wave rectifier. Such an arrangement is represented in Fig. 241, whilst Fig. 242 shows the variation of the voltage across the reservoir condenser as a function of time with full-wave rectification. This wave-form can be obtained from the wave-forms in the case of two single-phase rectifiers whose input voltages are displaced 180° in phase with respect to each other. From this diagram it appears that the ripple at the condenser has a frequency twice as high as that obtained with single-phase rectification. As a consequence the ripple can be smoothed more easily. Moreover, the ripple is smaller than that with single-phase rectification, because the discharge time of the condenser is not so long, the

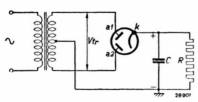

Fig. 241
Diagram of a full-wave rectifier.

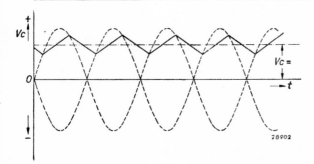

Fig. 242
Full line: Ripple voltage across the reservoir condenser of a full-wave rectifier as a function of time.
Dot-dash line: Mean d.c. voltage across the reservoir condenser as a function of time.

next successive charging pulse beginning much earlier. Since the discharge time is so much shorter the current peaks in full-wave rectification during the charging of the reservoir condenser are much smaller. Consequently not such heavy demands are made on the cathode of full-wave rectifying valves. Seeing that full-wave rectification offers important advantages this is almost exclusively employed in modern receiving sets. Single-phase rectifying circuits are still used occasionally for producing the negative d.c. voltage required by output valves and in D.C./A.C. receivers working without a mains transformer.

111. Voltage Doubling

It is often desired to produce from an alternating voltage a higher d.c. voltage than is normally possible with a single- or double-phase rectifying circuit. This is of importance, for instance, if there is no transformer for stepping up the alternating voltage and the rectifier has to be connected directly to the a.c. mains. In this case a system for doubling the voltage can be employed. There are two methods available. The so-called Greinacher circuit, which is the better known, is shown in Fig. 243, where two rectifiers are connected in parallel with each other and to the alternating voltage source in such a way that the

anode of one rectifier and the cathode of the other are connected to the same mains terminal. Across the condenser C_1 there is a d.c. voltage which is positive and across the condenser C_2 there is a voltage which is negative with respect to the other mains terminal. The two condensers C_1 and C_2 are connected in series between the points A and B, and the d.c. voltages across these two condensers are additive. Between A and

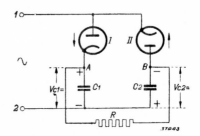

Fig. 243
Diagram of the Greinacher circuit for voltage-doubling.

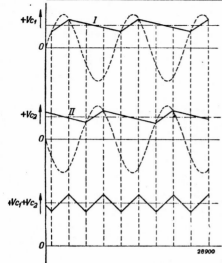

Fig. 244
Top: Wave-form of the d.c. voltage across the reservoir condenser C_1 of rectifier I in the voltage-doubling system of Fig. 243.
Middle: Wave-form of the d.c. voltage across the reservoir condenser C_2 of rectifier II in the voltage-doubling system of Fig. 243.
Bottom: Sum of the voltages across the condensers C_1 and C_2. The saw-tooth line represents the wave form of the ripple voltage across the load resistance.

B, therefore, there is a d.c. voltage twice as great as the voltage between A and the lower mains terminal 2. Hence the load R is connected between A and B. By plotting the wave-form of the voltage across the two condensers C_1 and C_2 we obtain the curves I and II of Fig. 244. Since curve II is displaced 180° in phase with respect to curve I, the resultant curve will have a wave-form like the lowermost saw-tooth line; this line was constructed by adding up the voltages across C_1 and C_2 taken from curves I and II. These curves show that the resultant ripple is smaller than that obtained with single-phase rectification. Moreover, it is to be noted that the frequency of the ripple is twice the mains frequency and is therefore the same as that obtained with full-wave rectification.

Another system of voltage doubling is shown in Fig. 245. Here the action is as follows:

Imagine, first, a half-cycle of a.c. with terminal 2 of the mains positive with respect to terminal 1. The condenser C_1 is then charged by rectifier I to a voltage V_{C1}. During the next half-cycle terminal 1 is positive with respect to terminal 2. This voltage reaches the anode of rectifier II, via the condenser C_1. The anode of rectifier II receives at the same time the d.c. voltage produced during the preceding half-cycle across C_1, so that the two voltages add up. As a consequence the d.c. voltage across the condenser C_2 is approximately doubled and is applied to the load R. In practice it will frequently be found that the voltage obtained in this way is slightly lower than that obtained by the Greinacher circuit. By employing a condenser of twice the capacity for C_1 and C_2, however, a load characteristic practically identical to that formed by the Greinacher circuit is obtained.

With this voltage-doubling system the ripple has the same frequency as occurs with a single-phase rectifier, that is to say a frequency equal to that of the mains. The amplitude is greater than that with the

Greinacher system, so that the smoothing should be better. The advantage associated with this system is that on one side the load is connected directly to the mains. If that mains terminal is earthed the load is also earthed. This is not the case with the Greinacher circuit, as may be seen from Fig. 243.

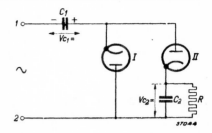

Fig. 245
Another system for voltage-doubling.

112. Calculating the Ripple Voltage

As is evident from Figs 237 and 242, the wave-form of the ripple voltage is saw-tooth shaped. This can be analysed into a sinusoidal fundamental wave and harmonics. The latter have higher frequencies and smaller amplitudes than those of the fundamental wave and with the smoothing components employed are much better suppressed, so that after this smoothing very little remains of the already smaller amplitudes. For calculating the smoothing element it is of importance to know the amplitude of the fundamental wave. This can easily be determined with the aid of a few simple rules of thumb

a) With a single-phase rectifier the r.m.s. value of the fundamental wave is approximately 4.5 V per mA of the d.c delivered per microfarad of the reservoir condenser.

b) With a double-phase rectifier the r.m.s. value of the fundamental wave is about 1.7 V per mA of the d.c. delivered per microfarad of the reservoir condenser.

c) With the Greinacher circuit for voltage-doubling the ripple is about two-thirds of that with a single-phase rectifier. Therefore the ripple per mA per microfarad of the condenser for the Greinacher circuit is about 3 V.

d) With the other system of voltage-doubling the ripple is about $1\frac{1}{2}$ times as great as that with the Greinacher circuit, i.e., about 4.5 V per mA per microfarad of the reservoir condenser.

Examples:

For a double-phase rectifier with a load of 60 mA and a 16 μF reservoir condenser it can be calculated from the foregoing rules that the r.m.s. value of the fundamental wave of the ripple amounts to approximately

$$1.7 \times \tfrac{60}{16} = 6.4 \text{ V}.$$

With the Greinacher system for voltage-doubling, a load current of 40 mA and 16 μF condensers the r.m.s. value of the fundamental wave is found to be approximately

$$3 \times \tfrac{40}{16} = 7.5 \text{ V.}$$

113. The Average, Effective and Peak Currents in the Rectifying Circuit

For a total d.c. supplied, I_u, the average current flowing through the rectifying valve per anode is equal to I_u in the case of a single-phase rectifier but only $\tfrac{1}{2} I_u$ in the case of a double-phase rectifier.

It is important to know, however, also the peak value and the effective (r.m.s.) value of the current. The latter determines, for instance, the heating of the transformer winding and consequently the minimum gauge of the wire needed. The former is of importance in determining the surface of the rectifying-valve cathode.

In Fig. 246 the shaded areas represent the current peaks with which the reservoir condenser in the rectifying circuit is charged. It will be evident that the peak value of the charging current pulses, I_m, is at the same time the maximum value of the current passing through the valve. The period T of the mains alternating voltage corresponds to 2π radians. If the time $t_2 - t_1$ during which a charging current is flowing corresponds to 2α radians, then current is flowing through the valve during the fraction α/π of the period T. The shaded current pulses in Fig. 246 are truncated peaks of a sine curve.

The shape of these current pulses, however, can be approximated with sufficient accuracy by assuming that they represent half-sine curves. The shaded area of such a current pulse represents the charge conducted per cycle to the reservoir condenser. If the peak value of the current pulse is I_m, it is found by integration that the charge acquired per cycle is equal to

$$Q = \frac{2\,\alpha\,T}{\pi^2} I_m. \tag{167}$$

This acquired charge should be equal to the discharge per cycle and per phase through the load resistance. If n is the number of phases of the rectifier (with full-wave rectification n is equal to 2) the acquired charge is equal to $\dfrac{I_u}{n}$ T.

Thus we have:

$$Q = \frac{2\,\alpha\,T}{\pi^2} I_m = \frac{I_u}{n} T \tag{168}$$

or

$$I_m = \frac{\pi^2\, I_u}{2\,\alpha\,n}. \tag{169}$$

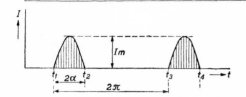

Fig. 246
Charging current pulses flowing through
the rectifying valve to the reservoir condenser as a function of time.

In practice a/π is usually found to
be equal to 0.25, so that the peak
value of the current pulse through
the rectifying valve is found to be

$$I_m \approx 6 \times \frac{I_u}{n}. \qquad (170)$$

With a double-phase rectifier
supplying 60 mA d.c., i.e. 30 mA

per phase, the peak current will be about $6 \times 30 = 180$ mA.

The effective current per phase is the current which during the period
T would develop the same heat as the current pulse shown in Fig. 246
develops during the time $t_2 - t_1$. A simple calculation shows that the
effective current for a current pulse having approximately the shape
of a half-sine wave (see Fig. 246) is equal to

$$I_{eff} = I_m \sqrt{\frac{a}{2\pi}}. \qquad (171)$$

If, with the aid of Equation (169), the effective current is expressed as
the d.c. I_u supplied to the load it is found that

$$I_{eff} = \frac{I_u}{n} \frac{\pi^2}{2a} \sqrt{\frac{a}{2\pi}} = \frac{I_u}{n} \sqrt{\frac{\pi^3}{8a}} = \frac{I_u}{n} \sqrt{\frac{3.9}{a}}. \qquad (172)$$

Assuming again that a/π is in most cases equal to 0.25, it is found that
the effective current per phase is approximately $2.2 \times$ the d.c. per
phase. If this is checked by measurement it will be found to be in
good agreement with the calculated result; therefore the wire gauge
of the transformer secondary winding must be chosen for 2.2 times the
d.c. per phase.

For voltage-doubling circuits as a rule a/π has a value of 0.22, so that
the maximum $I_{eff} = 2.36 I_u$. The permissible current quoted in publications for Philips rectifying valves always relates to the total average
d.c. supplied by the rectifier. This corresponds, therefore, to the
current indicated above by I_u.

CHAPTER XX

Generation of Oscillations

In the technique of radio reception the generation of oscillations is employed in superheterodyne reception and in measuring instruments. As already explained in Chapter IX, the principle of superheterodyne reception is based on the fact that the modulated R.F. signal received from the aerial and an unmodulated H.F. signal produced in the receiver itself are combined in a valve. In telegraphy-receiving apparatus oscillators are also employed for other purposes. The unmodulated H.F. oscillations are generated by means of a radio valve.

114. Regeneration by an Amplifying Valve

We will consider the case of an oscillatory circuit consisting of a condenser C and a coil having inductance L and ohmic resistance r (see Fig. 247). If through any cause an oscillation at the resonant frequency of the circuit commences and the cause of that oscillation is suddenly removed, the oscillation amplitude will very soon drop to zero, because the a.c. circulating through the circuit becomes transformed into heat owing to the unavoidable resistance of the coil (represented in Fig. 247 by r) and consequently power is dissipated. The lower the resistance the longer it will take for the oscillation to stop. This is expressed by saying that with a small resistance the **damping** is small. Fig. 248 gives an example of a damped a.c. voltage.

An oscillatory circuit comprises capacity, inductance and resistance. If the condenser is given a certain charge and the circuit is then left to itself, the condenser discharges itself through the inductance and the resistance, after which the e.m.f. of inductance within the coil still maintains the current through the coil, which results in the condenser being charged with the opposite polarity. The voltage of opposite polarity between the condenser plates reaches its maximum as soon as the current flowing through the coil is reduced to zero. At that moment the condenser discharges itself again through the inductance and the resistance. In this manner an oscillation is generated the amplitude of which, however, as already indicated, decreases gradually to zero (see Fig. 248), because the

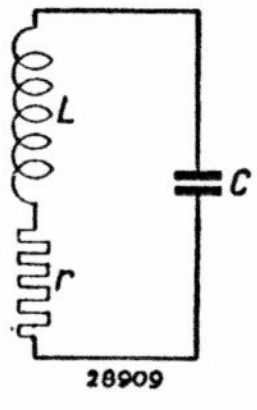

Fig. 247
Oscillatory circuit consisting of a coil of inductance L and resistance r and a condenser of capacity C.

266

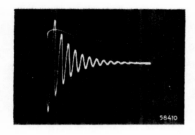

Fig. 248
An alternating voltage which due to damping dies out in an oscillatory circuit.

energy originally stored in the condenser is lost in the resistance as thermal energy (damping). The resistance thereby manifests itself by the voltage drop taking place across it. Owing to the a.c. flowing through the resistance this voltage drop is an alternating voltage in phase with the current (this a.c. voltage will be termed **degenerating voltage**). If now in some way or other a voltage is applied to the circuit which is in series with the degenerating voltage and just neutralizing the latter (being of equal amplitude and in antiphase to the degenerating voltage), then the oscillation in the circuit will continue. If this **compensating voltage** were greater than the voltage across the resistance the oscillation amplitude would increase until equilibrium was restored between the compensating voltage and degenerating voltage in the circuit. The compensating voltage can be fed to the oscillatory circuit by connecting the latter, for instance, between grid and cathode of a valve and introducing in the anode circuit a coil coupled inductively to the grid-circuit coil. An oscillation in the grid circuit produces an alternating current in the anode circuit which in turn induces an a.c. voltage in the grid circuit. This is called **back-coupling** the anode circuit to the grid circuit. In order to excite in the grid circuit an oscillation that maintains itself, the back-coupling voltage must have the right phase and amplitude with respect to the degenerating voltage in the circuit. Fig. 249 indicates a circuit as described above.

115. Automatic Grid Bias

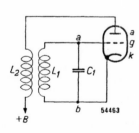

Fig. 249
Fundamental circuit diagram of an oscillator operating with an amplifying triode.

In the circuit of Fig. 249 the grid acquires no negative voltage in respect to the cathode, so that the point a in respect to the cathode (point b) is alternately positive and negative. During the positive half-cycles of the grid a.c. voltage grid current is flowing and the oscillatory circuit L_1C_1 is heavily damped. This damping makes it essential to apply a tight back-coupling in order to get a proper alternating-voltage amplitude. Moreover, the efficiency with Class-A operation, represented by the circuit of Fig. 249, is comparatively small. By efficiency is to be understood the ratio of the

267

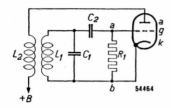

Fig. 250
Oscillator circuit with automatic negative grid bias obtained by means of a grid-leak resistor and condenser.

useful a.c. power in the anode circuit to the total power input. This would be much better for a Class-B operation, but the greatest efficiency is attained with Class-C operation. By **Class-C oper-ation** is understood adjustment of the working point in such a way that the negative d.c. voltage between grid and cathode is greater than the grid voltage at which the anode current is practically zero. The greater the efficiency the smaller the anode dissipation for a given a.c. power in the anode circuit. When Class-C operation is chosen, in consequence of the better efficiency a much smaller valve can be used than with Class-A operation.

In Class-C operation the oscillatory circuit is damped only during a very small part of the a.c. voltage cycle. This damping is analogous to the diode damping. On the other hand with Class-C operation the effective transconductance is less, which practically neutralizes the advantage of less damping. Usually the negative grid bias for Class-C operation is automatically produced, by connecting the grid to the oscillatory circuit via a condenser and to the cathode via a leak resistor (see Fig. 250). Since the electron current can flow only from the cathode to the grid, the condenser C_2 is charged merely by the positive half-cycles, the condenser plate connected to the grid receiving the negative charge from the electrons. During the negative half-cycles the condenser C_2 partially discharges itself via the resistance R_1; consequently a current flows from b to a, because the plate of C_2 connected to the grid is negative. In this manner a constant negative direct voltage is obtained at the grid, as in most cases the RC constant of the resistance-condenser combination is very great compared with the duration of the cycle (see also Chapter XVIII, Section 102). The grid-cathode path of the valve thus acts as a rectifier. Fig. 251 shows how the voltage of the grid adjusts itself to a mean negative value. On the assumption that grid current begins to flow at —1 volt, this current will flow as soon as the amplitude in the positive direction exceeds —1 V. From this instant, t_1, the condenser will be negatively charged until at t_2 the voltage drops again below —1 V and the charging of the condenser ceases. The condenser discharges itself via the resistance until charging begins again at the instant t_3. If Q is the charge applied during the time $t_2 - t_1$, the negative voltage V_g built-up across the leak resistor is equal to $\dfrac{Q}{t_3 - t_1} \times R$.

$\dfrac{Q}{t_3 - t_1}$ is the direct current flowing through the leak resistor from b to a. This can be measured and by multiplying it by the resistance value we obtain the value of the negative direct voltage. This negative direct voltage is somewhat lower than the amplitude of the oscillation. By determining the direct current we can find the approximate amplitude of the oscillatory voltage.

The slope of the I_g/V_g curve with respect to the horizontal axis has, of course, some influence upon the magnitude of the negative grid bias built-up. The smaller the slope the farther the amplitude penetrates into the positive grid-voltage range.

In order to determine accurately the amplitude of the oscillator voltage from the d.c. measurement, curves are often published for oscillator valves used in superheterodyne receivers; the grid current is represented as a function of the oscillator voltage for a certain leak resistance (see Fig. 283).

The advantage of automatic grid bias is that the negative grid voltage always adjusts itself to the value corresponding to the alternating voltage. If a fixed bias is applied which, in order to give Class-C operation, is greater than the grid voltage where the anode current is equal to zero, the transconductance of the valve for low alternating grid voltages is also equal to zero. A small voltage impulse on the grid is then unable to cause the valve to oscillate. With automatic grid bias, on the other hand, the transconductance without an alternating

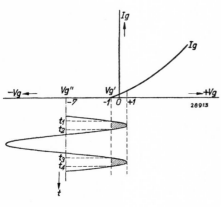

Fig. 251
Adjustment of the grid-alternating-voltage amplitude and the negative grid bias of an oscillator valve with automatic negative grid bias.

voltage on the grid is large enough and a small voltage impulse causes the valve to oscillate, provided that the back-coupling is sufficient to induce a voltage in the grid circuit which is greater than that required to compensate for the loss of power in the circuit at the moment oscillation begins, i.e., when $V_g = 0$. Oscillation then starts in such a way as to cause an increase in the alternating voltage amplitude, accompanied by an increase of the negative grid direct voltage. Consequently both increase, and this continues until the average transconductance of the oscillator

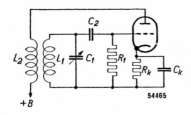

Fig. 252
Circuit of a triode oscillator with automatic negative grid bias. The valve also receives an automatic negative grid bias via a by-passed cathode resistance so as to start oscillation of the valve at a greater slope of the characteristic.

valve (see Section 118) has fallen, owing to the increased automatically induced negative grid bias, to the point where the damping is just neutralized. The oscillatory voltage has then reached its final value.

The automatic production of the oscillator grid bias has the great advantage that a very constant oscillator amplitude is obtained. Any deviation of the amplitude results in a variation of the effective transconductance due to the accompanying shifting of the negative bias. By this means the amplitude regains approximately its original value.

With automatic bias there is a high negative voltage between grid and cathode and consequently the anode current and the space-charge density are small. The advantage of this is that variations in the space-charge density, for instance, owing to mains-voltage fluctuations, have little influence upon the grid capacity and thus upon the resonant frequency of the oscillatory circuit connected to it, so that frequency changes are appreciably reduced. Therefore in this respect, too, automatic bias is favourable.

Sometimes it is advantageous to apply a negative starting voltage to the grid, for instance by interposing a resistance in the cathode lead (see Fig. 252). This is of importance if for example the transconductance of the valve at $V_g = 0$ is smaller than that at a low negative grid voltage of, say, —2 V and if, in consequence of this, the valve does not so easily begin to oscillate.

116. Interposing the Oscillatory Circuit in the Anode Lead

The oscillatory circuit can also be introduced in the anode lead, in which case the back-coupling coil is connected to the grid circuit. The arrangement of such a system is represented in Fig. 253.

In essence the working of this system is the same as that of Fig. 250. Owing to the a.c. flowing through L_1 an alternating voltage is induced in L_2, which is applied to the grid. This a.c. voltage produces a negative direct

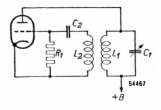

Fig. 253
Oscillator circuit where the oscillatory circuit is interposed in the anode circuit and the back-coupling coil in the grid circuit.

270

voltage on the grid and an alternating current in the anode circuit.

In consequence of this alternating current an alternating voltage is obtained across the oscillatory circuit and, if the coils are correctly connected, damping is thereby neutralized.

In this manner an undamped oscillation is again generated. The drawback of this system is that the oscillatory circuit has the anode direct voltage of the oscillator valve applied to it, unless the anode of this valve is fed via a resistance and condenser (see R_2

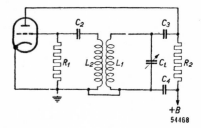

Fig. 254
Oscillator circuit where the top of the oscillatory circuit L_1C_1 is capacitively coupled (by means of C_3) to the anode and the latter is fed from the direct-voltage source $+ B$ via a resistance R_2. The negative grid bias is automatically produced by means of C_2 and R_1.

and C_3 in Fig. 254); this has the disadvantage that the oscillatory circuit is damped by the resistance R_2, which is effectively in parallel with the oscillatory circuit so far as high frequencies are concerned.

117. Conditions for Oscillation

In the preceding sections of this chapter the underlying principles of the production of oscillations have been given and the basic circuit for automatic grid bias has been shown.

We shall now consider in more detail the conditions for oscillation in a valve circuit. The fundamental circuit of a back-coupled valve as represented in Fig. 255 is used as a basis. In the circuit enclosed by the rectangle it is assumed that an oscillation arises. This oscillation results in an alternating voltage V at the grid, which in turn induces an anode alternating current I_a, according to

$$I_a = g_m V. \qquad (173)$$

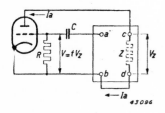

Fig. 255
Representation of an oscillator where the oscillatory circuit with back-coupling is drawn schematically by means of a rectangle.

This a.c. maintains the oscillation by setting up the voltage V between the terminals a and b. Now for a given circuit as represented by the rectangle in Fig. 255 there will always be a certain relation between the a.c. I_a flowing through the rectangle from terminal c to terminal d and the a.c. voltage V between the terminals a and b of the rectangle. Between the terminals c and d there must be an impedance Z, as otherwise the anode current I_a would have no effect. This impedance includes the internal resistance of the valve, which is in parallel with the circuit between the terminals c and d. Owing to the anode current I_a an alternating voltage

$$V_z = - Z I_a \qquad (174)$$

occurs across the impedance Z.

If now owing to the circuit inside the rectangle an a.c. voltage is produced between the terminals a and b the value of which equals a fraction t, the **back-coupling ratio,** of the a.c. voltage V_z between c and d, then we have

$$V = t V_z = - t Z I_a. \qquad (175)$$

In order to maintain an oscillation the alternating voltage V—which according to (173) causes an alternating current I_a that in turn induces an a.c. voltage V between a and b—must be of equal value and of the same phase as the last-mentioned a.c. voltage V. Substituting in (175) the value found for I_a from (173), we get

$$-t Z g_m V = V \qquad (176)$$

or

$$-t Z g_m = 1. \qquad (177)$$

Provision has now to be made that at the beginning of the oscillation —t Z g_m is much greater than 1. As the amplitude V increases, the negative d.c. voltage between grid and cathode likewise increases, with the result that the effective transconductance of the valve decreases until an amplitude is reached at which —t Z g_m = 1 and the amplitude of the oscillation is constant.

The phase of the a.c. voltage V which occurs, owing to the current I_a, across the circuit inside the rectangle, given by Equation (175), must be in phase with the a.c. voltage V given by Equation (173). If there is no phase displacement due to the transit time of the electrons, as is the case for frequencies below about 30 Mc/s, then the anode current occurring in (173) is in phase with the a.c. voltage V. This means that the induced a.c. voltage given by (175) must likewise be in phase with I_a. The factor —t Z indicating the relation between I_a and V for the circuit within the rectangle of Fig. 255 must therefore be real. The phase of the voltage V according to (175) is determined not only by the circuit itself but also by the frequency of the alternating current I_a and the alternating voltage V. A stable situation can exist only when the frequency has so adjusted itself that the anode current I_a and the a.c. voltage V are actually in phase. All normal back-coupling circuits

contain an oscillatory circuit, from which a grid a.c. voltage is obtained
in such a way as to bring about the correct phase at a frequency equal
to the resonant frequency of the oscillatory circuit.

As g_m and Z are positive quantities it follows from (177) that the back-
coupling ratio t must be negative, which implies that the back-coupling
must cause a phase shift of $180°$. This can be attained in a simple
manner by using an inductive back-coupling (see Figs 249 and 250).
When the impedance Z consists of an oscillatory circuit connected to
the anode then by means of an inductive back-coupling a smaller
voltage is induced at the grid. In that case therefore t is smaller than
unity. The impedance Z may, however, also be the back-coupling coil
connected in the anode lead (see Fig. 250), whilst the grid circuit contains the oscillatory circuit. In that case the voltage from the anode
to the grid is stepped up, so that t is then greater than unity. Equation
(177) gives the transconductance g_m required for oscillation in a certain
circuit. In the case of an oscillator whose frequency has to be variable
over a very wide range, e.g., oscillators for superheterodyne receivers,
as a rule the required transconductance will vary with the selected
frequency, because the impedance Z depends upon the frequency.

In Equation (177) t and Z are quantities dependent upon the circuit, whilst
g_m is a quantity determined by the
valve. In the next section it will be
shown how this transconductance depends upon the amplitude of the grid
alternating voltage and upon the valve
transconductance where $V_g = 0$.

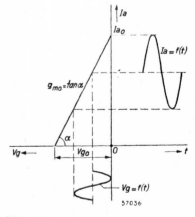

118. The Effective Transconductance and the Mean Anode D.C. of the Oscillator Valve

We will first consider a rectilinear valve
characteristic. Further it will be assumed that the grid condenser with leak
resistance yields a negative grid bias
equal to the peak value of the grid-
cathode a.c. voltage.

Fig. 256 gives the anode a.c. as a
function of time (right-hand upper half)
when the grid has imposed upon it an
alternating voltage whose peak-to-peak

Fig. 256
Top left: Rectilinear transfer characteristic of a valve with transconductance $g_{mo} = \tan \alpha$.
Bottom left: Sinusoidal grid voltage
as a function of time. The peak-to-
peak value of this voltage is smaller
than the grid voltage V_{go}, where the
anode current is zero.
Top right: Anode alternating current
resulting from the grid alternating
voltage, as a function of time.

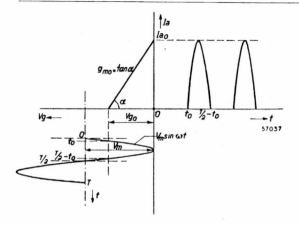

Fig. 257
Top left: Rectilinear transfer characteristic of a valve with transconductance $g_{mo} = \tan \alpha$.
Bottom left: Sinusoidal grid voltage as a function of time. The amplitude is greater than the grid bias V_{go}, where the anode current is zero.
Top right: Anode current resulting from the grid alternating voltage as a function of time.

value is less than the negative grid voltage V_{go}, where the anode current is zero. In that case the anode a.c. as a function of time is sinusoidal. With large grid alternating voltage, on the other hand, the anode current consists of current pulses and the anode current as a function of time is considerably distorted.

In the former case the relation between the anode a.c. and the grid a.c. voltage equals the transconductance of the valve. In the latter case only the fundamental wave of the distorted anode a.c. is of importance for maintaining oscillation, as this fundamental wave has the same frequency and phase as the grid a.c. voltage.

The harmonics of the anode a.c. will cause practically no a.c. voltages across the oscillatory circuit owing to the selectivity of the latter. We must therefore know the amplitude of the fundamental wave. The useful or **effective transconductance** ($g_{m \, eff}$) is the ratio of the amplitude of this anode alternating current to the amplitude of the grid alternating voltage. If the latter is V_m then in the anode circuit the fundamental component of the anode current I_1 will have an amplitude equal to

$$I_1 = g_{m \, eff} \, V_m. \tag{178}$$

When, as represented in Fig. 257, the grid alternating-voltage amplitude is greater than half the grid voltage V_{go}, where the anode current is equal to zero—as is the case with most oscillators—then for the transconductance g_m in Equation (177) we have to substitute the effective transconductance $g_{m \, eff}$ and choose such a magnitude of $-t \, Z$ that the condition for oscillation determined by this equation is satisfied. The effective transconductance is dependent upon the amplitude of the grid alternating voltage V_m and upon the grid voltage V_{go}, where the anode current is zero (determined by the amplification factor of the valve and the anode direct voltage applied).

The current pulses occurring when the grid alternating-voltage amplitude is greater than half the grid voltage V_{go}, where the anode current is zero, represent a certain average direct current. This **mean direct current** $\bar{I}_a$ is supplied by the source of the anode current, and therefore constitutes a load of this source. It is dependent upon the transconductance g_{mo} of the transfer characteristic where $V_g = 0$ and upon the magnitude of the grid alternating voltage V_m with respect to the grid voltage V_{go} where the anode current is zero. The **current efficiency** η is defined as the ratio of the amplitude of the fundamental component I_1 of the anode current to the mean d.c. value $\bar{I}_a$.

For practical oscillators one can now measure the effective transconductance $g_{m\,eff}$, the mean direct current $\bar{I}_a$ and the current efficiency η for different values of the amplitude of the grid alternating voltage V_m and plot these data in the form of curves. In this way an idea is obtained of the behaviour of the valve as an oscillator. In order to be able to compare easily the behaviour of different oscillator valves it is more practical to plot the ratio $g_{m\,eff}/g_{mo}$ of the effective transconductance to the slope of the transfer characteristic for $V_g = 0$ and also the ratio $\bar{I}_a/I_{ao}$ of the mean anode direct current to the anode current value for the transfer characteristic for $V_g = 0$ as functions of the amplitude V_m of the grid alternating voltage.

Fig. 258 shows such curves for the triode part of the valve ECH 3 (triode-hexode).

From the curve for the effective transconductance $g_{m\,eff}/g_{mo}$ it is seen that the oscillatory voltage

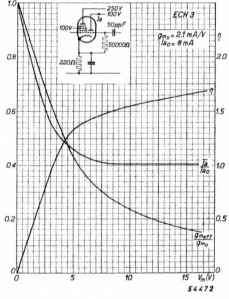

Fig. 258
The ratio $g_{m\,eff}/g_{mo}$ between the effective transconductance and the transconductance of the transfer characteristic at the point on this characteristic where $V_g = 0$, the ratio $\bar{I}_a/I_{ao}$ between the mean anode d.c. and the anode current value for the transfer characteristic, where $V_g = 0$, and the efficiency η ($I_1/\bar{I}_a$) as a function of the amplitude V_m of the grid alternating voltage for the triode part of the ECH 3 valve, when automatic grid bias is obtained with the aid of a 50,000-ohm leak resistance and a 50-$\mu\mu$F condenser.

275

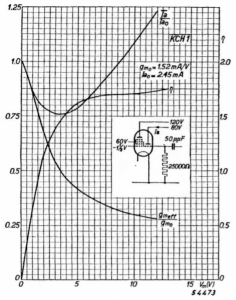

Fig. 259
The ratio $g_{m\,eff}/g_{mo}$ between the effective transconductance and the transconductance of the transfer characteristic where $V_g = 0$, the ratio $\bar{I}_a/I_{ao}$ between the mean anode d.c. and the anode current value for the transfer characteristic where $V_g = 0$ and the efficiency η ($I_1/\bar{I}_a$) for the triode part of the triode-hexode KCH 1 (a battery-operated valve), as a function of the amplitude V_m of the grid alternating voltage when automatic grid bias is obtained with the aid of a 25,000-ohm leak resistance and a 50-$\mu\mu$F condenser.

V_m is approximately inversely proportional to the effective transconductance. (In superheterodyne receivers this voltage is of great importance.) If, therefore, the effective transconductance required in a certain wavelength range increases in the proportion of 1 to 3 (due to the decrease of the impedance of the oscillatory circuit) the oscillatory voltage will fall to approximately one-third (when V_m remains greater than half V_{go}). With the aid of the curve $g_{m\,eff}/g_{mo}$ plotted as a function of V_m for a given valve —so that g_{mo} is known—it can be determined how large the product —t Z has to be, since according to the foregoing,

$$g_{m\,eff} = \frac{1}{-t\,Z}. \qquad (179)$$

On the other hand with the aid of the curve $g_{m\,eff}/g_{mo} = f(V_m)$ the amplitude of the oscillator voltage can also be calculated for a given circuit impedance and back-coupling ratio. If, for instance, the impedance of a tuned anode circuit in the short-wave range amounts to 5,000 ohms and the absolute value of the back-coupling ratio t = 0.6 (as a rule for practical reasons this will not be chosen greater than unity), then, according to Equation (177), the effective transconductance during oscillation will be

$$g_{m\,eff} = -\frac{1}{t\,Z} = \frac{1}{3000} = 0.33 \text{ mA/V.}$$

Now, according to the data for the ECH 3 given in Fig. 258, g_{mo} equals 2.1 mA/V, so that $g_{m\,eff}/g_{mo}$ is equal to 0.33/2.1 = 0.16. According to the curve of Fig. 258 this ratio corresponds to an amplitude $V_m = 15$ V, or an r.m.s. value of 10.5 V.

From the curves for $\bar{I}_a/I_{ao}$ and η it appears
that in a certain range of oscillator voltages
the mean anode current and the efficiency
remain fairly constant. From the curves for
the triode part of the directly-heated valve
KCH 1, however, (see Fig. 259) it is noticed
that after a certain value of V_m has been
reached (assuming that V_m increases from
zero) the ratio $\bar{I}_a/I_{ao}$ increases rapidly. At
the same time Fig. 259 shows that the ratio
$g_{m\,eff}/g_{mo}$ is relatively greater than the corresponding value for the indirectly-heated valve
ECH 3. The increase of the average anode d.c.
$\bar{I}_a$ leads one to suppose that at the positive
peaks of the grid alternating voltage the grid
becomes very positive, as is confirmed by
measurements of the grid current. Whereas
with high oscillatory voltage values ($V_m = 15$
V) of the ECH 3 positive values of the grid
voltage of about 2.5 V could be measured,
for the KCH 1 a positive voltage of 7.5 V
was found in the peak of the grid alternating
voltage. This is due to the fact that with directly-heated valves the grid-current charac-

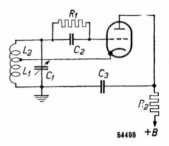

Fig. 260
Hartley oscillator circuit

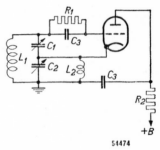

Fig. 261
Colpitts oscillator circuit (L_2
is a h.f. choke coil).

teristic $[I_g = f(V_g)]$ is less steep than that of indirectly-heated valves.
If these curves are now plotted again to show the relation between the
effective transconductance and the transconductance at the point of
the transfer characteristic where the positive peak of the grid alternating voltage lies, and also the relation between the mean anode d.c.
and the value of the anode current at the same point of the transfer
characteristic, then better agreement is found between the curves for
directly-heated valves and those for indirectly-heated valves. Therefore
g_{mo} has to be defined as the slope of the transfer characteristic at the
point where the positive peak of the oscillatory voltage lies, and I_{ao}
as the anode current at that point.

119. Special Oscillator Circuits

Figs 260 and 261 show possible circuits for oscillator valves other than
those consisting of an oscillatory circuit connected to the grid or anode
with a back-coupling coil in the anode or grid circuit. That in Fig. 260
is the so-called Hartley circuit and that in Fig. 261 the Colpitts circuit.

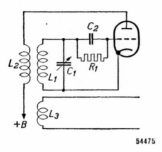

54475

Fig. 262
Oscillator circuit with a coupling coil L_3 for obtaining the alternating voltage induced in the oscillator circuit L_1C_1.

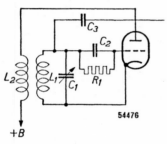

54476

+B

Fig. 263
Use of a capacitor C_3 for obtaining the oscillatory voltage.

Owing to the cathode being connected to a tapping point on the oscillatory circuit the voltages between grid and cathode and between anode and cathode are in antiphase. As already explained, this is necessary for maintaining oscillation. There is, therefore, no back-coupling coil in these circuits.

The drawback of the systems depicted in Figs 260 and 261 is that there is a H.F. voltage between cathode and earth, which is inadmissible in the case of indirectly-heated valves if the heater-current winding of the mains transformer is directly earthed. It is therefore advisable, where this is practicable, to earth the cathode and to disconnect from earth the earthed points of the diagram of Figs 260 and 261.

In the diagram of Fig. 260 back-coupling takes place owing to the anode a.c. flowing to the cathode via the chassis (earth) through the coil L_1 and thus inducing a regenerating a.c. voltage in the oscillatory circuit.

In Fig. 261 the anode a.c. flows through C_2, so that across this condenser is an a.c. voltage in the correct phase, which via C_1 is applied across the entire circuit.

In the diagrams of Fig. 260 and those following, the leak resistance R_1 is connected in parallel with the grid condenser instead of between grid and cathode. These two possibilities always exist.

120. Making Use of the Induced Oscillator Voltage

The a.c. voltage produced in the oscillator circuit can be taken off in different ways. First of all it is possible to couple a coil to the oscillatory circuit (see Fig. 262). The a.c. voltage induced in it can then be applied to the grid of the mixer valve or of an amplifying valve.

Another method is to employ a capacitive coupling (Fig. 263).

These coupling methods are often used but have the drawback that variations in the constants of the circuits coupled with the oscillator circuit affect the characteristic frequency of that circuit.

So-called "electronic coupling" is also often employed. The electronically-coupled oscillator corresponds, in fact, to a Hartley oscil-

278

lator with a tetrode or a pentode connected as a tetrode (see Fig. 264).

The screen grid of a tetrode is used
as the anode of the oscillation generator, which consists of cathode, grid
and screen grid. An alternating voltage
occurs at the control grid. The anode
current is thus alternating and if the
anode circuit contains an oscillatory
circuit tuned to the oscillator frequency an alternating voltage is produced across it, and this can be used
to advantage. This type of circuit is
much recommended on the grounds
that the reaction of the anode circuit
upon the oscillatory circuit is very
small, due to the electronic coupling,
so that this circuit is extremely stable.
Further it is maintained that mains-
voltage fluctuations have less effect
upon the oscillator frequency than
is the case with other types of circuit. From a closer investigation of
the conditions obtaining and from
measurements made, however, it is
to be seen that these assertions may
not always be correct, as is shown

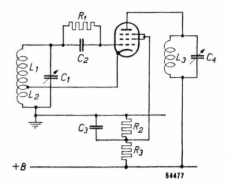

Fig. 264
Circuit of an electron-coupled oscillator.

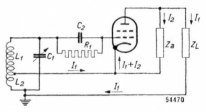

Fig. 265
Representation of an electron-coupled
oscillator, indicating the impedance
Z_L interposed in the anode circuit
and the internal impedance Z_a in the
valve between anode and cathode,
also the currents through the impedance Z_L and Z_a.

also by Fig. 265. In addition to the electronic coupling between the
anode circuit and the input circuit there is also a coupling across the
internal valve impedance Z_a. This internal valve impedance is formed
mainly by the anode-cathode capacity C_{ak}. Owing to the coupling
across Z_a any variation in the anode circuit may result in a perceptible
change in the grid circuit.

As stated at the beginning of this chapter, the generation of oscillations
in radio receivers is utilised principally in superheterodyne sets. The
oscillations produced across an oscillatory circuit are fed to the control
grid of a mixer valve, and the electron current flowing through this
valve is therefore controlled by that alternating voltage (see Chapter
XXI). With multiple valves composed of an oscillator triode and a mixer
hexode or heptode the grid of the oscillator triode is often directly
connected within the bulb to a grid of the mixer part.

279

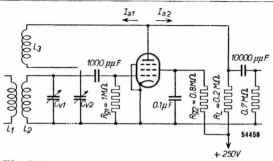

Fig. 266
Circuit diagram of a regenerative grid detector.

121. Regeneration with Grid Detection by Means of Back-coupling

In Chapter XVIII, Section 107, it was pointed out that with grid detection the H.F. alternating voltage between control grid and cathode results in a H.F. anode alternating current, which has to be filtered out, so as to avoid a H.F. alternating voltage reaching the control grid of the next A.F. amplifying valve. This H.F. anode alternating current, however, may induce in the tuned input circuit an alternating current in phase with that induced in the grid circuit by the signal received. This gives rise to regeneration in the grid oscillatory circuit and consequently a considerable coil magnification factor, resulting in high sensitivity of the detection circuit and great selectivity.

As appears from Fig. 266, the anode alternating current divides itself between the anode series resistance R_L and the back-coupling coil L_3 in series with the variable condenser C_{v2}. Care has to be taken that L_3 is connected up correctly, i.e. it must induce in the tuning circuit an alternating voltage in phase with the anode alternating current. By means of C_{v2} the degree of back-coupling and consequently the sensitivity of the valve circuit can be regulated. The sensitivity can be increased until the point of oscillation is reached. The circuit must not, however, oscillate freely in its own frequency, because then, owing to the detection, the H.F. oscillation set up in the circuit would interfere with the input signal (it would mix with it) and one then gets in the anode circuit a strong signal with the differential frequency, causing a loud howling note in the loudspeaker.

The regenerative detector is frequently employed in single-tuned-circuit receivers, where the aerial signal is induced in the tuned grid circuit without previous amplification (L_1 in Fig. 266 then forms a part of the aerial circuit).

CHAPTER XXI

Frequency Conversion

122. Fundamentals of Frequency Conversion

As explained in Chapters XV and XVI, if two signals of different frequencies are applied to the grid of an amplifying valve having a curved characteristic, alternating-current components with frequencies equal to the difference and the sum of the frequencies at the grid will be produced in the anode circuit.

Superheterodyne reception is based on this principle, for if, in addition to the R.F. signal received from the aerial, the grid receives a signal generated in the set itself, and means are provided for keeping the difference in frequency between the two signals always the same, then in the anode circuit of the valve there will be a signal with a fixed frequency. This constant-difference frequency is called the **intermediate frequency.** The advantage of this method is that for the intermediate frequency a relatively low frequency can be chosen which can be easily amplified with the aid of circuits which are tuned to the fixed frequency and which have a high impedance at the resonant frequency. The process of converting a R.F. carrier wave into an I.F. carrier wave by modulation with a locally generated oscillation is called mixing, or **frequency conversion.**

If, for example, the R.F. signal received has a frequency of 1,500 kc/s and an alternating voltage with a frequency of 1900 kc/s is generated in the set itself, in the mixer anode circuit there will be an a.c. component having a frequency of 400 kc/s. If then a signal of, say, 1,000 kc/s is received and provision is made for a signal with a frequency of 1,400 kc/s to be locally generated, there will likewise occur in the anode circuit an a.c. component having a frequency of 400 kc/s. If in the anode circuit there is an oscillatory circuit tuned to 400 kc/s, an alternating voltage will be produced across it. This voltage can be conducted to the grid of a following amplifying valve.

In Chapter XV it was stated that in addition to the difference frequency a sum frequency also occurs. In the example just given the sum frequency at a radio frequency of 1,500 kc/s is 1,500 + 1,900 = 3,400 kc/s, and at a radio frequency of 1,000 kc/s it equals 1,000 + 1,400 = 2,400 kc/s.

The sum frequency, and also multiples of the sum frequency, of the signal frequency, of the oscillator frequency and of the difference frequency, differ considerably from the resonant frequency of the anode oscillatory circuit and consequently no appreciable alternating voltages of these frequencies are produced across this circuit. The frequency generated in the set is usually termed the **oscillator frequency.** Generally the chosen value of this frequency is higher than the frequency of the received signal by an amount equal to the I.F., so that, given a signal of 1,500 kc/s and an I.F. of 400 kc/s, the oscillator frequency amounts to 1,900 kc/s. The valve in which the two signals are combined in order to obtain the intermediate frequency is called the **mixer valve.** Various types of mixer valves will be described below.

The data and characteristics published for mixer valves have to be adapted to the particular requirements of such valves. In the first place the amplification obtainable is, of course, of importance. If there is a R.F. signal on the grid then, after mixing with the oscillator signal, there will be an alternating voltage of intermediate frequency across the I.F. circuit connected to the anode. The relation between the I.F. voltage across the I.F. circuit and the R.F. voltage at the grid is called the **conversion amplification.** The relation between the I.F. anode alternating current and the R.F. grid alternating voltage is called the **conversion conductance** and is generally indicated by the symbol g_c. The conversion conductance depends upon the amplitude of the oscillator voltage and therefore can only be quoted for a definite oscillator voltage. The anode direct current of a mixer valve is likewise dependent upon the oscillator voltage.

Given a certain oscillator voltage and direct voltage on the control grid and the screen grids, the anode voltage also has a certain influence upon the anode current. This influence can be indicated by quoting the ratio of a very small anode voltage change to the small anode current change resulting from it. This ratio is termed the **internal resistance** of the mixer valve. This internal resistance, denoted by the symbol R_a as in the case of other valves, therefore represents the differential resistance between anode and cathode inside the valve for a given oscillator voltage and given voltages on the electrodes. It is in parallel with the anode oscillatory circuit and thus reduces the resultant impedance of that circuit, as is the case also for the internal resistance of a R.F. valve, which reduces the resultant anode impedance.

As has already been explained in Chapter XV, the action of frequency-converting valves is based upon the fact that the transconductance of the grid to which the R.F. signal is applied fluctuates at the oscillator frequency. In the case, for instance, where the oscillator voltage is applied to the grid to which the R.F. voltage is fed, the variation of the transconductance at the oscillator frequency is due to the curvature of the characteristic showing the variation of anode current with control-grid voltage. The change of the transconductance with time will produce a sinusoidal curve only if the transconductance varies linearly with the voltage on the grid to which the oscillator voltage is applied, i.e., in the case quoted above, when the grid-voltage-anode-current characteristic follows the square-law. This is not usually the case, so that the curve showing the change of transconductance with time is not sinusoidal. If $V_h \cos \omega_h t$ represents the oscillator voltage, then with the aid of Fourier analysis the transconductance as a function of time may be written as:

$$g_m = g_{m0} + g_{m1} \cos \omega_h t + g_{m2} \cos 2 \omega_h t + g_{m3} \cos 3 \omega_h t + \dots \quad (180)$$

in which ω_h is the angular frequency of the oscillator signal. Since transconductance is an even function of time, that is to say it does not change if t is replaced by —t, only cosinal terms occur.

In the expression for transconductance as a function of time only the fundamental component $g_{m1} \cos \omega_h t$ is of importance, as after mixing with the R.F. signal the terms involving a multiple of ω_h yield sum and difference frequencies greatly deviating from the intermediate frequency.

If a R.F. signal $V_i \cos \omega_i t$ (in which $\omega_i = 2\pi f_i$, the angular frequency of the RF. signal) is applied to the grid whose transconductance fluctuates periodically due to the oscillator signal, then if we consider only the fundamental wave $g_{m1} \cos \omega_h t$ of the transconductance variation we obtain the following anode alternating current:

$$I_a = V_i \cos \omega_i t \times g_{m1} \cos \omega_h t. \quad (181)$$

As is well known, this is also equal to:

$$I_a = \tfrac{1}{2} V_i g_{m1} \cos (\omega_h - \omega_i) t + \tfrac{1}{2} V_i g_{m1} \cos (\omega_h + \omega_i) t. \quad (182)$$

Thus the resulting alternating current has two components, one of which has the difference frequency and the other the sum frequency of the signal alternating voltage and the oscillator voltage.

By definition, the conversion conductance is equal to the ratio of the anode current of frequency $(\omega_h - \omega_i)$ to the input alternating voltage V_i of frequency ω_i. According to Equation (182) the amplitude of the anode alternating current of frequency $\omega_h - \omega_i$ is equal to $\tfrac{1}{2} V_i g_{m1}$ and the amplitude of the R.F. signal is V_i, so that the conversion conductance g_c must be equal to

$$g_c = \tfrac{1}{2} g_{m1}. \quad (183)$$

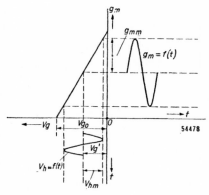

Fig. 267
Top left: Rectilinear transconductance characteristic (transconductance as function of grid voltage) for the grid of a valve to which the oscillatory voltage is applied.
Bottom left: Oscillatory voltage as a function of time. The amplitude V_{hm} is less than half the value of the negative grid potential, at which the transconductance g_m equals zero. The negative grid bias is so chosen that the transconductance during the whole grid alternating voltage cycle is greater than zero.
Top right: Transconductance g_m as a function of time.

Thus it follows from this equation that the conversion conductance equals half the amplitude of the fundamental component of the transconductance variation with time, where this transconductance variation is due to the oscillator voltage on one of the valve grids. It will be apparent that the amplitude g_{m1} of the fundamental component of the transconductance-time curve is dependent upon the amplitude of the oscillator voltage and also upon the negative bias of the grid to which the oscillator voltage is applied.

Fig. 267 depicts the case of a rectilinear transconductance characteristic where the negative grid bias V_g' and the amplitude of the oscillator voltage V_{hm} are so chosen that the transconductance is always greater than zero. In this case the variation of transconductance with time is sinusoidal and consequently the conversion conductance is simply equal to half the amplitude g_{mm} of that curve.

If, with variable amplitude of the oscillator voltage, the situation is always such that during the whole alternating voltage cycle the transconductance remains greater than zero, it is clear that the amplitude g_{mm} must be proportional to the amplitude of the oscillator alternating voltage. In that case, therefore, the relation between oscillator voltage and conversion conductance is very simple.

With the frequency-converting valves used in practice, however, the transconductance curve $g_m = f(V_g)$ is not a straight line (here we are concerned still with the transconductance of the grid to which the R.F. voltage is applied); neither is the condition satisfied that the grid bias is such as to keep the transconductance greater than zero during the whole a.c. cycle. The transconductance-time curve produced by a sinusoidal oscillator voltage is then greatly distorted (see Fig. 272); the relation between the amplitude of the fundamental component of the transconductance curve and the oscillator voltage is in that case no longer a simple one. This effect is more pronounced where, as happens in most cases, due to a resistance-condenser coupling, the negative bias is automatically dependent upon the oscillator-voltage amplitude. With small oscillator-voltage amplitudes and automatic bias the negative bias on the grid controlled by the oscillator voltage is still so small that the transconductance is greater than zero during the whole alternating-voltage cycle. With such a voltage, therefore, the amplitude of the fundamental component of the transconductance-time curve increases practically linearly with the oscillator amplitude. This holds good until amplitudes are reached where the negative grid voltage is so high that the transconductance is no longer

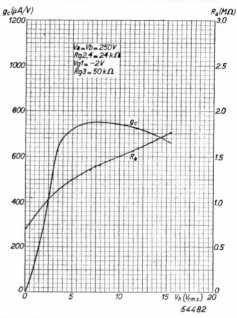

Fig. 268
Curves showing how, for valve ECH 4, the conversion conductance g_c and the internal resistance R_a depend upon the oscillatory voltage V_h.

greater than zero during the whole alternating-voltage cycle. The curve showing the conversion conductance as a function of the oscillator amplitude then bends downwards and shows a certain maximum. Fig. 268 represents such a curve, $g_c = f(V_h)$, for the triode-heptode ECH 4. This curve clearly shows the effect described above; the conversion conductance reaches its maximum at an r.m.s. value of the oscillator voltage of about 8 V. Fig. 268 also shows the internal anode resistance R_a as a function of the oscillator voltage. Now it is known that the oscillator voltage varies considerably in a wave-range of the receiver owing to several factors. This may result in the conversion conductance not remaining constant in a wave-range, so that the sensitivity of the receiver will not be the same for all wavelengths. As this is considered undesirable it is fortunate that the conversion-conductance curve is fairly flat over a wide range of oscillator-voltage values. In Fig. 268 the conversion conductance varies from 750 to 670 μA/V when the oscillator voltage changes from 8 to 15 V. This flat tendency of the conversion conductance can be favourably influenced by the shape of the curve connecting the transconductance of the grid to which the R.F. voltage

285

is fed and the voltage on the grid to which the oscillator voltage is applied.

The transconductance variation of the R.F. grid by means of the oscillator voltage can be obtained in two ways:

1) by applying the oscillator voltage to the same grid as the R.F. voltage, and

2) by applying the oscillator voltage to another grid introduced in the valve for that purpose.

On this basis frequency-converting valves can be classified under two groups, viz:

A) frequency-converting valves in which the oscillator signal and the R.F. signal are applied to the same electrode (grid);

B) frequency-converting valves in which the oscillator signal and the R.F. signal are applied to different electrodes (grids).

123. Frequency-converting Valves with Oscillator Voltage and R.F. Signal Applied to the Same Electrode

If the oscillator signal, which is usually generated in a separate valve, is applied together with the R.F. signal to the grid of an amplifying valve, for instance a pentode, then in consequence of the curvature of the characteristic an alternating current with the difference frequency will occur, among others, in the anode circuit [1]. Fig. 269 shows the essentials of a stage in which mixing is obtained in this manner. The oscillator voltage is fed between chassis (earth) and cathode via a coupling coil in the cathode lead. Thus the cathode voltage changes with respect to earth at the oscillator frequency. The R.F. tuned circuit is between grid and earth. Between grid and cathode we therefore have the R.F. voltage and the oscillator voltage in series.

With this arrangement the oscillator voltage, which is usually fairly large (several volts), is, apart from other factors, not present between grid and earth. Even when there is no

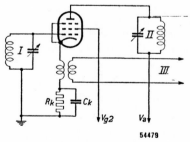

Fig. 269
Circuit diagram of a mixer valve with separate oscillator, where the R.F. signal and the oscillator signal are applied to the same grid. I = R.F. oscillatory circuit, II = I.F. oscillatory circuit, III = leads to the local oscillator.

54479

---

[1] See Chapter XVI, Section 85.

R.F. amplifying valve preceding the mixing stage, radiation from the aerial due to the oscillator voltage may thus be avoided. Such radiation causes disturbances in the neighbouring receivers and is therefore undesirable. Consequently if mixing is to be obtained by introducing the R.F. signal and the oscillator voltage between grid and cathode, the oscillator voltage has to be applied between cathode and earth.

Since the filament is usually earthed so far as high frequencies are concerned, the oscillator voltage between cathode and earth exists also between cathode and filament, and owing to the nature of the insulation between cathode and filament irregular currents occur which can be observed in the loudspeaker as rustling or crackling.

With this arrangement, notwithstanding the application of the oscillator voltage between cathode and earth, it is possible that a considerable part of this voltage exists between grid and earth, due to the grid-cathode capacity. Therefore radiation from the aerial at oscillator frequency is still possible.

Another and very considerable drawback is the great dependency of the conversion conductance upon oscillator voltage in the arrangement given in Fig. 269; in consequence provision has to be made in the receiver to keep this voltage very constant throughout the whole frequency range.

Finally there is the disadvantage of a considerable influence of the alternating voltage in the R.F. circuit upon the voltage in the oscillator circuit.

For all these reasons mixer valves with a common grid for R.F. signal and oscillator signal are very seldom used nowadays in receiving sets.

Diode mixer valves are among the group of mixer valves in which the input signal V_i and the auxiliary or oscillator signal V_h are applied to the same electrode. Diodes are now also used for the mixing stage of superheterodyne receivers for ultra-short waves. When mixer valves are used for normal broadcast receivers for wavelengths of the order of a metre all sorts of difficulties arise (see also Chapter XXV) owing to the long transit time of the electrons, and it is advantageous to use diodes, in which the distance between cathode and anode may be very short.

Fig. 270 shows the essentials of a

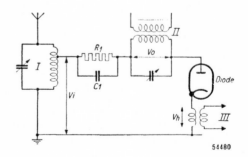

Fig. 270
Circuit diagram of a mixing stage using a diode. I = R.F. input circuit, II = intermediate-frequency transformer, III = leads to the local oscillator.

diode mixer circuit. The R.F. voltage V_i is in series with the oscillator signal V_h, which is applied between cathode and chassis (earth).

In series with the diode is also an I.F. circuit tuned to the frequency difference between ω_h and ω_i. Further, the negative bias of the diode anode with respect to cathode is obtained automatically by means of a leak resistance R_1 in combination with a condenser C_1, the impedance of which is small at the intermediate frequency ω_0. In consequence of the oscillator voltage applied between anode and cathode, due to the curvature of the plate characteristic of the diode, the transconductance of the diode (the ratio of the very small anode current increase to the corresponding very small anode voltage increase, i.e., the reciprocal of the internal resistance) for the H.F. voltage V_i changes in sympathy with the oscillator frequency, and in this manner the desired mixing is obtained.

The conversion amplification of the diode mixing stage is given by:

$$\frac{V_o}{V_i} = \frac{\frac{1}{2} g_{m1}}{g_{mo} + 1/R_o} \text{ }^1),\tag{184}$$

in which g_{m1} is the fundamental component of the curve indicating the transconductance as a function of time, g_{mo} the transconductance of the diode at the automatically adjusted bias (see e.g. Eq. 180) and R_o the impedance of the I.F. circuit at the resonant frequency. For high oscillator voltages $g_{mo} = \frac{1}{2} g_{m1}$. When $1/R_o$ is small compared with g_{mo}, as is the case when impedance of the I.F. circuit is large, the conversion amplification V_o/V_i equals unity. This, therefore, is the limiting value of the conversion amplification of a diode mixing stage, and is generally achieved approximately in practice.

124. Frequency-converting Valves with Oscillator Voltage and R.F. Signal Applied to Different Electrodes

If the oscillator voltage and the R.F. signal are applied to two diffferent grids of a valve an alternating current with the difference frequency of these signals is likewise produced in the anode circuit. The electron current to the anode is then controlled by the two grids successively. In principle there are now two possibilities:

a) the electron current is first controlled by the R.F. signal and then by the oscillator signal;

---

1) See J. Haantjes and B. D. H. Tellegen, The diode as mixer valve and as detector, Tijdschrift Ned. Genootschap, 1943; M. J. O. Strutt, Moderne Kurzwellenempfangstechnik, publishers Julius Springer, p. 159, and M. J. O. Strutt, Diode frequency changers, Wireless Engineer 13, 1936, pp. 73—80. Further see M. J. O. Strutt, Empfänger und Verstärker, publishers Julius Springer, 1943, pp. 221 et seq.

b) the electron current is first controlled by the oscillator signal and then by the R.F. signal.

To group b belong, for instance, the self-oscillating mixer valves, such as the octodes, whilst to group a belong the hexodes and heptodes, which work with a separate oscillator valve or with an oscillator built into the valve.

(a) Frequency-converting Valves where the Electron Current is First Controlled by the R.F. Signal and then by the Oscillator Voltage

The working of this kind of mixer valve will be described by taking as an example a heptode, i.e. a valve with five grids and a separate oscillator. The R.F. input signal is applied to the first grid and the oscillator signal to the third grid. Fig. 271 illustrates the principle of a heptode connected as a mixer valve, in which g_1 is the first control grid to which the R.F. signal is applied, g_2 and g_4 are screen grids and g_3 is the control grid for the oscillator signal. The electron current flows through the first control grid with the aid of g_2. The electrons thereby attain such a velocity that they shoot through the meshes of the screen grid and arrive in the vicinity of the second control grid g_3. Due to the negative voltage of g_3, in front of this grid a space charge is formed whose density depends on the number of electrons passing through g_1. The screen grid g_4 draws electrons from this pulsating space charge through the control grid g_3. The alternating voltage at this grid further controls the electron current, which then continues its way to the anode. The screen grid between g_1 and g_3 serves the purpose of reducing as much as possible the capacity between those grids, in order to minimize the influence of the input-circuit voltage upon the oscillator circuit. Owing to the screen grid g_4, which acts in a similar manner to the screen grid in a screen-grid valve, the valve has a high internal resistance.

The grid g_5, between g_4 and the anode, is a suppressor grid and its object is to render the secondary emission of the anode harmless. The first grid has a certain trans-conductance g_{ag1} with respect to the anode, and this is apparently dependent upon the voltage on the

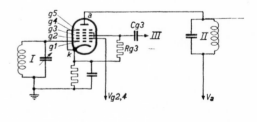

54481

Fig. 271
Circuit diagram of a mixer heptode. I = R.F. input circuit, II = intermediate-frequency circuit, III = lead to local oscillator.

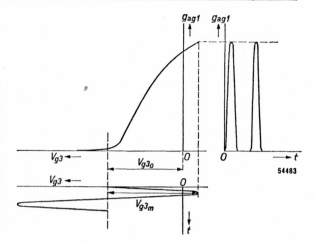

54483

Fig. 272
Transconductance g_{ag1} of the first grid of a heptode with respect to the anode, plotted as a function of the voltage V_{g3} on the third grid, with a certain fixed negative bias potential on the first grid. On the right is shown the transconductance g_{ag1} of the first grid as a function of time when a sinusoidal alternating voltage with amplitude V_{g3m} and a negative grid bias V_{g3o} are applied to the third grid.

third grid; where the negative bias on that grid is low more electrons will flow through the grid to the anode than is the case with a high negative bias. The controlling action of the first grid is naturally greater for a large anode current than for a small one. The transconductance g_{ag1} can be plotted as a function of the voltage on g_3, giving a curve as shown in Fig. 272.

If the oscillator voltage is applied to grid 3 via an RC link in accordance with the diagram of Fig. 271, a negative bias with a superimposed alternating voltage is produced across the leak resistance. The transconductance then varies with the grid voltage, in the manner indicated in Fig. 272.

At the positive amplitude of V_{g3} the transconductance of the first grid and the amplification of the alternating voltage on that grid are maximum. Owing to the alternating voltage on g_3 the transconductance of g_1 varies between a maximum and zero.

On the right-hand side of Fig. 272 the transconductance is set out as a function of time. This transconductance-versus-time curve is not sinusoidal. It is possible, however, to determine the fundamental component of this curve by graphical means and to find from that the value of the conversion conductance. According to Section 122 this is equal to half the amplitude of the fundamental component of the transconductance-versus-time curve. If this curve is constructed for different oscillator voltages and the corresponding negative grid voltages, the relation between the conversion conductance and the oscillator voltage can be found. Usually this relation is determined by measurement and published in the form of a curve (see Fig. 268).

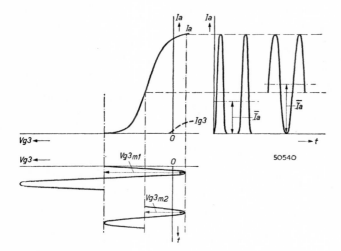

Fig. 273
Anode current of the heptode of Fig. 272 as a function of the voltage on the third grid, with a certain fixed negative voltage on the first grid. On the right are shown the anode currents as functions of time when large and small sinusoidal alternating voltages with the corresponding biases are applied to the third grid and the direct-current components $\bar{I}_a$ of these currents (dot-dash lines).

The anode direct current depends also upon the oscillator voltage at the third grid. Fig. 273 gives the anode current of the same heptode as a function of the voltage at the third grid for a given negative voltage at the first grid. From the figure it is clearly seen that the anode direct current is greater with a small grid alternating voltage than with a large alternating voltage. The current peaks of the current-versus-time curves become narrower as the alternating voltage on the grid increases, and consequently the anode direct current drops. The conversion conductance increases simultaneously until a certain maximum is reached.

This figure 273 applies when automatic negative grid bias is derived from a leak resistor and condenser. If a fixed negative grid voltage is used it must have such a value that in the absence of oscillator voltage the anode current is practically equal to zero. The amplitude of the oscillator voltage is thereby chosen of such a value that the maximum conversion conductance is obtained. For smaller oscillator voltages the anode direct current will be lower than that with automatic negative grid bias.

Modern mixer valves generally take the form of a **hexode** or **heptode** combined with a triode in a common bulb and with a common cathode.

Examples of such combinations are the Philips ECH 3, ECH 4 and ECH 21. The modulating or mixing system is then built together with the oscillating system. Fig. 274 shows the essentials of a triode-heptode circuit (using the ECH 4 or the ECH 21).

In the ECH 3 (triode-hexode) the grid of the oscillator system is connected directly with the third grid of the hexode system inside the valve. In the ECH 4 and the more recent ECH 21 (both are triode-

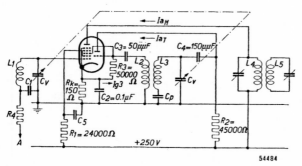

Fig. 274
Circuit diagram for the operation of types ECH 4 and ECH 21 in a frequency-conversion stage. A = lead to the automatic-volume-control-voltage source, see Ch. XXIII.

heptodes) these grids are not interconnected inside the bulb but have to be connected up outside it (the advantage of this is that the valves can then be used also for other purposes). With this arrangement the oscillator voltage set up on the triode grid owing to the oscillation of the feedback circuit is applied directly to the third grid of the modulator part and thus influences its transconductance in the manner desired. Furthermore both grids automatically adjust themselves to the same negative grid-bias voltage.

The conversion conductance of mixer hexodes and heptodes can be regulated by varying the voltage on the first grid (see Chapter XXII), because a separate oscillator is used which is mounted either in the same bulb or in a separate one. Therefore in these valves the electron current through the oscillator valve bears no relation to the voltage on the first grid of the mixer part, and consequently regulating of the mixer part has hardly any effect upon the transconductance of the oscillator triode and the oscillator voltage.

(b) Frequency-converting Valves where the Electron Current is First Controlled by the Oscillator Voltage and then by the R.F. Alternating Voltage

Another group of mixer valves are those in which, as previously mentioned, the electron current is first controlled by the oscillator signal. To this group belong the so-called **pentagrid** valves and **octodes,** in which a part of the electron current is used for generating the oscillation in the valve itself. Fig. 275 shows the essential circuit components of an octode (valve with 8 electrodes). In order to give an idea of the action of an octode the EK 2 will be described as typifying this kind of valve.

An octode may be regarded as being a valve consisting of a "lower" part with which the oscillation is produced and an "upper" part in which the electron current, fluctuating at the frequency of the auxiliary

oscillation, is con-
trolled by the R.F.
signal. Grids 1 and 2
act respectively as
control grid and
anode of a triode
oscillator. The second
grid absorbs part
of the pulsating
electron current flow-
ing through the first
grid and conducts it
to the feedback coil.
This second grid (ac-
tually we can no

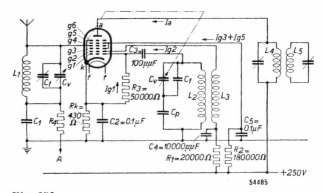

Fig. 275
Circuit diagram of an octode frequency converter (type EK 2).
A = lead to the automatic-volume-control-voltage source,
see Ch. XXIII.

longer strictly speak of it as a grid) is constructed in such a way that
its controlling action upon the rest of the electron current passing
to the "upper" part is limited to a minimum. For this reason this
electrode consists merely of two parallel rods. The electron current
is drawn off from the "lower" part by means of a screen grid g_3 and
the electrons are concentrated in front of the negative control grid g_4,
where another pulsating space charge is created. From this space charge
electrons are attracted by the screen grid g_5 via the control grid g_4,
and after being further controlled by the alternating voltage on g_4
these electrons rush through the screen grid g_5 and the suppressor grid
g_6, finally reaching the anode.

The advantage of this arrangement is that with a **self-oscillating mixer
valve** it is possible to regulate the conversion transconductance
(Chapter **XXII**) by varying the negative voltage on the fourth grid,
without appreciably affecting the oscillator transconductance. The
oscillator voltage therefore remains practically constant even when the
negative voltage on the fourth grid is high.

125. Secondary Phenomena with Frequency-converting Valves [1])

(a) Transit-time Current

Transit-time current occurs particularly with frequency converters
of the hexode or heptode type. Fig. 276 shows the principle of measuring
the transit-time current with a heptode. In the case of very high fre-

---

[1]) See the respective publications by **M. J. O. Strutt** quoted in the bibliography appended
to this book.

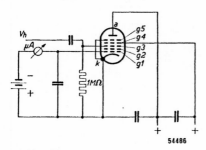

Fig. 276
Basic circuit used for measuring the transit-time current on a heptode.

quencies in the short-wave range it will be found that a considerable current flows through the microammeter. This current causes damping of the connected oscillatory circuit or such a negative grid voltage across the leak resistor that the working point is shifted considerably and consequently the amplification of the valve is very much reduced. This current is called **transit-time current** because it has to be ascribed to the finite transit time of the electrons in the valve.

Owing to the high-frequency alternating voltage on the third grid, some electrons which turn back in front of the third grid may, on their relatively long path, reach such a velocity that they are able to overcome the negative potential of the first grid, reach this grid and set up a grid current. With a hexode or heptode the transit-time current is roughly proportional to the square of the frequency of the alternating voltage on the third grid and to the square of the magnitude of that voltage. Further, the transit-time current is proportional to the square of the distance between grids 2 and 3 and inversely proportional to the voltage on grid 2. With hexodes the transit-time current is usually much larger than with octodes. In a certain hexode and on a wavelength of 10 m, for instance, it amounts to 22 μA, and in an octode to 3 or 4 μA. In hexodes and heptodes this current can be considerably reduced by shortening the distance between grids 2 and 3 and lowering the oscillator voltage; this results in a smaller space charge and thereby reduces the transit time of the electrons between these grids.

(b) Induction effect

In the case of mixer valves built on the octode principle an **electronic coupling** is observed between grids 1 and 4, especially in the lower bands of the short-wave range, as a consequence of which the conversion amplification is reduced. Owing to the induction effect there occurs across the input circuit an alternating voltage with a frequency equal to that of the oscillator circuit, which under certain circumstances may amount to several volts. This voltage is dependent upon the properties of the input circuit, the I.F. employed and the oscillator alternating voltage.

If the frequency of the oscillator is higher than that of the signal received then the voltage on grid 4 will be in antiphase to the oscillator voltage on grid 1. In that case the disturbing voltage on grid 4 has the same effect as a lowering of the oscillator voltage on grid 1, this resulting in the reduction of the conversion amplification already mentioned. If the mixer stage is not preceded by a R.F. stage, owing to the induction voltage on the fourth grid there will also be interference in neighbouring receivers. If this voltage assumes too high a value grid current may also occur, and this entails damping of the input circuit.

The existence of the induction voltage on the fourth grid can be explained as follows. The density of the electrons forming a space charge in front of the fourth grid changes under the influence of the oscillator voltage on the first grid. This electron cloud has a certain capacity with respect to the fourth grid; thus a variation in the charge of this cloud also results in a variation of the fourth-grid charge. If the resonant frequency of the oscillator circuit is higher than that of the input circuit the impedance of the latter circuit will be capacitive for the oscillator voltage. The space charge in front of the fourth grid together with that grid itself forms a condenser. If now the fourth grid and the cathode form a capacity owing to the capacitive impedance of the oscillatory circuit between these electrodes, we have a capacitive voltage divider (see Fig. 277) between space charge and cathode. When the instantaneous value of the alternating voltage on the first grid increases in a positive direction the density of the space charge increases, that is to say the negative charge becomes greater, and the same applies to the induced charge on the fourth grid. Consequently between the fourth grid and cathode an alternating voltage occurs which is in antiphase to that between the first grid and cathode.

If, on the other hand, the frequency of the oscillator is lower than that

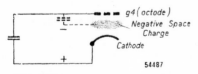

Fig. 277
Representation of the capacitive voltage divider between the space charge in front of the fourth grid of an octode and the cathode, which occurs when the circuit between the fourth grid and cathode has a capacitive impedance at the frequency of variation of density of the space charge.

of the signal received the impedance of the input circuit will have, practically speaking, an inductive character for the oscillator voltage, so that the voltage induced on the fourth grid is in antiphase to the voltage set up across the oscillatory circuit when it has a capacitive impedance. It is consequently then in phase with the oscillator voltage. Therefore it is advantageous to choose an oscillator frequency in the short-wave range lower than the fre-

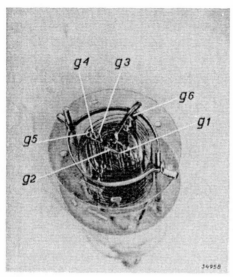

Fig. 278
Photograph of the electrode system of the battery octode DK 21. The two rods of g_3 lie parallel to the supporting rods of grid g_4 and are connected with the first (oscillator) grid. Owing to the capacity between g_3 and g_4 the induction effect is neutralized.

quency to which the input circuit is tuned. Considering the phase of the voltage that is induced on the fourth grid, the electronic coupling between grids 1 and 4 may be compared to a **negative capacity** between these grids. This capacity, however, does not exist in the reverse direction, i.e., from grid 4 to grid 1, since the fourth grid has little influence upon the electrons around the first grid. For an r.m.s. value of the oscillator voltage of about 8.5 V in the octode AK 2 the value of this capacity is about 2 $\mu\mu$F. If, now, a small condenser of the same value placed outside the valve is connected up between grids 1 and 4, the electronic coupling in the medium-wave range will practically disappear, because then via the same condenser the oscillator voltage also is applied to the fourth grid, but in antiphase. In this way a neutralization is obtained by these two voltages on the fourth grid and there is no reduction of the conversion conductance.

The following factors affect the magnitude of the disturbing oscillator voltage on the fourth grid:

1) The induction voltage decreases according as the capacity of the tuning condenser in a wave range increases. The electronic coupling in a given wave range is inversely proportional to the third power of the frequency.

2) With the tuning condenser across the input circuit set in a given position the electronic coupling is proportional to the frequency if this is varied by a change of the wave range (by means of the wave-range switch).

3) The electronic coupling is inversely proportional to the value of the intermediate frequency, i.e., to the difference between the frequency of the signal received and that of the oscillator.

In the case of the octode EK 2 the unilateral negative capacity between grids 1 and 4 is very much reduced by a judicious choice of voltages on the electrodes. Further, in the EK 2 the electronic coupling is neutralized by means of a small condenser mounted in the valve between grids 1 and 4. In the battery octode DK 21 the negative unilateral capacity between grids 1 and 4 is compensated by setting up two small rods (g_3 in Fig. 278) close to the fourth-grid support rods and connecting them to the first grid carrying the oscillator voltage (this octode has no screen grid between the oscillator anode g_2 and the control grid g_4).

When the oscillator voltage varies the induction voltage changes only slightly. From this it follows that the unilateral negative capacity from grid 1 to grid 4 must vary considerably when the oscillator alternating voltage is changed. Therefore the positive capacity inserted between grids 1 and 4 in order to neutralize the induction effect can only do so at a determined value of the oscillator voltage. It is of importance that this neutralization point should be adjusted accurately at the bottom end of a wave range, because the induction voltage decreases with increasing rapidity when turning away the tuning condenser from the position corresponding to that wavelength (see above under factor 1).

The inference that the induction effect in the medium-wave range is to be regarded as a unilateral negative capacity from grid 1 to grid 4 no longer holds for the short-wave range, for there the induction effect can no longer be compensated by a positive capacity and an induction voltage always remains on the fourth grid. This phenomenon is to be ascribed to the finite transit time from grid 1 to grid 4, as a result of which at high frequencies the electrons are somewhat delayed in reaching the fourth grid. The induction voltage is retarded with respect to the alternating voltage on the first grid, so that it becomes displaced in phase. In this case the inductive action can be compared to a unilateral negative capacity in series with a unilateral negative resistance between grids 1 and 4, and it can be neutralized by interposing a capacitor in series with a resistor between those grids.

When a fixed resistor is connected in series with a fixed condenser, however, there is only compensation of the induction effect at a fixed value of the oscillator voltage. It is advisable so to adjust the oscillator part that this value of oscillator voltage lies at the bottom end of a wave range.

In one of the latest methods of construction of mixer valves on the octode principle, viz. the octode EK 3 with 4 electron beams, this

28584

Fig. 279
Top view of the internal construction of the octode frequency-converting valve EK 3 showing clearly the means used to compensate the induction effect. A points to the condenser and B to the resistor.

compensation is achieved by means of a resistor mounted in the valve in series with a condenser (see Fig. 279). By this means the induction effect of an average valve is compensated at an r.m.s. value of the oscillator voltage of about 12 V.

Induction effect also occurs in valves constructed on the hexode or heptode principle. The electrons moving through the first (control) grid and the second (screen) grid to the third (modulation) grid partly turn back in front of the third grid when this is considerably negative during the negative half-cycle of the oscillator voltage. These returning electrons pass again through the second grid towards the first and, since this is likewise negative, partly turn back again towards the second grid. In this manner a negative space charge is set up between grids 1 and 2 which varies periodically at the frequency of the oscillator voltage (see Fig. 280).

This varying space charge induces a fluctuating charge on the first grid.

When the circuit between cathode and grid g_1 has a capacitive impedance at the frequency at which the space charge fluctuates we again have a capacitive voltage divider between space charge and cathode. Contrary to the case of an octode, however, the space charge between the first and second grids increases in density when the voltage on the third grid changes in a negative direction (with the octode the space charge in front of grid 4 increases when the voltage on the first grid changes in a positive direction). The charge induced on the first grid therefore increases when the voltage on the oscillator grid g_3 becomes more negative; the induction voltage on the first grid is consequently in phase with the oscillator voltage and thus more or less assists the mixing. However, this induction voltage of oscillator frequency is across the R.F. input circuit and may cause interference in neighbouring receivers.

The induction effect of a hexode or heptode may be represented by a positive capacity between grids 3 and 1 acting in one direction (not from grid 1 to grid 3). In hexodes and heptodes this quasi-capacity is about 10 times smaller than that in the case of octodes, because the space charge formed in the vicinity of the first grid due to the returning electrons is of much smaller density than the space charge in front of the fourth grid of an octode.

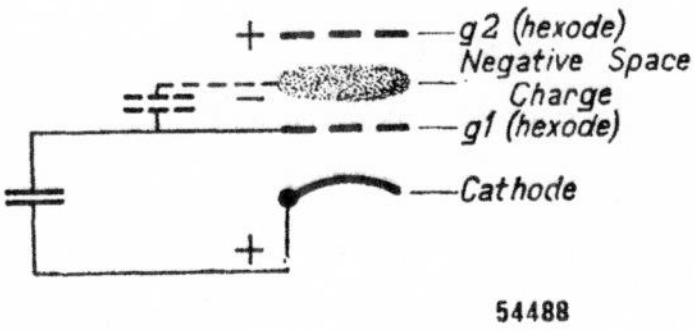

With triode-hexodes and triode-heptodes there are capacitive couplings between oscillator anode and R.F. input grid g_1, and also between oscillator grid and

Fig. 280

Representation of the capacitive voltage divider between the space charge outside the R.F. grid g_1 of a hexode (or heptode) and cathode. This capacitive voltage divider is formed when the circuit between g_1 and cathode has a capacitive impedance at the frequency at which the density of the space charge varies.

g_1 (see Fig. 281). If the R.F. oscillatory circuit between g_1 and the cathode has a capacitive impedance at the oscillator frequency ($\omega_h > \omega_i$) the coupling by the capacity C_{g1aT} will produce an alternating voltage at the grid g_1 in antiphase to the oscillator alternating voltage on the grid g_3 and thus having a tendency to reduce the conversion conductance. The coupling via the capacity C_{g1gT}, on the other hand, produces at the grid g_1 a voltage which is in phase with the oscillator voltage at grid g_3.

It depends upon the magnitude of the capacities C_{g1aT} and C_{g1gT} and upon the alternating voltages at the anode and the grid of the triode part whether the resulting voltage at the grid g_1 of the modulator part is in phase with or in antiphase to the oscillator voltage at g_3 and whether consequently the conversion conductance is increased or decreased.

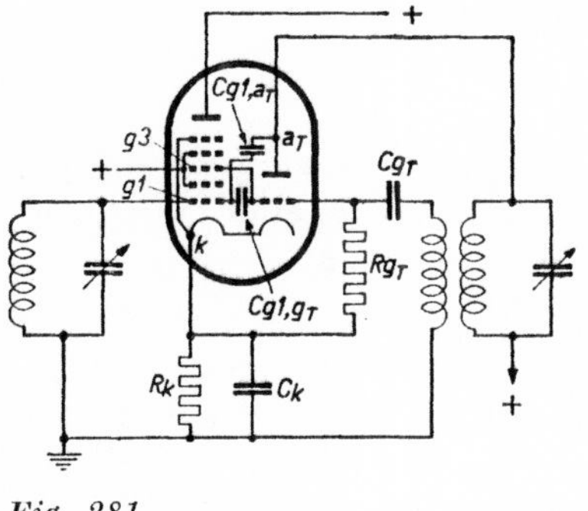

Fig. 281

Representation of the capacitive coupling between the R.F. input grid and the oscillator anode, and also between oscillator grid and R.F. input grid of a triode-heptode.

(c) Frequency Drift

By frequency drift is understood the undesired changing of the oscillator frequency that may occur after tuning. This also causes the intermediate frequency to change and perhaps to fall outside the pass-band of the I.F. amplifier. A small fluctuation of the oscillator frequency causes the intermediate frequency to lie beside the peak of the I.F. resonance curve and the side-bands are then no longer equal-

299

ly amplified. This leads to distortion in the detector. If there is a large oscillator-frequency drift the intermediate frequency produced by the mixer will be displaced so far outside the I.F. resonance curve that it can hardly be amplified at all by the I.F. amplifier. The signal will then have disappeared almost entirely and the oscillator tuning has to be readjusted to bring it back again. Let us take first the case of a separate oscillator valve. Here frequency drift may occur through a variation in the space charge before the first grid. This space charge may be regarded as an increase of the grid-cathode capacity, and when that capacity is directly in parallel with the tuning capacity of the connected oscillatory circuit the change in the space charge will affect the resonant frequency of that circuit. A variation of the anode voltage of the oscillator valve and of the heating voltage, as may be caused for example by a fluctuation of the mains voltage, influences the density of the space charge and gives rise to frequency drift. If the oscillatory circuit is connected to the anode and the feedback coil to the grid, then the capacity change is modified by transformer action between the oscillatory circuit and the feedback coil. If there is a step-up in the transformer from the feedback coil this will reduce the capacity change in the oscillatory circuit. Therefore, as regards frequency drift, there is an advantage in connecting the oscillatory circuit to the anode if the drift is not neutralized in some other way. The drawback, however, is that it makes the connections more complicated.

Obviously a capacity change due to a variation in density of the space charge will have the greatest effect upon the tuning of the oscillatory circuit when the capacity of that circuit is smallest. It will also be apparent that the absolute change in frequency will be greater for high frequencies than for low frequencies, so that at high frequencies frequency drift may assume inadmissible proportions, especially in the short-wave range.

If the oscillator valve and the mixer valve are built in one common bulb and both use the same electron current the space charge in front of the control grid of the oscillator valve may be affected not only by the voltage of the oscillator anode but also by the direct voltage variations on the other electrodes. Thus, for instance, the space charges in front of the first grid in the octodes AK 2 and EK 2 are affected by a negative gain-control voltage applied to the fourth grid, due to which frequency drift takes place. The electrons driven back in front of the fourth grid return partly to the space charge in front of grids 1 and 2 and thus affect the density of the space charge and the tuning of the oscillatory circuit. On wavelengths of 200 m and higher very little frequency drift is noticed with these octodes. In the short-wave range, however, frequency drift assumes inadmissible proportions. With the AK 2 for instance a capacity change of 0.2 to 0.3 $\mu\mu$F was measured when the bias on the fourth grid was varied from —2 to —20 V. At a wavelength of 12 metres and with an oscillatory circuit having a total capacity of 50 $\mu\mu$F a capacity change of 0.2 $\mu\mu$F causes a frequency drift of 50 kc/s. (This frequency drift increases in a given wavelength range in proportion to the third power of the frequency.)

In view of the magnitudes that frequency drift may assume in the short-wave range when using the AK 2 the negative grid bias of this valve must not be altered in that range.

In the case of hexodes and heptodes with separate oscillator valve the capacity change of the third grid can hardly have any effect upon the tuning of the oscillatory circuit. This capacity is little influenced by mains-voltage variations, and even if a negative gain-control voltage is applied to the first grid the capacity change of the third grid can be kept small by adopting a suitable valve construction. By keeping the coupling of the third hexode grid with the oscillatory circuit loose, for instance by connecting the oscillatory circuit to the anode of the oscillator valve and connecting the feedback coil to the grid of that valve, the influence of capacity variation can be greatly reduced.

If, when using the ECH 3 (triode-hexode), the oscillatory circuit is connected to the anode and the feedback coil to the grid of the triode part, the frequency drift for a mains voltage variation of 10% and for a wavelength of 15 metres is less than 1 kc/s. At full down-control of the conversion conductance, a wavelength of 15 metres and 50 $\mu\mu$F oscillatory-circuit capacity the frequency drift is less than 2 kc/s. Under these conditions the frequency drift with the ECH 4 (triode-heptode) is less than 3 kc/s, due to the full down-control of the conversion conductance.

In the case of the octodes AK 2 and EK 2 the frequency drift in the short-wave range is increased owing to the fact that a voltage variation at the electrodes and the resultant variation of current to the second grid causes a variation in transconductance and consequently an apparent variation of oscillator capacity. Owing to the transit time of the electrons the oscillator-anode alternating current in the short-wave range is no longer in phase with the alternating voltage on the oscillator grid. This delay of the anode current causes the feedback coil to induce an alternating voltage in the oscillatory circuit which is not in phase with the alternating current in the oscillatory circuit. This voltage has two components, one of which is in phase with the oscil-latory-circuit current and thus produces regeneration, whilst the other is phase-shifted 90° with respect to that current. The latter component, however, would also be the result of a correspondingly increased capacity of the circuit. Consequently the effect of the transit-time delay of the anode a.c. is an apparent increase of the circuit capacity. Therefore a mains-voltage variation that brings about a change in the transconductance of the oscillator part of an AK 2 or EK 2 causes a variation of this quasi-capacity and consequently fre-quency drift.

In the case of the octode EK 3 frequency drift is greatly reduced by:

1) eliminating the influence of the fourth-grid bias potential on the space charge in the oscillator part of the valve;

2) shortening the transit time of the electrons in the oscillator part.

This is achieved by applying the principle of electron-beam concen-tration and screening the oscillator part almost completely [1]). As a consequence the frequency drift is so small that the EK 3 can also be fully gain-controlled in the short-wave range. At 15 metres and with an oscillatory circuit capacity of 50 $\mu\mu$F the drift is 4.5 kc/s for a variation of the bias on the fourth grid from —2 to —20 V.

(d) **Whistles** [2])

If in a superheterodyne receiver there is, in addition to the inter-mediate frequency, a second signal of a frequency differing slightly from that of the I.F. signal applied to the detector, the presence of these two signals causes a whistle in the loudspeaker. The pitch of the whistle depends upon the difference between the frequencies of the two signals reaching the detector. This undesired signal may arise in all sorts of ways. It is possible, for instance, that in addition to the selected signal on the R.F. grid of the mixer valve there is an interfering signal of a frequency almost equal to the intermediate frequency. The interfering signal is amplified directly by the modulator part and sets up an I.F. alternating current in the anode circuit, whilst through mixing with the oscillator signal the selected R.F. signal likewise sets up in the anode circuit an alternating current of intermediate fre-quency. Due to the two I.F. signals there will be a difference-frequency

---

[1]) See also J. L. H. Jonker and A. J. W. M. van Overbeek, Wireless Engineer 15, 1938, pp. 423–431.
[2]) See also M. J. O. Strutt, Wireless Engineer, 12, 1935, pp. 194—197, and M. J. O. Strutt, Verstärker und Empfänger, Julius Springer, 1943, pp. 224 et seq.

component in the detector circuit which manifests itself in the loud-speaker as a whistle.

Whistles may also arise if, for instance, a harmonic of the oscillator—which is likewise applied to the modulator grid of the mixer valve—together with a harmonic of the input signal caused by the curvature of the mixer-valve characteristic, forms a signal whose frequency deviates slightly from the intermediate frequency.

Further it is possible for whistles to arise if, for instance, a strong undesired signal penetrates to the grid of the mixer valve and this has such a frequency that together with an oscillator harmonic it sets up a frequency almost equal to the intermediate frequency. If, for example, the I.F. is 125 kc/s and the receiver is tuned to a transmitter with a frequency of 600 kc/s (= 500 metres) the oscillator frequency is 725 kc/s. The second harmonic of the oscillator frequency is then 1,450 kc/s and with a signal of 1,324 kc/s (= 226 m) may cause a frequency of 126 kc/s. The whistle will then have a frequency of 1,000 c/s.

In the first place, however, we have to consider the disturbances due to the so-called **image frequency.** If a receiving set is tuned to 1,000 kc/s and the I.F. is 125 kc/s, the oscillator signal has a frequency of 1,125 kc/s. Now if we have a signal that differs from the tuned signal by about twice the amount of the I.F.—usually called the **image signal**—this will also produce the intermediate frequency in the anode circuit. Suppose the frequency of this signal is 1,249 kc/s, then in the anode circuit we get a frequency of 1,249 — 1,125 = 124 kc/s. As the frequency of the 1,249 kc/s signal does not differ too much from that of the 1,000 kc/s signal it may happen in certain circumstances that a very strong signal of that frequency will be applied on the grid of the mixer valve and give rise to an intense whistle. As the I.F. is increased, the difference in frequency between the desired signal and the image signal becomes greater and the more easily can the disturbing signal be filtered out by the preceding tuned circuits. In this respect, therefore, a high I.F. offers great advantages.

There is a large variety of possible causes of whistles, a brief summary of which is given below. The frequency of the desired signal will be represented by f_1 and that of the interfering signal by f_2.

1) The interfering frequency is equal to the I.F.:

$$f_2 = \text{I.F.}$$

2) Interference due to the image frequency:

$$f_2 = f_1 + 2 \times \text{I.F.}$$

3) Due to mixing with the tuned signal the interfering frequency forms the I.F.:

$$f_2 = f_1 \pm \text{I.F. or}$$

$$f_2 = \text{I.F.} - f_1.$$

4) A harmonic of the oscillator together with the interfering frequency or a harmonic thereof forms the I.F.:

$$n (f_1 + \text{I.F.}) \pm mf_2 = \text{I.F.}$$

Here n and m are whole numbers.

5) A harmonic of the oscillator together with a harmonic of the tuned signal forms the I.F.:

$$n (f_1 + \text{I.F.}) \pm mf_1 = \pm \text{I.F.}$$

It is obvious that in order to keep the number of whistles as low as possible care has to be taken that no interfering signals reach the grid of the mixer valve and that any signals that cannot be avoided cause the weakest possible whistles. In the first place, therefore, the tuned circuits preceding the mixer valve must give the highest possible selectivity. For this reason it is advisable to use two tuned R.F. circuits.

By selecting a high I.F. there is less chance of interference being caused by the image frequency. The I.F. should be so chosen that the minimum number of whistles occur on the tuning dial, taking into account the effect of any local stations.

Under certain circumstances a

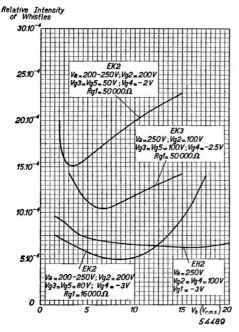

Relative Intensity of Whistles

EK2
Va = 200-250V; Vg2 = 200V
Vg3 = Vg5 = 50V; Vg4 = -2V
Rg1 = 50000Ω

EK3
Va = 250V; Vg2 = 100V
Vg3 = Vg5 = 100V; Vg4 = -2.5V
Rg1 = 50000Ω

EK2
Va = 200-250V; Vg2 = 200V
Vg3 = Vg5 = 80V; Vg4 = -3V
Rg1 = 16000Ω

EH2
Va = 250V
Vg2 = Vg4 = 100V
Vg1 = -3V

54489

Fig. 282
Curves for the valves EH 2, EK 2 and EK 3 indicating the relative strength of whistles (vertical axis) caused by the second harmonic of the input signal, related to the strength of the desired signal as a function of the oscillator alternating voltage V_h. The relative strength of the whistle is measured after the detector and applies to linear rectification of the detector valve with a modulation depth of the carrier wave of 30% and a R.F. signal of 3 mV at the grid of the frequency-converter.

whistle can be eliminated at a certain place on the tuning dial by introducing in the set a wavetrap for the interfering frequency.

The selectivity of the preceding tuned R.F. circuits or a wavetrap has no effect upon the fifth possible cause of whistles given above. It is therefore of importance that the mixer valve should generate the fewest possible harmonics of the input signal, and this can be achieved when the characteristic has a satisfactory shape.

Fig. 282 gives some curves for the valves EH 2, EK 2 and EK 3 indicating the strength of the whistle in the loudspeaker due to the second harmonic of the input signal (which is most prominent) compared with the strength of the desired signal in the loudspeaker, as a function of the oscillator voltage. In plotting these curves it has been assumed that the rectification of the detector valve of the receiver in which the measurement is made on the mixer valve is linear, that the modulation depth of the R.F. signal is 30% and the strength of the signal at the grid of the mixer valve is 3 mV. From these curves it appears that the whistle depends mainly on the magnitude of the oscillator voltage. In this respect, therefore, it is preferable to keep the oscillator voltage as small as possible, so long as this does not lead to any other unpleasant phenomena. such as valve noise.

(e) Valve Noise

With mixer valves the background noise produced is stronger than with valves used for direct R.F. amplification. The nature of this noise will be considered in greater detail in Chapter XXIV. Apart from a factor subject to slight variation for different types of valves, the noise of a mixer valve is roughly proportional to the ratio of the square-root of anode current to the conversion conductance. Since the anode current increases as the oscillator voltage decreases, this ratio increases as the oscillator voltage is reduced, so that the noise becomes stronger. A compromise has to be sought between the weakest possible noise and the least possible interference by whistles.

(f) Cross-modulation, Modulation Distortion and Hum-modulation

As in the case of R.F. and I.F. valves, with mixer valves phenomena are observed which are to be attributed to the curvature of the characteristic and manifest themselves as cross-modulation, modulation distortion and hum-modulation. Cross-modulation is of particular importance, as in most cases the mixer stage constitutes the input stage of the receiver.

Modulation distortion is only to be expected when the receiver is tuned in to very high-powered transmitters. In this respect it is necessary that the characteristic of the mixer valve should have a favourable shape. For mixer valves curves are published indicating also the admissible signal voltage at the grid as a function of the conversion conductance for 1% cross-modulation or for 1% hum-modulation. (For the older types the curves were published for 6% cross-modulation and 4% hum-modulation.)

126. Grid-current Curves

Usually the bias on the oscillator grid is automatically produced by means of a leak resistor (self bias). The current passing through the leak resistor is an indication of the amplitude of the alternating voltage on the oscillator grid.

Since it is easier to measure the current I_R passing through the leak resistor than the alternating voltage across the oscillatory circuit—

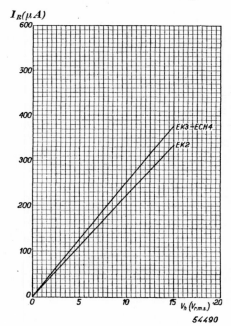

because in the latter case the effect of the measuring instrument on the voltage source has to be avoided— for every mixer valve curves are published showing the oscillatory voltage as a function of the grid current for a given leak resistance. Such curves are shown in Fig. 283 for the valves EK 2, ECH 4 and EK 3.

Fig. 283
Curves for the valves EK 2, ECH 4 and EK 3 indicating how the current I_R through the leak resistor depends upon the oscillator voltage V_h.

CHAPTER XXII

Gain Control (Variable-transconductance Valves)

127. Object and Achievement of Gain Control

A large number of signal voltages differing in frequency and magnitude are induced in the aerial of a receiving set. When a set is tuned in to a local transmitter the voltage across the oscillatory circuit coupled to the aerial will be very large and may amount, for instance, to 2 volts. If, the on other hand, the set is tuned to a distant station the signal from which is very weak when picked up by the set, the voltage across that circuit is very small, often of the order of only a few microvolts. With such widely divergent signal strengths the alternating voltage supplied to the loudspeaker should be as constant as possible, whilst at the same time the volume of sound from the loudspeaker must be adjustable. For the loudspeaker to be able to reproduce very weak signals with sufficient volume it is essential that the amplification in the set should be adequate. When a strong signal is picked up the amplification and thereby the volume of sound from the loudspeaker must be reduced, in order also to avoid overloading the valves, which leads to distortion.

In the first place it is necessary that the R.F. and I.F. amplification is such that the signal voltage at the detector is kept practically constant and, if possible, of sufficient strength to drive the output valve fully. This makes it essential that the R.F. amplification, the conversion amplification and the I.F. amplification or one or more of these can be controlled. Separate measures to control the volume of sound from the loudspeaker must then be taken in the A.F. part of the circuit. Usually the A.F. amplification or gain is controlled with the aid of a potentiometer operated by hand and reducing the signal strength applied to the output valve.

In simple receiving sets with direct R.F. amplification (T.R.F. receivers) the R.F. signal is sometimes also attenuated by a potentiometer on the input side of the set; this then serves at the same time to adjust the sound volume. In simple sets with one circuit and a regenerative grid detector gain control is obtained by adjustment of reaction.

In modern superheterodyne receivers the R.F., conversion and I.F. amplification are attenuated by applying negative voltages to the control grid of each amplifying valve. This gives a very wide range of control, particularly where several valves are controlled. In most cases

the gain-control voltage is automatically adjusted by the signal voltage
and this is given the name of **automatic volume control** (A.V.C.).
A study of the transfer characteristic of a valve shows that when the
negative grid bias is increased the transconductance and consequently
the amplification by the valve is reduced, owing to the curvature of
the characteristic. The same applies in the case of frequency-converting
valves. This affords the possibility of controlling the amplification.

128. Requirements to be met by the Gain-controlled Valves

When a strong signal is picked up by the aerial a high-frequency valve
has to cope with a large alternating voltage at the grid. This makes it
desirable to reduce the R.F. amplification or gain, which is attained
by raising the negative bias of the valve until the amplification is
sufficiently limited. In the case of a valve with a normal characteristic, i.e. not one whose shape is artificially modified, control by increasing the negative bias will cause the operating point to be shifted
towards the cut-off point [1]), where the curvature of the transconductance characteristic is much greater than over the part of the characteristic corresponding to low grid bias. As shown in Chapter XV,
modulation distortion (and cross-modulation) depends upon the curvature of the transconductance characteristic $g_m = f(V_g)$. Any displacement of the operating point towards the cut-off point of the characteristic therefore results in a considerable increase of modulation distortion
even with a constant signal strength. If, moreover, the signal on the
grid increases, which is the cause of the operating point being shifted,
then the modulation distortion is still greater (it is proportional to
the square of the grid alternating voltage).
In view of modulation distortion, therefore, for gain-control purposes
a valve characteristic which shows reduced curvature of the transconductance characteristic for increasing negative grid bias is desired. In
that case modulation distortion is not increased so much by gain
control as when the curvature becomes greater due to increased grid
bias.
The risk of modulation distortion will generally be great in the first
valve of a receiver, since on the grid of that valve the signal from a
local station may be fairly strong. This is also the case with valves
preceded by several amplifying stages and having to cope with strong
grid signals, particularly so with the I.F. valve preceding the detector
valve, especially when only small A.F. amplification is employed. The
signal on the grid will then have to be of considerable strength if the

[1]) The point where the anode current becomes practically equal to zero.

final stage is to be fully driven. There is then a very great chance of modulation distortion, against which the necessary counter measures have to be taken.

Apart from modulation distortion there is also the possibility of cross-modulation. From Chapter XV, Section 82b, it follows that cross-modulation does not depend upon the carrier-wave strength of the desired station but is proportional to the square of the amplitude of the interfering carrier wave. The greatest risk of cross-modulation occurs in the first valve of the receiver (usually the R.F. or the mixer valve) when the signal of adjacent carrier-wave frequency reaches the grid of this valve in great strength. This risk is equally great in the reception of both strong and weak stations. Like modulation distortion, cross-modulation depends upon the curvature of the transconductance characteristic at the operating point. If with larger grid bias the curvature increases then also the risk of cross-modulation becomes greater. With regard to cross-modulation, therefore, it is also desirable that the transconductance characteristic should show a uniform and gentle curvature for the whole control range, so that even for the strongest interfering carrier wave there will be no trouble from cross-modulation. To this may be added the fact that in the reception of a powerful station (a local transmitter) there is little probability of another station of adjacent frequency coming through also in such strength as to cause troublesome cross-modulation. This possibility is much greater in the reception of weak stations, so that it would be desirable for the curvature of the transconductance characteristic to be less with low bias than with higher bias; this is contrary to what is required of the transconductance characteristic of a gain-control valve to avoid modulation distortion.

Now the shape of the transconductance characteristic, which is derived from the transfer characteristic, depends upon the shape of the latter. From this it follows that there are exceptional requirements for the shape of the transfer characteristic of a valve in which the amplification is to be controllable.

It is not, however, only cross-modulation or modulation distortion that has to be taken into account for a valve that is intended for gain-controlling purposes. There are other requirements to be considered and these are usually of a conflicting nature. In a gain-controlled valve a combination of the following properties would be desirable:

1) largest possible transconductance change with the smallest possible gain-control voltage;

2) small non-linear distortion at all points of the transfer characteristic;

3) largest possible initial transconductance (the transconductance at the lowest bias to be applied in view of grid current);

4) not too large an anode current with the lowest bias, i.e., in the uncontrolled state.

Not a single one of these requirements can be satisfied without adversely influencing at least one of the others. A large transconductance change with limited control voltage requires a pronounced curvature of the characteristic, which results in excessive cross-modulation and modulation distortion. High initial transconductance accompanied by a not too pronounced curvature of the characteristic means that in order to reduce the transconductance to a certain extent a very high control voltage is needed. Moreover, the anode current for the minimum bias will then be very large. Consequently for a gain-controlled valve a choice has to be made from these properties and it depends upon their relative importance which is to be most favoured.

129. Means of Influencing the Shape of the Characteristic

In the case of a pentode the anode current as a function of the negative grid voltage is represented in the simplest case by the formula

$$I_a = k \left(V_{g1} + \frac{1}{\mu_{g2g1}} V_{g2} \right)^{3/2} \tag{185}$$

in which k is a constant and μ_{g2g1} the amplification factor of the control grid with respect to the screen grid.

Here the influences of anode and suppressor-grid voltages are disregarded. This formula represents a curved characteristic departing from the square-law shape. This departure implies that modulation distortion or cross-modulation may arise. The characteristic represented by Formula (185) applies to the case where the equipotential planes between grid and cathode are parallel to the cathode, i.e., as explained in Chapter XIII, when the mutual distances between the grid wires is small compared

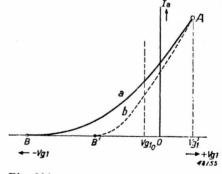

Fig. 284
Curve a: Transfer characteristic of a valve with island effect.
Curve b: Transfer characteristic of a similar valve without island effect.

with the distance between the grid and cathode surfaces. If, however, the distance between the grid wires is about equal to or greater than the grid-cathode distance then, as explained in Chapter XIII, with high negative grid voltages the **island effect** arises. As a consequence the anode current for a certain direct negative grid voltage is greater than would be obtained if the $^3/_2$ power law were obeyed. Furthermore the transconductance is less.

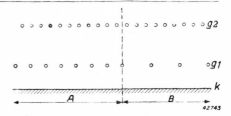

Fig. 285
Diagrammatic representation of a valve where the control grid of one half (A) has a small pitch and the other half (B) a large pitch.

Instead of curve b in Fig. 284, representing a transfer characteristic which conforms to the $^3/_2$ power equation of Formula (185), we get the full-line curve a. From the shape of that curve it appears that the curvature of the characteristic for corresponding anode currents is much less pronounced throughout; the transconductance, however, is also at all points much less than that of curve b. Moreover, the "cut-off" point of anode current (the point where the anode current is practically zero—see the points B and B' in Fig. 284) lies at a much higher negative grid voltage in curve a than in curve b. The anode current is also much greater at all points, except at the grid voltage corresponding to point A. (This grid-voltage value corresponds to the potential of the potential-distribution diagram between cathode and anode which would occur with the given anode voltage at the position of the grid if the grid were left out.) In such a case the characteristic a is said to have a long tail. Owing to the fact that the curvature of characteristic a is smaller for all grid voltages than

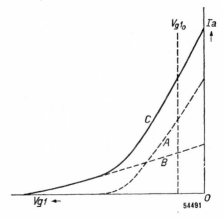

Fig. 286
Curve A: Transfer characteristic showing normal transconductance and normal anode current in the non-controlled condition (for a grid voltage V_{g1o}).
Curve B: Characteristic showing a long tail and low anode current in non-controlled condition.
Curve C: Sum of the curves A and B. In this manner a characteristic is obtained of which the transconductance is normal and the anode current in the non-controlled condition is not too great, whilst for large negative grid voltage there is a long tail with slight curvature.

311

that of characteristic b, the former is more favourable with regard to avoidance of modulation distortion and cross-modulation. Characteristic a, however, has a great drawback in that for the smallest direct grid voltage at which grid-current flow is avoided (generally —2 to —3 V) the transconductance is much less than that of curve b (compare the slopes at the grid voltage V_{g1o} in Fig. 284). If we now consider a valve of which one half (A in Fig. 285) has a small control-grid pitch and the other half (B in Fig. 285) a large control-grid pitch, it is found that the two halves have different transfer characteristics (see curves A and B in Fig. 286); whereas curve B has a long tail with small slope at point P (at the grid voltage V_{g1o}, which is the least required for amplification without grid current), curve A has a large slope at V_{g1o} and a short tail.

The resultant characteristic is represented by curve C, obtained by adding the currents of the two valve halves for each grid voltage. The valve-half A gives a large amplification for small direct grid voltage, but with smaller transconductances is unfavourable as regards modulation distortion. The valve-half B, on the other hand, contributes hardly anything at all towards amplification in the uncontrolled condition, but less modulation distortion is associated with the tail of the characteristic; the valve-half A is then inoperative. Curve C thus shows two slightly curved parts which are favourable from the point of view of avoiding modulation distortion and cross-modulation, and a greater curvature over the transition part. In practice valves to be used for gain control are made in such a fashion that the shape of their characteristics is determined by a large number of valve segments connected in parallel, each having a characteristic of its own. In order to achieve this a control grid with variable pitch is employed. A valve having such a grid may be considered as consisting of a number of valves connected in parallel, having different amplification factors; they can be successively switched off as the negative grid voltage increases (so-called variable-μ valves). The shape of the characteristic can be influenced by the law of the change of pitch. With a grid having a variable pitch, if the pitch and variation of pitch are suitably chosen, a characteristic is obtained that has properties favourable for gain control of the valve.

130. Logarithmic Transconductance Characteristic and Non-linear Distortion

Owing to the grid-transconductance change of a valve used for gain-control purposes, it is possible to give the characteristic of such a valve the shape most appropriate to definite requirements. However, no rule that is of general application can be given for the most favourable shape of the characteristic. A certain shape of curve that may be optimum for one case may be unsuitable for another. In practice, therefore, a valve has to be constructed in such a way that it best serves the purpose to which it is most commonly applied. Consequently the aim will be to get a certain shape of characteristic that offers the best possible compromise between the conflicting factors mentioned in Section 128. These factors can be most readily evaluated with the aid of the transconductance characteristic of the valve $[g_m = f(V_{g1})]$, plotting the transconductance on a logarithmic scale and the grid voltage on a linear scale (see Fig. 287). This characteristic indicates in the first place

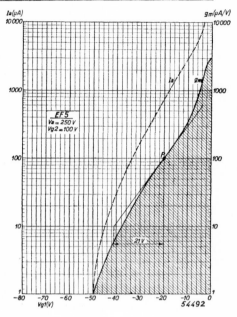

Fig. 287

The transconductance g_m of valve EF 5 as a function of the negative grid voltage (transconductance on logarithmic scale and grid voltage on linear scale). The broken-line curve represents the anode current I_a on a logarithmic scale as a function of the grid voltage on linear scale.

the rate of transconductance control for a given variation of grid bias (point 1 of Section 128). Further, from that characteristic the initial transconductance (point 3 of Section 128) can be read off.

Also the non-linear distortion can be determined from this transconductance characteristic. By computation it is found that the distortion due to the second harmonic and the hum-modulation are proportional to the slope of the logarithmic transconductance curve. If m_b represents the modulation depth due to hum (see Chapter XV, Section 81), V_{s1} the amplitude of the interfering hum voltage and S' the slope of the logarithmic transconductance curve, we obtain the equation:

$$m_b = \frac{S'}{0.434} V_{s1}$$

or

$$V_{s1} = 0.434 \, m_b \, \frac{1}{S'}. \tag{186}$$

313

The slope of the curve of Fig. 287 at point P where $g_m = 100\ \mu A/V$ equals $\dfrac{\log 10}{21} = \dfrac{1}{21} = 0.0476$ logarithmic units per volt grid bias. According to Equation (186), therefore, the peak value of the interfering alternating voltage admissible on the grid for 1% hum-modulation is equal to

$$V_{s1} = 0.434 \times 0.01 \times \frac{1}{0.0476} = 0.91\ V$$

and the r.m.s. value is

$$V_{s1\,eff} = \frac{0.91}{\sqrt{2}} = 0.605\ V.$$

Since the hum-modulation and, therefore, also the distortion due to the second harmonic are directly proportional to the slope of the logarithmic transconductance curve, it is desirable to aim at a curve with the smallest possible slope. A small slope of the logarithmic g_m curve, however, means that the transconductance changes only slightly per volt of grid-voltage variation. For small distortion, therefore, as already indicated, it is necessary that the transfer characteristic should have a long tail.

According to Equation (81) in Chapter XV the cross-modulation equals $m_k = \frac{8}{3} F V_{s2}^2 m_2$, in which V_{s2} is the amplitude of the voltage of the interfering carrier wave and m_2 its modulation depth. It can now be proved that in the straight parts of the logarithmic transconductance characteristic the factor F is equal to

$$F = S'^2, \tag{187}$$

in which S' is again the slope of the logarithmic transconductance characteristic. Now, from Equation (82) in Chapter XV it follows that the cross-modulation factor K is given by $K = \frac{8}{3} F V_{s2}^2$, so that with the aid of Equation (187) we find for the straight parts of the logarithmic g_m characteristic that

$$K = 2.65\ S'^2 V_{s2}^2 \tag{188}$$

or

$$V_{s2} = \frac{1}{S'} \sqrt{\frac{K}{2.65}}. \tag{189}$$

Thus the latter equation gives the amplitude of the interfering H.F. signal that causes $K\%$ cross-modulation. This is also the value of the admissible H.F. signal for $\frac{1}{2} K\%$ modulation rise and $\frac{3}{8} K\%$ modulation distortion. If in Equation (189) we substitute 1% for K we obtain

$$V_{s2} = \frac{0.0615}{S'}. \tag{190}$$

The r.m.s. value of the interfering H.F. signal causing 1% cross-modulation is consequently

$$V_{s2\,eff} = \frac{0.0433}{S'}. \tag{191}$$

From Equation (189) it follows that for the straight parts of the logarithmic transconductance characteristic a small slope, i.e., a long tail, is favourable also for the avoidance of the third harmonic, cross-modulation, modulation rise and modulation distortion.

For the curved parts of the logarithmic transconductance characteristic the rule is, as may also be proved, that a curve with the concave side underneath produces greater modulation distortion, cross-modulation, etc. than is evident from the slope of that characteristic at the point in question. A curvature with the convex side downwards, on the other hand, reduces these effects, but in practice this cannot be turned to advantage because with a high negative grid voltage the slope of the logarithmic transconductance characteristic would be too steep, and this leads to excessive cross-modulation and modulation distortion.

As regards the anode current at the minimum negative grid bias the following is to be noted. The transconductance $\dfrac{dI_a}{dV_g}$ of the valve is the first derivative of the function $I_a = f(V_g)$, so that in order to know the anode current for a certain grid bias V_{go} we have to integrate the function $\dfrac{dI_a}{dV_g} = f(V_g)$. Here we have to introduce as the limits of integration $V_g = -\infty$ and $V_g = V_{go}$. The function $\dfrac{dI_a}{dV_g} = f(V_g)$ can be represented by a curve which then gives the transconductance plotted on a linear scale as a function of V_g. The integral of the function $\dfrac{dI_a}{dV_g} = f(V_g)$ from $V_g = -\infty$ to $V_g = V_{go}$ is the area enclosed by the transfer characteristic, the V_g axis and the vertical line through $V_g = V_{go}$. The larger this area the greater is the anode current for $V_g = V_{go}$. If we compare two valves which have the same transconductance for the same negative grid bias V_{go}, one of which has a short tail and the other a long one, it will be found that the transconductance characteristic of the valve with a short tail encloses a smaller area than the

315

valve with a long tail, and consequently at $V_g = V_{go}$ the former valve will have a lower anode current than the latter valve. Therefore, in order to obtain favourable cross-modulation properties a valve which for gain-control purposes must have a long tail will usually have a high anode current at the minimum bias.

According to the foregoing the most suitable logarithmic characteristic is one that is approximately rectilinear and has not too great a slope. Therefore one would select, for instance, in Fig. 288 the line abcd as logarithmic transconductance characteristic. There are two reasons, however, why one has to deviate from this characteristic, viz.:

1) In order to have a large transconductance at a small negative grid bias, the part ab is given a steeper slope (e.g. fb). In this range, therefore, the effects in question will be greater; this is usually permissible, since this part of the curve is used only for small alternating grid voltages and therefore modulation distortion will not assume any serious proportions.

2) In order not to make the tail of the transfer characteristic too long, which implies also a high anode current with small negative grid voltage, the part cd of the logarithmic transconductance characteristic is also given a steeper slope, i. e. cg. Therefore in this range cross-modulation and modulation distortion will be great. For this reason the automatic volume control (see Chapter XXIII) will be arranged in such a way that the point c is not passed.

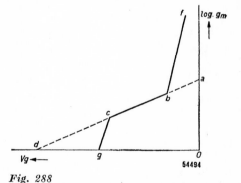

Fig. 288
Sketch representing the logarithmic transconductance characteristic of a valve with variable transconductance. The line abcd would yield the best properties as regards cross-modulation or modulation distortion. In order, however, to get a larger initial transconductance the slope of the part ba is made greater and we get a part as represented by bf. To get a not too long tail of the transfer characteristic and a not too high anode current at the minimum bias, for large values of grid voltage the transconductance characteristic will be given a shape corresponding to cg instead of cd.

The logarithmic transconductance characteristic of a normal valve with variable transconductance is therefore as a rule of the shape represented in Fig. 288 by fbcg, with the three more or less rectilinear parts merging into each other.

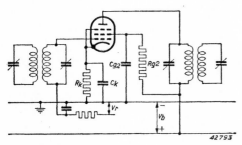

Fig. 289
Basic circuit diagram of an I.F. amplifying valve with variable screen-grid voltage. $V_r =$ control voltage of the automatic volume control.

131. Variable Screen-grid Voltage

Valves with variable transconductance, the characteristics of which are favourable as regards freedom from crossmodulation, will have, for the minimum negative grid voltage, either a relatively large anode current or a relatively small transconductance when a constant voltage is applied to the screen grid. By applying the principle of variable screen-grid voltage it is possible to obtain a smaller anode current and also greater transconductance at the operating point corresponding to the minimum negative grid voltage, with practically unchanged properties as regards crossmodulation.

Fig. 289 illustrates the circuit of a pentode with variable screen-grid voltage. The screen grid is fed via a series resistor, and the value of that resistance must be chosen so that with the minimum negative grid voltage the correct value of screen-grid voltage is obtained.

As the negative grid voltage increases, the screen-grid current will decrease and the voltage drop across the resistor will consequently diminish. From this it follows that the screen-grid voltage increases while the negative control-grid voltage increases, and when the cut-off point of the anode current is reached it approximates the value of

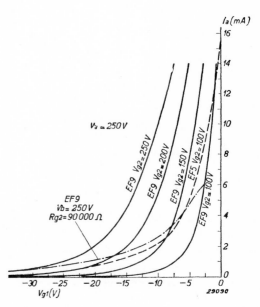

Fig. 290
Some transfer curves of the EF 9 valve for different screen-grid voltages. The transfer curve of the EF 5 for $V_{g2} = 100$ V is represented by the broken line, whilst the dot-dash line gives the variation of the anode current of the EF9 as a function of the negative voltage when the screen grid is fed via a 90,000-ohm series resistance from the 250-V supply.

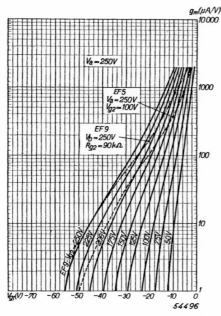

Fig. 291
Full lines: Logarithmic transconductance characteristics of the EF 9 valve for different screen-grid voltages.
Dot-dash line: Transconductance of the EF 9 as a function of the negative grid voltage when the screen grid is fed via a 90,000 ohm series resistance from the 250 V anode supply source.
Broken line: Logarithmic transconductance characteristic of the EF 5 valve with fixed screen-grid voltage of 100 V.

the supply voltage. Usually the supply source of the anode is used for this.

If such a valve has a control grid of which the greatest pitch is less than that of a valve for a fixed screen-grid voltage, then the initial anode current will be smaller and the transconductance greater. In that case a characteristic is obtained as shown in Fig. 290 for valve type EF 9 at $V_{g2} = 100$ V; for the sake of comparison the transfer characteristic of the EF 5 at $V_{g2} = 100$ V is shown by a broken line. From these curves it is evident that the cross-modulation of the EF 5 will be much less than that of the EF 9 with fixed screen-grid voltage.

Fig. 291 shows the logarithmic transconductance curves of the valve EF 9 corresponding to these curves (full lines) for different screen-grid voltages, from which the following may be deduced.

With increasing screen-grid voltage the slope of the logarithmic transconductance characteristic is reduced. In Fig. 291 the broken line represents the logarithmic transconductance characteristic of the EF 5, whilst the dot-dash line indicates the transconductance as a function of the grid bias when the valve EF 9 is fed via a series resistance of 90,000 ohms from an anode-supply source of 250 V.

As the screen-grid voltage increases, the transfer characteristic of the EF 9 and thus also the logarithmic transconductance characteristic are displaced to the left, and the curvature of the transfer characteristic and the slope of the logarithmic transconductance characteristic are reduced. It might be supposed that, although the cutting-off of the anode current is deferred, the steep transfer characteristic with which we started is not changed, so that after all no improvement is obtained as regards cross-modulation. Such a parallel displacement would take

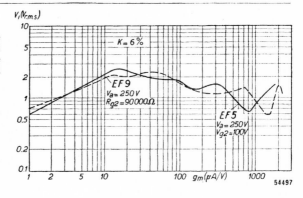

Fig. 292
Cross-modulation curve of the EF 9 with variable screen-grid voltage (broken line) and of the EF 5 with fixed screen-grid voltage (full line).

place if there were no island effect. In practice, however, this displacement is not parallel, and certainly not in the case of valves designed for the application of a variable screen-grid voltage. This is evident from Fig. 290, which shows that with increasing screen-grid voltage and constant anode current the transconductance decreases. As already stated, this phenomenon is to be ascribed to island effect. The dot-dash line of Fig. 290 indicates the shape of the anode current of the EF 9 as a function of the negative grid voltage when the screen grid is fed via a 90,000-ohm series resistance from the supply source with a voltage of 250. This line, however, is not to be regarded as the transfer curve of the EF 9. For a certain negative grid voltage the amplification takes place according to the transfer curve corresponding to the respective screen-grid voltage (the screen grid being short-circuited to the cathode so far as alternating currents are concerned). Fig. 292 shows, further, that by applying the principle of variable screen-grid voltage practically the same results are obtained, as regards cross-modulation, as are obtained with a fixed-screen-grid-voltage valve, whilst in the case of the EF 9 the initial anode current is less, namely 6 mA instead of 8 mA for the EF 5, and the initial transconductance is greater, viz. 2.2 mA/V instead of 1.8 mA/V for the EF 5.

132. Application of Valves with Variable Transconductance to A.F. Amplification

As pointed out in Section 127, the A.F. amplification of a receiver is usually varied by means of a potentiometer operated by hand. In some cases, however, it is desirable to effect this gain control by varying the grid bias of an audio-frequency amplifying valve. This is the case, for instance, when the A.F. amplification has to be automatically regulated together with the R.F., I.F. and conversion amplification by means of a negative control voltage.

In Chapter XV we have seen that the curve for 1% hum-modulation (see Fig. 176) also indicates as a function of the transconductance the

319

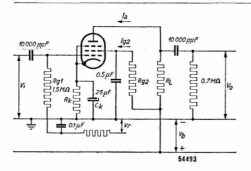

Fig. 293
Essentials of the circuit of a pentode with
variable transconductance used as A.F.
amplifier with resistance coupling.
V_i = input alternating voltage.
V_o = output alternating voltage.
V_r = gain-control potential at the grid.

magnitude of the grid alter-
nating voltage at which $\frac{1}{4}\%$ distortion occurs due to the second
harmonic. Likewise, the curve
for 1% cross-modulation indicates also the magnitude of
the grid alternating voltage at
which 0.083% distortion due to
the third harmonic is produced.
The higher harmonics may be
disregarded because, generally
speaking, these are very small
compared with the second and
third harmonics resulting from
curvature of the characteristic.
It will be obvious that a valve
that is satisfactory as regards hum-modulation and cross-modulation
from small to large grid biases will give rise to little distortion due to
the second and third harmonics. The behaviour of a pentode used as
an A.F. amplifier with variable gain cannot, however, be judged merely
from the published hum-modulation and cross-modulation curves
directly.
These curves are intended to apply when the pentode is used as R.F.
and I.F. amplifier, with the anode circuit, incorporating an impedance
having a low d.c. resistance, so that a normal screen-grid voltage can
also be used (e.g., 100 V). With A.F. amplification, on the other hand,
in order to get adequate amplification a high d.c. resistance will be
introduced in the anode circuit (resistance coupling), so that a low
screen-grid voltage will have to be employed (e.g., 40 V). The published
curves for hum-modulation and cross-modulation, however, apply only
to the screen-grid voltage quoted (for R.F. and I.F. amplification), and
in order to get an idea of the behaviour of the valve as A.F. amplifier
with variable gain it would be necessary to know the hum-modulation
and cross-modulation curves for the screen-grid voltage corresponding
to resistance coupling. Favourable hum-modulation and cross-modulation curves do, however, give an indication that the behaviour of a valve
as A.F. amplifier may be satisfactory.
It is more practicable to determine directly by measurement how a
valve behaves as a variable-gain A.F. amplifier.
Fig. 293 shows a circuit diagram where a pentode is used as A.F.
amplifier with resistance coupling. A study of the cross-modulation

and hum-modulation curves of existing H.F. valves with variable transconductance shows that the third harmonic, bearing a certain relation to the cross-modulation factor, as a rule is less prominent than the second harmonic, which bears a certain relation to the percentage of hum-modulation. Consequently in the designing of special valves for variable A.F. amplification it is of importance to obtain characteristics yielding a low distortion due to the second harmonic. It is then that recourse is had to the principle of variable screen-grid voltage. Such a valve, for example, is the pentode part of the EFM 1.

With variable A.F. amplification it should, however, be borne in mind that the range of the control is not so great as with R.F. and I.F. amplification. Generally one has to be content with a control ratio of 1 : 5 to 1 : 10.

CHAPTER XXIII

Automatic Volume Control

133. Object of Automatic Volume Control

The object of automatic volume control is to adjust automatically the amplification in a receiving set in such a way as to obtain, if possible, always the same volume of sound regardless of the strength of the incoming signal. A perfect volume control or fading compensation should therefore make it possible to compensate completely the differences in strength of signals induced in the aerial, and also the variations in strength which occur in the reception from a given station due to fading. In practice, however, automatic volume control is never perfect and in fact only capable of converting large differences in strength into small variations in volume of sound.

Theoretically automatic volume control should operate in such a way that the signal at the detector remains constant and of sufficient amplitude to drive fully the output stage with the A.F. amplification available, no matter what the strength of the signal may be. With a hand-operated volume control it is, in fact, possible to adjust the A.F. amplitude and, consequently, the volume of sound at will. Perfect automatic volume control, however, would require a very great maximum amplification in the set for the signals from the least powerful transmitters to come through in sufficient strength. In practice amplification in receiving sets is limited, so that it is only possible to reproduce with sufficient volume in the loudspeaker the signals from stations which produce a certain field strength at the place of reception.

134. The Principle on which Automatic Volume Control is Based

As already stated in connection with the detection of the R.F. or I.F. signals, across the leak resistor of the detector diode a direct voltage is produced, one end of the resistor becoming negative with respect to the other; the magnitude of this is dependent on the voltage of the carrier wave applied to the diode. In the case of carrier waves having an amplitude of more than one volt the direct voltage is practically proportional to the R.F. or I.F. voltage. If the cathode of the diode is earthed via the chassis of the set and the negative direct voltage is applied to the grid of a R.F. or I.F. amplifying valve, the amplification of the latter is dependent on the strength of the H.F. voltage fed to the diode. As the H.F. voltage on the grid of the controlled valve

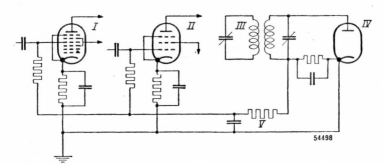

Fig. 294
Principle of automatic volume control where the negative
direct voltage produced across the leak resistor of the detector
diode is used as control voltage.
  I = frequency converter,
 II = I.F. amplifier,
III = I.F. band-pass filter,
 IV = detector diode and
  V = filter circuit for modulation.

increases, so the alternating voltage applied to the diode and hence
the negative grid voltage of the gain-controlled valve will increase. The
amplification of the gain-controlled valve decreases until a state of
equilibrium is reached between that amplification and the increase of
the I.F. voltage on the diode. It is on this principle that the systems
of automatic volume control usually applied in receiving sets are
based. In Fig. 294 a diagram is given showing the circuit for providing
automatic volume control.

It must be borne in mind that the I.F. voltage on the detector diode
must necessarily increase in order to reduce the amplification in the
R.F. and I.F. stages; given a constant A.F. amplification, the volume
of sound will in consequence also increase, although possibly to a much
smaller extent than would be the case without automatic volume
control. If, owing to a slight rise of the diode voltage, the amplification
in the preceding stages drops appreciably, the increase of sound volume
will be less than would occur if the same rise in voltage resulted in less
reduction of the amplification. In order to obtain effective automatic
volume control in receiving sets provision has to be made for the application of the gain-control voltage obtained from the diode resistor to
several valves. There is a drawback, however, in a highly effective
automatic volume control, in that the amplification increases very
greatly as the tuning is changed from one transmitter to another,
so that all the attendant noises are reproduced in great strength, between
the transmitter frequencies, which is very annoying when adjusting

323

the tuning.[1]) With a less effective automatic volume control, adjustment is made for an appreciable attenuation by the hand-controlled volume potentiometer when a transmitter signal of normal intensity is received; if the set is tuned to any other transmitter the maximum amplification in the set cannot be attained between the two transmitter frequencies because of the lower A.F. amplification.

For automatic volume control it is advisable to use valves with variable transconductance designed for this purpose, because the cross-modulation and modulation distortion over almost the entire gain-control range are less with these valves than with others designed for constant amplification, i.e. for a fixed operating point.

135. Delayed Automatic Volume Control

If the control voltage for varying the amplification of valves in the R.F., mixing and I.F. stages is obtained from the leak resistor of the detector diode (on which direct voltage is superimposed the A.F. modulation, which should be smoothed by means of a filter circuit—see Fig. 295), then the amplification of the set will undergo a reduction even with very weak incoming carrier waves. Should these carrier waves be so weak that the superimposed modulation is unable to drive fully the output stage, then a situation arises where the already inadequate sound volume is still further reduced by the automatic volume control.

For this reason the point at which automatic volume control is brought into operation is often shifted to a certain signal level, so that it does not begin to work until carrier waves are received which are of such an intensity that with a given modulation depth the output valve can be fully driven. (Usually the standard for modulation depth is taken as 30%.) In such a case one refers to **delayed automatic volume control.**

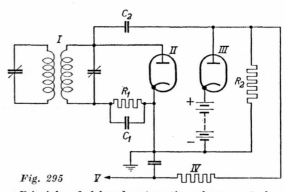

Fig. 295

Principle of delayed automatic volume control.
  I = I.F. band-pass filter,
 II = detector diode,
III = diode for A.V.C.,
 IV = filter circuit,
  V = lead to the gain-controlled valves,
C_1 and C_2 = reservoir condensers,
R_1 and R_2 = leak resistors.

---

[1]) In certain receiving sets a so-called silent-tuning or muting device is incorporated in order to prevent hiss between station tunings.

This so-called **delayed action** can be brought about, for instance, by applying a negative bias to the diode supplying the control voltage. For example, the anode is maintained at chassis potential, the bias then being obtained by applying a positive voltage to the cathode (see Fig. 295). As long as the greatest peak value of the I.F. signal is lower than the bias there will be no negative direct voltage across the leak resistor. Not until the positive peaks of the alternating voltage exceed the value of the negative bias will a current start to flow through the diode and leak resistor, thereby setting up across the latter (see Fig. 295) a negative direct voltage and bringing the automatic volume control into action.

It is not possible to apply a negative direct voltage to the detector-diode anode itself to attain the desired delayed action, because then with weak signals no detection would take place and the set would be inoperative. Moreover, with carrier waves exceeding the **delay voltage** there would be considerable distortion. For delayed automatic volume control it is therefore necessary to use a second diode solely for supplying the gain-control voltage, and since delayed automatic volume control is commonly applied nowadays valves are usually employed having two diode systems in one bulb.

The delay voltage must be such that with a modulation depth of 30% the modulation peaks on the carrier wave just exceed the bias of the diode at the moment that the I.F. voltage applied to the detector diode reaches a strength which allows the full grid drive of the output stage at that modulation depth. If such is the case with a carrier wave having a voltage of 10 V then at a modulation depth of 30% the delay voltage must therefore be equal to $10 \sqrt{2} \times 1.3 = 18.4$ V.

136. Connecting the Diode for Automatic Volume Control to the Preceding Band-pass Filter

Usually there is an I.F. band-pass filter in front of the detector. The diode for automatic volume control can be connected either to the primary circuit or to the secondary circuit of this filter (see Fig. 296), thereby influencing:

a) the selectivity of the band-pass filter,

b) the accuracy or sharpness of the tuning to a station either by ear or by means of a tuning indicator, and

c) the modulation distortion occurring as a result of the periodic variation of the load on the preceding band-pass filter at the frequency of the A.F. modulation or introduced by the distorted A.F. voltage across the leak resistor of the A.V.C. diode, both due to the delaying of the automatic volume control.

(a) Influence on the Selectivity of the Band-pass Filter

Selectivity of the set is influenced by the automatic-volume-control diode because, this being connected to the last I.F. band-pass filter, the damping of the diode influences the resonance curve (see Chapter XVIII).

With regard to the magnitude of that damping, the following is to be noted. Modern receivers are usually so designed that the voltage of the I.F. signal at the detector diode (d_2 of Fig. 296) is relatively high (giving the advantage of linear rectification). The a.c. equivalent resistance of this diode with its leak resistor then equals half the value of the leak resistor R_2 (see Fig. 296); if the leak resistance is, say, 0.5 megohm then the damping resistance for the diode circuit is 0.25 megohm. Since a delay voltage is applied to the automatic-volume-control diode d_1 the damping action of that diode is dependent on the strength of the incoming I.F. signal. If in the circuit of Fig. 296 the peak value of the modulated carrier wave applied to d_1 is lower than the delay voltage, then the oscillatory circuit to which d_1 is connected is damped only by the resistors R_1 and R_3 connected in parallel. For signal voltages several times higher than the delay voltage V_d the equivalent resistance due to damping by the diode d_1 can be calculated approximately in the following manner. The direct voltage on the resistor R_1 is approximately equal to the peak value of the unmodulated carrier wave V_o less the delay voltage V_d. Therefore the d.c. power absorbed by the resistor R_1 is equal to $\dfrac{(V_o - V_d)^2}{R_1}$. If, therefore, the diode circuit is to be replaced by a resistance R_{d1} the latter must absorb the same power.

Hence R_{d1} must satisfy the following conditions:

$$\tfrac{1}{2} \frac{V_o{}^2}{R_{d1}} = \frac{(V_o - V_d)^2}{R_1} \qquad (192)$$

or

$$R_{d1} = \tfrac{1}{2} R_1 \left(\frac{V_o}{V_o - V_d} \right)^2. \qquad (193)$$

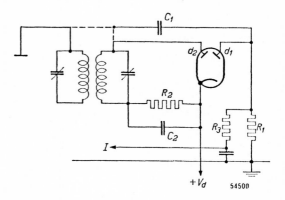

Fig. 296
Connection of the detector diode and the delayed-automatic-volume-control diode to the preceding I.F. transformer. The automatic-volume-control diode can be connected either to the primary or to the secondary circuit, both possibilities being indicated by broken lines.
I = lead to gain-controlled valves.

In addition to the d.c. power dissipated by R_1, the resistors R_1 and R_3 connected in parallel absorb also some a.c. power.

If, now, R_1 and R_3 each have, for instance, a value of 1 megohm, then with small I.F. voltages (less than the delay voltage V_d) the circuit of diode d_1 can be replaced by a resistance $R_1R_3/(R_1 + R_3) = \frac{1}{2}$ megohm, and with I.F. voltages very much higher than the delay voltage, where $\left(\dfrac{V_o}{V_o - V_d}\right)^2$ is about equal to 1, it can be replaced by a resistance R_{d1} which is the resultant of $\frac{1}{2}$ megohm (determined above) and $\frac{1}{2}R_1$ in parallel, i.e., $\frac{1}{4}$ megohm.

Theoretically, therefore, the damping produced by the automatic-volume-control diode may vary by a factor 2. For the sake of simplicity we may consider an average value of $^1/_3$ megohm when a value of 1 megohm is assumed for R_1 and R_3.

If the diode d_1 is connected to the primary tuned circuit then the secondary circuit of the band-pass filter is damped only by the detector diode d_2, the primary circuit being damped by the automatic-volume-control diode d_1 and the internal resistance of the preceding I.F. valve. If, on the other hand, the diode d_1 is likewise connected to the secondary circuit of the band-pass filter, the primary circuit will be hardly damped at all (only by the internal resistance of the preceding I.F. valve), while the secondary circuit will be damped by both the diodes d_1 and d_2. It will be evident that in the latter case the average quality factor of the two band-pass filter circuits, and thus also the selectivity, is better than when the diode for automatic volume control is connected to the primary circuit. This may be explained by the fact that in any case the secondary circuit is damped by the detector diode d_2, so that without connecting to it the diode for automatic volume control the quality factor of the secondary circuit is less than that of the primary circuit. The connection of the automatic-volume-control diode has less effect upon the reduced quality factor of the secondary circuit than on the higher quality factor of the primary circuit.

It may therefore be concluded that the selectivity of the band-pass filter is better when the automatic-volume-control diode is connected to the secondary circuit than when it is connected to the primary circuit.

(b) Influence on the Sharpness of Station Tuning

The accuracy with which a station is tuned-in is reduced by automatic volume control. This may be explained by the fact that when there is a certain detuning of the resonant circuits with respect to the signal frequency, the attenuation of the carrier-wave voltage on the detector that should occur according to the resonance curve is partly neutralized by the automatic volume control adjusting the amplification to a higher value. If the automatic-volume-control diode is connected to the primary circuit it has the tendency to keep the voltage across that circuit constant.

The primary circuit is followed by the secondary, as a result of which, when there is detuning, the I.F. voltage on the detector diode is attenuated according to the selectivity of that circuit, consequently the sharpness of tuning with this circuit will be better than that obtained when both diodes are connected to the secondary circuit. This applies not only to sets tuned by ear, but also to sets with visible tuning, since the tuning indicator, too, is usually controlled by the voltage on the secondary circuit.

(c) Influence on the Modulation Distortion

If the modulation depth of the I.F. voltage applied to the automatic-volume-control diode has such a value that the smallest amplitude of the carrier wave occurring during an audio-frequency cycle is lower than the delay voltage, then the damping of the connected oscillatory circuit due to the diode is periodically zero. As a result the I.F. amplification during an A.F. cycle will not be uniform, hence giving rise to modulation distortion. This distortion is greatest when the amplitude of the carrier wave is roughly equal to the delay voltage, because then during one half-cycle of each modulation cycle no current is flowing through the automatic-volume-control diode. In practice distortion up to 3% has been measured.

The modulation distortion occurring when the automatic-volume-control diode is connected to the secondary circuit is less than that occurring when the connection is made to the primary circuit, because the I.F. amplification is less affected by a variation of the damping of the secondary circuit by that diode, since that circuit is already damped by the detector diode. In that case, however, in addition to modulation distortion, another form of distortion occurs which is apt to assume far more serious proportions. Across the resistor R_1 (see Fig. 296), in addition to a direct voltage, there is an A.F. voltage, and when the amplitude of the I.F. voltage during part of the A.F. cycle is less than the delay voltage the wave form of that A.F. voltage will not correspond to the modulation but will be severely distorted. Part of this distorted A.F. voltage is applied via the condenser C_1 to the resistor R_2 (the impedance of the tuned circuit is negligible for low

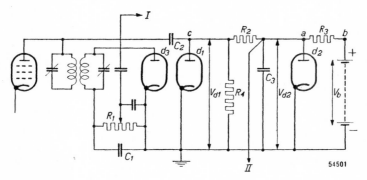

Fig. 297
Essentials of a triplex-diode circuit for delayed automatic volume control.
  I = lead to the grid of the A.F. amplifier,
 II = lead to the gain-controlled valves

frequencies). As a result the A.F. voltage across R_2 is distorted. This distortion may prove serious. When the automatic-volume-control diode d_1 is connected to the primary circuit the distorted A.F. voltage is not transferred to the secondary circuit and to R_2.

If the amplitude of the carrier wave applied to the automatic-volume-control diode is several times greater than the delay voltage then the detection by that diode will be interrupted only during part of the A.F. cycle when great modulation depths occur. Both kinds of distortion described here are much less under those circumstances, and for that reason it is of importance not to make the automatic volume control too effective. Otherwise after the delay voltage has been exceeded the carrier-wave voltage applied to the automatic-volume-control diode would increase very slowly indeed with an increasing signal and will be only a few times greater than the delay voltage with a very strong signal.

137. The Triplex-diode Circuit

The employment of a diode for delayed automatic volume control has certain drawbacks, some of which have been explained. From the view-point of quality of reproduction it would therefore be preferable to utilise a diode without delay voltage, but this would be attended by the disadvantage that the output stage is only fully driven when very powerful signals are induced in the aerial.

The ideal solution would be:

1) to connect the diode for automatic volume control to the primary circuit of the last I.F. band-pass filter;

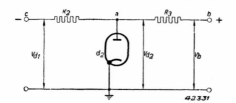

Fig. 298
Voltage-divider between positive potential V_b and negative potential V_{d1}, loaded by the diode d_2.

2) not to apply any bias to that diode, having regard to the distortion which would otherwise arise;

3) to obtain, notwithstanding, a delayed automatic volume control.

These conditions may be satisfied by employing three diodes in the circuit, termed the **triplex-diode circuit** (see Fig. 297, in which for the sake of clarity three separate diodes have been drawn).

The diode d_3 is used in the normal way as detector diode. The diode d_1 serves for rectification of the I.F. voltage for obtaining the automatic volume control; it is connected to the primary of the last I.F. transformer and has no delay voltage. The delay of the automatic volume control is effected by means of the diode d_2 [1]); the anode of this diode receives via the resistor R_3 a positive voltage of suitable magnitude, which in the diagram is represented by a battery voltage V_b.

Owing to this positive voltage on the anode of diode d_2 a current flows through that diode. The d.c. resistance of the diode being small, the anode (point a) is only slightly positive with respect to the cathode (or the earth lead to which the cathode is connected).

When a carrier voltage is applied to the diode d_1, as a result of the rectification in that diode a negative d.c. voltage is developed across the leak resistor R_4 (point c therefore becomes negative). Point a is then the junction point of a voltage divider R_3R_2 loaded by the diode d_2, connected between a positive potential V_b and a negative potential V_{d1} (see Fig. 298). As long as the latter potential is small the no-load voltage of point a will be positive and consequently current will flow through the diode d_2; if, on the other hand, that potential is large the no-load voltage at point a will be negative, and no current can flow through the diode d_2. In the latter case the potential at point a is directly dependent on the potential V_{d1}.

The voltage V_{d2} at point a is then determined by the equation

$$V_{d2} = V_b \frac{R_2}{R_2 + R_3} + V_{d1} \frac{R_3}{R_2 + R_3}. \tag{194}$$

If, now, the control voltage applied to the grids of the valves to be

---

[1]) These diodes are numbered in agreement with the purpose for which the three diodes of the valve EAB 1 are intended.

controlled is obtained from point a, automatic volume control is achieved. Therefore, as long as the negative voltage on the anode of d_1 is only small, with weak carrier waves, due to the positive voltage V_b, a current flows through d_2 and the automatic volume control does not operate, but when a certain value of the negative voltage V_{d1} is surpassed, that control comes into operation. Thus we have an automatic volume control that begins to function when the voltage across diode d_1 reaches a certain value, that is to say when the carrier wave has attained a certain voltage. Consequently this volume control is delayed.

The choice of the resistances R_2 and R_3 and of the voltage V_b in Equation (194) is governed by the amplitude of the carrier wave on the detector diode at which the automatic volume control is to come into operation. When this voltage is known then V_{d1}, being the amplitude of the unmodulated carrier wave across diode d_1, is known. Assuming that current is just prevented from flowing through d_2 when the negative voltage on that valve is 0.8 V, then Equation (194) may be written:

$$-0.8 = V_b \frac{R_2}{R_2 + R_3} + V_{d1} \frac{R_3}{R_2 + R_3}.$$

The triplex-diode circuit has yet another advantage over the application of a diode with a delaying bias. The control voltage obtained in the latter case with the aid of a diode is dependent to a rather large extent upon the modulation depth of the signal. This is particularly the case in sets having a very effective automatic volume control and in which only a small increase in the signal strength at the detector diode is possible after the automatic volume control starts to operate[1]). The result in the first place is that with a greater modulation depth, i.e., with greater sound volume, the control voltage is greater, thereby reducing the contrast between the loudest and softest sounds.

Much more disturbing, however, is the fact that the deflection of a fluorescent-screen indicator (for visible tuning), being operated by this control voltage, is dependent on the modulation depth. Consequently the indicator will begin to flicker, especially when there are strong

---

[1]) This can be explained as follows: Assuming, for instance, that the amplitude of the unmodulated carrier wave is roughly equal to the delay voltage, then current flows through the diode only during that part of the modulation period in which the amplitude of the carrier wave is greater than the delay voltage. Consequently the voltage across the leak resistance of the diode adjusts itself to an average value depending on the difference between the highest peak value of the modulated I.F. carrier wave and the delay voltage, which difference depends on the modulation depth.

passages in the music or speech. This flickering also takes place when the indicator is connected to the detector diode, because the strength of the carrier wave at the detector also depends on the modulation depth, due to the dependency of the control voltage on modulation depth.

Since in the triplex-diode circuit the control voltage is obtained without a delay voltage, the drawbacks mentioned do not arise.

Finally, attention is drawn to the purpose of the resistor R_4 in this circuit. At first glance it might be thought that there was no need for this resistor, considering that the direct current through the diode d_1 should be able to flow back via R_2 and R_3 in the external circuit to the cathode. If, however, this resistor were omitted the following difficulty would arise. If the amplitude of the carrier wave applied to d_1 increases, for instance when the set is tuned to a station, the condenser C_3 becomes charged by the diode d_1 via the resistor R_2, but when the strength of the carrier wave decreases, for instance upon detuning the set, C_3 has to discharge itself, through R_3. Since the resistance value of R_3 is always a multiple of that of R_2 (otherwise too small a part of the control voltage V_{d1} would be used effectively) C_3 can only discharge itself slowly. As the amplitude of the carrier wave decreases the control voltage drops too slowly and for a short time the set continues to be adjusted to the small sensitivity that corresponded to the original, greater amplitude of the carrier wave. This is all the more troublesome when the strength of the carrier wave changes quicker. If, therefore, a set is tuned to a powerful station and the tuning knob is quickly turned further, some less powerful adjacent stations will not be heard. This phenomenon is avoided by introducing the resistor R_4, for then C_3 is able to discharge itself sufficiently rapidly through R_2 and R_3.

138. Practical Version of Triplex-diode Circuits

Obviously a triplex-diode should be used for a triplex-diode circuit, and for this purpose Philips made a special valve, the EAB1, which can be connected up in the circuit as indicated in Fig. 297. The battery for supplying the voltage V_b, however, is replaced by a voltage-divider which reduces the voltage of the H.T. supply to the desired value. In many cases, however, it is desired to use diodes incorporated in amplifying valves (e.g., the diodes contained in the double-diode-triode EBC3 and/or the double-diode-output-pentode EBL1 or EBL21). We shall give some examples of circuits for these cases. Fig. 299 shows a diagram in which the valves EBC3 and EBL1 are employed, with the

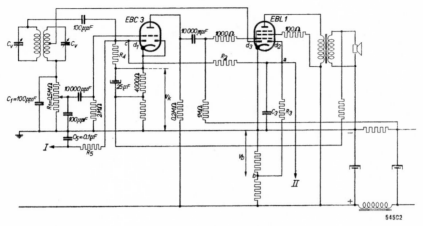

Fig. 299
Practical version of a triplex-diode circuit using a double-diode-triode and a double-diode-output-pentode. Of the four diodes available one is not used. This diagram shows also an example of negative feedback (see Chapter XXVII).
I = A.V.C. lead to the I.F. valve,
II = A.V.C. lead to the frequency-converting valve.

diodes, condensers and resistors indicated in accordance with the triplex-diode circuit of Fig. 297; here the control voltage for the mixer valve is taken from point a and that for the I.F. valve from point c (the latter is filtered by R_5 and C_5).

In this way one obtains a combination of a delayed and a non-delayed automatic volume control, since the control voltage to the I.F. valve is not delayed. This circuit also includes a negative feedback from the loudspeaker circuit to the grid of the A.F. preamplifying valve EBC 3 (see Chapter XXVII). Owing to the fact that for proper functioning the cathode of the EBL 1 has to be earthed, the negative bias of the pentode part of this valve is obtained from the voltage drop in the resistor in the negative H.T. line.

In analysing the action of this system it has to be borne in mind that the cathode of the diode d_1 is about 3 volts positive with respect to earth, which has to be taken into account when determining the carrier-wave voltage at which the automatic volume control begins to operate. If, for instance, the carrier-wave voltage at the instant when automatic volume control starts to operate is 10 V, the voltage between point c and earth is $3 - 10\sqrt{2} = -11$ V; therefore this is the value that has to be substituted for V_{d1} in Equation (194).

In the absence of a carrier-wave voltage on the diode d_1 the point c is positive with respect to earth. The potential of this point is then

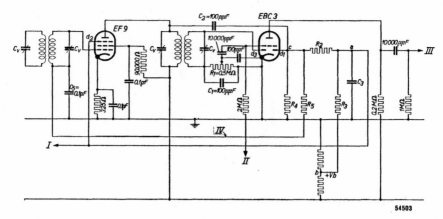

Fig. 300
Triplex-diode circuit where the suppressor grid of the I.F. valve is employed as third diode.
 I = A.V.C. lead to the gain-controlled frequency converter,
 II = lead to negative bias source of valve EBC 3,
 III = lead to grid of output valve,
 IV = A.V.C. lead to I.F. valve.

equal to $V_k \dfrac{R_2}{R_2 + R_4}$, because the cathode of the EBC 3 is positive
and point a is practically at earth potential (the diode d_2 is then
conductive and consequently its anode is nearly at cathode potential).
The grid of the I.F. valve is connected to point c via the resistance
R_5 and thus has the corresponding potential, so that in order to get the
correct negative grid bias the cathode voltage of this valve has to be
chosen 2 or 3 V higher than the potential of the grid (i.e., approximately 5 — 6 V).
Another practical version of the triplex-diode circuit is offered by
using as the third diode the suppressor grid of the I.F. amplifying
pentode. Fig. 300 shows a diagram for this purpose, where the pentode
EF 9 is used as I.F. amplifier and valve EBC 3 (double-diode-triode) as
A.F. amplifier.
Since in this case the cathode of valve EBC 3 has to be earthed, the
negative bias for this valve has to be taken from a resistor in the H.T.
negative line. The diodes of valve EBC 3 are used for detection and
automatic volume control (d_3 and d_1 of Fig. 297), whilst the suppressor
grid of the EF 9 serves as diode for delaying the automatic volume
control (d_2 in Fig. 297). In Fig. 300 the resistors and condensers are
again indicated in accordance with Fig. 297; the working of this system
therefore needs no further explanation. As shown in the diagram, to
the suppressor grid only a direct voltage and no I.F. or A.F. voltage

is applied. The control voltage of the I.F. valve EF9, as in the previous example, is again obtained from point c and thus not delayed, whilst the control voltage of the mixer valve and the R.F. valve, if one is used, comes from point a. In this way the I.F. valve receives an increased grid bias for even the weakest aerial signal, a factor that is favourable to reduction of noise.

Since the cathode voltage of the EF9 is 2.5 V positive with respect to the earth line, without a carrier-wave voltage on the diode d_1 the potential of point a equals that cathode voltage.

When analysing the action of the circuit this has to be taken into account, and for V_{d2} in Equation (194) we have to substitute 2.5 — 0.8 = 1.7 V. Furthermore it must be borne in mind that until the automatic volume control comes into operation point a is positive, and as this voltage is also applied to the grid of the mixer valve the cathode voltage of this valve has to be chosen correspondingly higher, in order to get the correct negative grid bias in the non-controlled state.

139. Control Curve of Automatic Volume Control

The action of the automatic volume control can be computed in advance fairly accurately. First it is necessary to know the amplification of the various gain-controlled valves as a function of the control voltage applied. For valves designed for this purpose curves are always published which give the transconductance on a logarithmic scale as a function of the negative grid bias. When the impedances of the coupling elements are known—as a rule, in the case of valves with high internal resistance, the amplification is equal to the product of the transconductance and the coupling impedance—the amplification of each stage can be calculated as a function of the control voltage. When this has been done then the amplification of the whole set—in so far as this precedes the diode for automatic volume control—can be plotted as a function of the control voltage. In addition to this curve for the stages preceding the diode for automatic volume control it is also necessary to have the curve indicating the manner in which the control voltage produced by the detector is dependent upon the carrier-wave voltage, whilst it must also be known to what extent the A.F. signal is influenced by the I.F. voltage on the detector; these data are given for all diode detectors in the form of characteristics.

With the aid of these three curves it is possible to construct point by point the curve representing the A.F. voltage at the loudspeaker as a function of the aerial R.F. voltage. It indicates the action of the

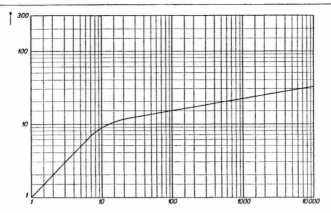

Fig. 301
Control curve of a set with delayed automatic volume control.
Vertical axis = times the audio-frequency voltage for an output power of 50 mW.
Horizontal axis = times the aerial voltage for an output power of 50 mW at a modulation depth of 30%.

automatic volume control and is therefore called the **control curve of the automatic volume control.**

Starting from a certain voltage on the loudspeaker and with the aid of the detector-diode curve indicating the relation between the A.F. voltage and the 30% modulated I.F. voltage, the corresponding I.F. voltage can be calculated. From the curve of the diode for automatic volume control, giving the negative direct voltage as a function of the I.F. voltage occurring simultaneously on that diode, it is possible to calculate this bias applied to the gain-controlled valves. Further the R.F. voltage on the aerial can be calculated from the curves for R.F. amplification, conversion amplification and I.F. amplification as functions of the direct voltage. Fig. 301 gives such a curve for a set with delayed automatic volume control, from which it can be seen that the loudspeaker voltage increases practically in proportion to the aerial voltage until the automatic volume control begins to operate, after which it increases much more slowly than the aerial voltage.

It is evident that with a large maximum amplification due to the part of the set preceding the detector the shape of the control curve is flatter than that where this amplification is small, at least if the same number of controlled valves with the same characteristics are used, because the signal at the detector is then amplified to a greater extent; in that case the variation in the direct voltage on the gain-controlled valves due to a given variation of the aerial signal will also be greater.

140. Amplified Automatic Volume Control

In cases where automatic volume control having a particularly effective action is desired it is possible to amplify the direct voltage supplied by the automatic-volume-control diode before feeding it to the grids of the gain-controlled valves. This is known as **amplified automatic volume control.** In normal receiving sets this is very seldom applied, because it not only makes the set more expensive but it is usually not desired. Fig. 302 gives an example of an amplified and delayed automatic-volume-control circuit, where the diodes of the EBC 3 are used for detection and to provide the delay of the automatic volume control, whilst the triode part of this valve serves for amplifying the control voltage set up across the leak resistance R_1 of the detector diode d_2 and used for the volume control (potential at point c). A variation of the direct voltage on the grid of the EBC 3 causes approximately 15 times as large a variation of the direct voltage across the cathode resistor R_3 (12,500 ohms) of that valve.

The potential difference between cathode and chassis (earth) is determined by the voltage drop across the cathode resistor R_3 and by the negative voltage with respect to earth at the junction point b between the resistors R_6 and R_7 in the negative H.T. line. In the absence of a carrier-wave voltage on the detector diode the cathode is positive with respect to earth. The anode of the diode d_1 is earthed via resistor R_4 and is moreover connected to the grids of the gain-controlled valves via a filter R_5C_5.

As long as the cathode is positive with respect to the chassis the automatic volume control is inoperative. As the carrier-wave voltage on the detector diode d_2 increases, the grid of the EBC 3 becomes more negative, the anode current drops and hence also the cathode voltage until, at a given instant, the positive voltage of the cathode with respect to the anode of d_1 has been so far reduced as to make that diode conductive. A current then starts to flow from the anode d_1 to the cathode and a voltage drop occurs across the resistance R_4. When the carrier-wave voltage applied to the detector diode increases still further, the cathode potential of the EBC3 becomes still more negative with respect to the chassis. The current flowing through the diode d_1 then increases still more and consequently the voltage drop across the resistance R_4 also increases. Point a thus becomes more and more negative, as also the grids of the gain-controlled valves.

Since the d.c. resistance of the diode d_1 is small compared with the resistance R_4, the potential at the point a will correspond approxi-

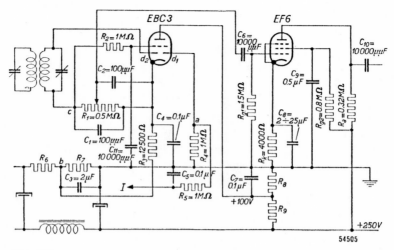

Fig. 302
Essentials of an amplified and delayed automatic-volume-control circuit.
I = A.V.C. lead to gain-controlled valves.

mately to the cathode potential. The voltage variation at the cathode, however, is a magnified reproduction of the voltage variation across the leak resistance R_1, so that the control voltage fed to the gain-controlled valves is an amplified reproduction of the direct voltage across the leak resistance of the detector diode.

As soon as the diode d_1 becomes conductive the automatic volume control comes into operation. In the absence of a carrier-wave voltage on the detector diode, the cathode voltage of the EBC 3 must be so chosen that the diode d_1 becomes conductive at that value of carrier-wave voltage for which the automatic volume control is required to operate. This voltage can be adjusted to the appropriate value with the aid of the negative voltage at point b.

141. Automatic Volume Control where the A.F. Amplification is also Controlled

The amplification of the control voltage for automatic volume control described above constitutes a means of obtaining a highly effective control. It will always be found, however, that when the control is confined to the valves preceding the automatic-volume-control diode a rise in the carrier-wave voltage is always accompanied by an increase, however small, of the A.F. voltage applied to the grid of the output valve (or an increase of the output power—see also the control curve of Fig. 301). If it is desired to keep the voltage on the grid of the output

338

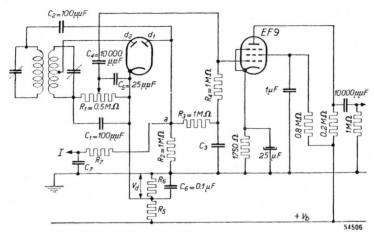

Fig. 303
Essentials of a circuit in which an A.F. amplifying valve, as well as the
mixer and I.F. amplifying valves, are controlled by automatic volume control.
I = A.V.C. lead to the I.F. and frequency-converting valves.

valve constant, it has to be arranged for the A.F. amplification to
decrease when the intensity of the aerial voltage increases.

If, for instance, with a constant A.F. amplification the A.F. voltage at
the grid of the output valve for a certain modulation depth of the
carrier wave increases by a factor 10 when the aerial voltage varies
from 100 μV to 1 V, then by reducing the A.F. amplification by a
factor 10 the same sound volume can be obtained with a strong signal
as with a weak one. This reduction of the A.F. amplification can be
attained automatically by using as an A.F. amplifier a valve with
variable gain and controlling that valve with the voltage supplied by
the automatic-volume-control diode.

By means of this A.F. gain control it is thus possible to obtain an
extremely effective automatic volume control, theoretically perfect (it
is also possible to obtain an excessive control which gives a reduction
of output power due to an increase of aerial voltage). The desirability
of having such a highly effective action, however, has to be investigated
for each individual case.

Fig. 303 shows the essentials of an automatically controlled A.F.
amplification circuit combined with a double diode for detection and
for delayed automatic volume control.

The A.F. alternating voltage across the leak resistor R_1 of the
detector diode d_2 is fed to the grid of the A.F. amplifying valve EF 9
via the condenser C_4.

The diode d_1 provides delayed automatic volume control; the delay voltage V_d is obtained by means of a voltage divider R_5R_6. The direct voltage produced across the leak resistor R_2 between point a and the earth line is fed to the grid of EF 9 via a filter circuit R_3C_3 (in order to smooth the A.F. and I.F. alternating voltages) and a leak resistor R_4 (necessary because otherwise the grid would be short-circuited by C_3 so far as the A.F. alternating voltage applied through C_4 is concerned).

The control voltage for the R.F. valve, the mixer valve and the I.F. valve can also be taken from point a, either by means of the same filter circuit R_3C_3 or via a separate filter as indicated in Fig. 303 by R_7C_7. The latter alternative may be necessary, for instance, when it is desired to make the gain control on the A.F. valve less effective than that on the R.F. valve, mixer valve and/or I.F. valve. In that case the control voltage for the A.F. valve will not be obtained from point a but from a junction on the leak resistor R_2, and each control voltage will have to be smoothed separately.

142. Note Concerning the Voltage across the Leak Resistor in the Absence of a Carrier Wave

Finally, it is to be noted that when employing a diode without delay voltage a negative voltage of about -0.8 V is obtained across the leak resistor of the diode before there is any signal. This voltage is also applied to the grids of the gain-controlled valves, thereby increasing their negative grid voltage in the non-controlled state. In order to utilise the maximum amplification obtainable with the valves it is then necessary to choose the cathode resistance proportionately lower.

CHAPTER XXIV

Noise of Amplifying Valves

143. General Considerations Regarding Noise and in Particular the Noise of Valves [1])

Noise is a well-known phenomenon which is often disturbing in radio reception and frequently has an adverse effect upon the sensitivity of the receiver. Noise arises mainly in the first stage of the set; the aerial and the circuits preceding the first valve, and also the first amplifying valve, all produce a share.

Noise can actually be ascribed to two causes:

1) arbitrary movements of the electrons (thermal agitation) in different current paths;

2) irregularity in the transition of the electrons from the cathode to the anode of the receiving valve.

At all points where an electric current may be produced under the influence of an electromotive force the electricity is always spontaneously in motion, even in the absence of such an e.m.f., because the electrons making conduction possible take part in the thermal agitation of the atoms (see Chapter II), that is to say, they possess a thermal velocity of their own (see Chapter IV). These thermal electron movements are quite irregular, with the result that on the average over short intervals there is a surplus of electron movement in one direction or another; one refers to these as **thermal current or voltage fluctuations.**

When, under the influence of a sinusoidal e.m.f., for instance, a regular alternating swarm movement of particles of electricity is superimposed upon the aforementioned, unavoidable, irregular movement, then, regardless of the amplification applied, that regular movement can only be distinguished if it is of sufficient magnitude compared with the irregular movement. Interference through noise therefore occurs in the first place when the signal is weak and the swarm movement due to the alternating e.m.f. is no longer great compared with the irregular movements causing the noise.

---

[1]) See also M. Ziegler, Philips Technical Review **2** (1937), p. 136, **2** (1937), p. 329 and **3** (1938), p. 189.

The intensity of the thermal fluctuations corresponds to the thermal energy kT [1]), in which k is Boltzmann's constant and T the absolute temperature, and is independent of the magnitude of the elemental charge, that is to say, it would be just as great if that charge were infinitely small.

As already stated in Chapter II, the particles of electricity (electrons) have a charge of 1.6×10^{-19} coulomb. When in a vacuum valve, due either to photo-emission or to thermal emission, electrons emerge from an electrode independently of each other and pass over to another electrode independently of each other, then the number of electrons passing over within a certain time is finite even when there is a large current flowing through the valve; that number, however, is not always the same but varies in a random manner around an average number. The corresponding current fluctuations are governed by the elemental charge e. The voltage arising in the two cases mentioned above are directly proportional to the irregular current and after amplification cause **noise**. The phenomenon is called **hail effect**, because the noise produced by it in a loudspeaker resembles that of falling hail-stones. It is mainly the circuit and the valve preceding the greatest amplification—i.e., the input circuit and the first valve of the set—that cause noise disturbance. The share of the first valve in this noise is particularly large in short-wave reception. In many cases the first valve in a set is a mixer valve, but as noise is greater with a mixer than with a H.F. pentode many of the more expensive sets are equipped with a R.F. pre-amplifying valve.

Within the scope of this book we are interested primarily in the noise caused by the valves. The simplest case of noise arising in a valve is found in the saturated diode, and we will therefore consider this first. In a heated cathode both the molecules and the electrons are in an excited state of motion and in consequence of the emission described in Chapter IV electrons emerge occasionally into the evacuated space surrounding the cathode. As soon as an electron has left the cathode it is attracted by the positive anode and travels in that direction. If the anode voltage is high enough this is repeated for every electron that happens to emerge from the cathode.

As long as the emerging electrons are drawn away, so that at any moment it is left to chance when other electrons emerge, the average magnitude of the fluctuations occurring can be determined from the theory of probability. The fluctuations will be greater when the average

[1]) See Eq. (199); the meaning of k is that the average kinetic energy of a free particle participating in the heat movement amounts to $^3/_2$ kT.

current is greater, and smaller as the current is of a "finer structure" (according as it shows more resemblance to a liquid). The fineness of structure of a current is determined by the charge transmitted per electron; this magnitude occurs in fact in the formula for noise fluctuations.

The conditions as described for the saturated diode change when the anode voltage is reduced, because then not all emerging electrons are drawn away. The negative space charge formed around the cathode in that case returns part of the electrons to the cathode, the extent to which this takes place being dependent on the magnitude of the space charge. Where the space charge is great more electrons are returned than is the case with a small space charge, and the more electrons pass from the space charge to the anode the greater is the number of electrons emitted from the cathode. In this manner a state of equilibrium is obtained which in a certain respect may be compared to the conditions in a steam boiler: as soon as a certain steam pressure is reached no more steam is released from the water until steam is allowed to escape from the boiler.

Owing to the joint action of the electrons in the space charge, chance is largely eliminated. As soon as a current of electrons slightly greater than the average emerges from the cathode and the space charge therefore becomes greater, further electrons are repelled to a larger extent. A momentary excessive emission is immediately compensated by a reduced intake by the space charge and vice versa, so that, to put it in other words, the effect of the fluctuations is compensated. From the foregoing it will be evident that in consequence of the space charge the noise with an unsaturated diode will be less than that with a saturated diode, as is confirmed by measurements.

Though noise is very much reduced by the space charge it is not entirely eliminated, because no control that is based on reaction can be perfect. For the control described a variation of the space charge is essential, though it may be exceedingly slight, and such variations naturally manifest themselves in the anode current.

In triodes, too, a negative space charge is produced around the cathode. With a negative control-grid voltage the cathode emission of a triode is not saturated, so that the noise with a triode is much less than that with a saturated diode (with equal current); the effective value of the noise component of the current is about five times as small, though of course this is only an approximation and in practice that value may vary considerably.

In the case of valves having more than one grid the noise may be fairly great; such valves are usually employed in the first stages of receiving sets. One would be inclined to think that the noise of a valve with more than one grid would be much about the same as that of a triode, but that is not so. It is true that the negative space charge in these valves also has a compensating effect, but chance again influences the electrons on their way to the anode. In the path of the electrons are the wires of one or more positive grids which absorb some of the electron current before the electrons reach the anode. It is again a matter of chance which and what number of electrons reach the positive screen grid of a pentode per unit of time. Reckoned over a longer period one may speak of an average current, the screen-grid current, but between successive equally short periods of time there are haphazard fluctuations. There is here no compensating action, for the potential of the screen grid remains unchanged, no matter whether a large or a small quantity of electrons has just been attracted to it. Consequently the screen grid of a pentode absorbs a varying amount of the almost constant current coming from the space charge, so that a varying amount finds its way to the anode.

In practice it is found that this so-called current-distribution noise (a result of the current distribution between the various electrodes in the valve) may be 8 to 9 times as strong as the noise already present in the space-charge-limited current. It is therefore of importance to suppress the current-distribution noise in a pentode, and to do this the screen-grid current evidently has to be kept small.

Consequently, with a view to producing low-noise R.F. amplifying valves, means were sought to reduce the screen-grid current, which may be done in various ways. In the Philips valve EF 8 between the control grid and the screen grid an extra grid with the same potential as the cathode has been introduced. This grid has the same pitch as the screen grid and is mounted in such a way that the windings of both grids lie exactly one behind the other. The electrons are concentrated by the second grid, so that they just shoot through between the windings of the third grid (screen grid). The third grid, it is true, produces a positive potential plane to draw the electrons through the first, negative, grid, but as the second grid concentrates the electrons in the meshes fewer of them collide with the screen grid than would otherwise be the case without that extra grid. By applying the principle of concentrating the electrons it has been possible to reduce the screen-grid current I_{g2} to 0.2 mA, or $1/40$th of the anode current (8 mA); with the EF 9 the screen-grid current is 1.7 mA, or $2/7$ths of the anode current (6 mA).

144. Influence of Reproduction Curve

Noise arises in a valve from the quite irregular fluctuations of the anode current, whilst in a resistor it is due to the random movements of the electrons in motion, which likewise produces an irregular current. In both cases the irregular fluctuations may be regarded as the sum of a large number of sinusoidal alternating currents of all possible frequencies.

Just as, in consequence of the selectivity of a receiver, only those signals picked up by the aerial which fall within a narrow frequency band are amplified, so only a part of the frequency spectrum of a noise source in the input stage of a receiver will be amplified. The R.F. part of the set allows only those R.F. noise components to pass through which have frequencies not differing too widely from the resonant frequency of the R.F. circuits. In the mixer valve these R.F. noise components are transformed into I.F. noise components, part of which are allowed to pass through the I.F. part of the receiver. When a carrier wave of a received signal is applied to the detector these I.F. noise components are detected and A.F. noise components occur, which after further amplification become audible in the loudspeaker; the A.F. noise components produced in the detector correspond to the difference between the frequencies of the I.F. noise components and the frequency of the carrier wave.

The effective value of a noise current or noise voltage can be measured, that is to say the root-mean-square value of that current or voltage can be found, for instance, by the heat produced by a noise current in a resistor. From the foregoing it follows that the magnitude of the noise depends upon the width of the frequency band passed through the circuits following the source of the noise.

Now from the theory of alternating currents it is known that if a current is composed of two components of different frequencies the square of the effective value of that current equals the sum of the squares of the effective values of the components. Therefore, if the frequency band passed through the circuits after the source of the noise is made twice as wide without altering the amplification, the square of the effective value of the noise will also become twice as great. Consequently as a rule the square of the effective value of the noise will be proportional to the width of the frequency pass-band. In practice, however, it is only rarely that the term "pass-band" may be employed, because actually a reproduction curve is not rectangular but follows a curved line. The curve can, however, be replaced in imagination by a somewhat narrower, purely rectangular band which, as regards noise, is equivalent to the actual curve (see Fig. 304). We shall call this band-width the **effective band-width.**

145. Intensity

It is not only in the addition of two noise currents of different frequencies that we have to add the squares of the effective current values of both components, but also in the addition of two mutually independent noise currents of equal frequency this has to be done. When two currents of equal frequency are in phase we have to add them directly, and when they are in antiphase one has to be deducted from the other. When, however, they have an arbitrary phase difference, as is the case with two mutually independent noise currents, we may regard the phase difference as being on the average equal to 90°, so that again we have to add together the squares of the currents in order to find their resultant. By this means we can determine in what manner the noise current is dependent upon the current intensity.

When considering the noise of valves it is usual to start from the simplest case, the saturated diode, and to compare other cases with that. Let us suppose that two diodes are connected in parallel with each other, one of which has an anode direct current I_1 and the other an anode direct current I_2. This parallel connection then yields a resultant anode direct current $I = I_1 + I_2$, which is the same as when the current through a diode is increased in some way or other. Corresponding to the anode direct currents I_1 and I_2 are noise currents I_{N1} and I_{N2} respectively. Now the squares of the noise currents of the two diodes connected in parallel have to be added up and from that sum the square root has to be calculated to find the total noise current I_{Nt}. Thus we obtain for the noise current of two diodes connected in parallel the equation:

$$I_{Nt} = \sqrt{I_{N1}{}^2 + I_{N2}{}^2} \quad (195)$$

If $I_1 = I_2$ then also I_{N1} will be equal to I_{N2}, which corresponds to the connecting of two similar diodes in parallel. In that case the

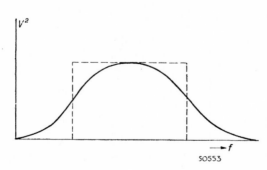

Fig. 304
Full line: Shape of a reproduction curve. The broken-line rectangle represents a curve with a somewhat narrower band, which, as regards noise, is equivalent to the actual curve.

anode current becomes twice as large, but the total noise current becomes:

$$I_{Nt} = \sqrt{2\,I_{N1}{}^2} \text{ or } \sqrt{2\,I_{N2}{}^2} = I_{N1}\sqrt{2} \text{ or } I_{N1}\sqrt{2} \qquad (196)$$

From this it follows that when the anode current is doubled the noise current is $\sqrt{2}$ times as large, or, in other words, the square of the noise current is proportional to the anode direct current. Usually the noise current in valves is reduced by the space charge around the cathode, so that in practice account has to be taken of a certain factor which depends on the construction of the valve; in low-noise valves this factor is smaller than in normal H.F. pentodes such as the EF 9 and EF 22. In this manner we arrive at the following simple formula for the noise current

$$I_N{}^2 = F^2 \cdot 2\,I_a eB, \qquad (197)$$

in which I_a = anode current,
I_N = noise current
e = charge on an electron,
B = band-width in c/s of the amplifier following the valve,
F = a factor related to the construction of the valve, which is always less than unity.

The noise currents in the anode circuit may also be regarded as arising from equivalent noise voltages on the grid. These equivalent noise voltages are equal to the noise currents in the anode circuit divided by the transconductance of the valve, thus $V_N = \dfrac{I_N}{g_m}.$.

Consequently

$$V_N{}^2 = F^2 \cdot 2eB\,\frac{I_a}{g_m{}^2} \text{ or } V_N = F\sqrt{2eB}\,\frac{\sqrt{I_a}}{g_m} \qquad (198)$$

The equivalent noise voltage V_N at the grid of a valve is mainly of importance for comparing the signal amplified in the valve with the noise arising in the anode current, which will be reverted to in Section 147.

146. Comparison between the Noise of Valves and that of Circuits

The noise of a receiving set is not caused by the valves alone. The thermal fluctuations in resistors and oscillatory circuits are likewise responsible in a large degree for the total noise. The irregular thermal movements of the electrons in a resistor may be regarded as being due to an electromotive force in series with the resistance (this e.m.f. is independent of the kind of resistance). The thermal fluctuations in the resistor therefore correspond to voltages between the extremities

of the resistor. The irregular fluctuating noise voltage becomes greater
as the value of the resistance is increased. As may easily be deduced [1]),
the square of the noise voltage is proportional to the value of the resistance. The noise caused by an oscillatory circuit has the same value
as that caused by a pure d.c. resistance of the same value as the impedance of the oscillatory circuit at resonance, and it is this noise particularly that is of importance in normal receiving sets. The formula
applying to the noise voltage at the terminals of a resistance is:

$$V_N{}^2 = 4kTRB, \tag{199}$$

where V_N = noise voltage in volts,
   k = Boltzmann's constant (1.38×10^{-23} joules/degree),
   T = absolute temperature in degrees,
   R = resistance in ohms,
   B = band-width in c/s.

By substituting for T the value 293 °K (20 °C) we obtain

$$V_N{}^2 = 1{,}600 \ RB \cdot 10^{-23}. \tag{200}$$

With a band-width of 10,000 c/s and a resistance of 100,000 ohms the
noise voltage amounts to 4 μV.
In order to compare easily or add together the noise due to valves
and that due to the circuits, the valve noise is often represented by
an equivalent resistance in the grid circuit having a noise voltage
equal to the equivalent noise voltage at the grid of the valve. A valve
causing strong noise can therefore be represented by a large resistance
and a valve causing little noise by a small resistance. The value of
this resistance is called the **equivalent noise resistance** of the valve
(symbol R_{eq}).
The value of the equivalent noise resistance is found from Equations
(198) and (199) by putting $V_N{}^2$ of Equation (198) equal to $V_N{}^2$ of
Equation (199) and replacing R in Equation (199) by R_{eq}, thus

$$R_{eq} = \frac{F^2 e}{2 \, kT} \cdot \frac{I_a}{g_m{}^2} = C \frac{F^2 I_a}{g_m{}^2}. \tag{201}$$

From this it is to be deduced that the equivalent noise resistance R_{eq}
is directly proportional to the anode current and inversely proportional
to the square of the transconductance. Consequently this resistance
depends on the operating conditions of the valve; it increases· with
rising negative grid bias.

---

[1]) Vide Philips Technical Review **2**, 1937, page 139.

When the noise resistances for different cases have been determined the noise voltages are found to bear the same relations to each other as the square roots of the relations between the corresponding noise resistances, since according to Equation (199) the noise voltage for a resistance is proportional to the square root of the value of that resistance.

147. Relation between Signal Strength and Noise

The interference from noise is governed by the relation between the magnitude of the noise and the strength of the signal picked up.

The simplest way to compare the noise with the signal is to substitute for the noise an equivalent voltage V_N on the grid of the valve and to determine the relation between that voltage and the signal voltage V_i. This is called the **noise-signal ratio** (V_N/V_i) and serves as a measure for judging the quality of reception with respect to noise.

From Equation (198) it appears that the equivalent noise voltage on the grid of a valve is directly proportional to the square-root of the anode current and inversely proportional to the transconductance. If the negative grid bias is increased then according to Equation (198) the equivalent noise voltage increases, since the transconductance decreases to a greater extent than the square-root of the anode current. This is also expressed by Equation (201), where it was already apparent that the equivalent noise resistance is dependent on the operating conditions of the valve. The relation between the noise resistance and the noise voltage is determined by Equation (200), and it is therefore obvious that both are dependent on the operating conditions of the valve.

Very often an increase of the negative grid bias is accompanied by an increase of the signal. Such is the case with variable-μ valves for automatic volume control, and then the noise-signal ratio may become more favourable, notwithstanding the increased equivalent noise voltage.

From Equation (198) it is also to be seen that, apart from the magnitude of the factor F, valves having a low anode current and high transconductance must consequently have a favourable noise-signal ratio. In the case of mixer valves the conversion transconductance is relatively low, so that with these valves a less favourable noise-signal ratio is to be expected than with normal H.F. pentodes.

Since the noise that arises in an amplifying stage is due not only to hail effect but also to thermal fluctuations in resistors and oscillatory circuits in the grid circuit, as explained in Section 146, a practical method of analysis is to represent the total noise by a noise resistance. In order, however, to determine the noise-signal ratio it is necessary to know the total noise voltage. This can be calculated from the total noise resistance with the aid of Equation (200) for a certain temperature (room temp.) and a given effective band-width (see Fig. 304).

The equivalent noise resistance of the valve EF 8, for instance, amounts to 3,000 ohms. If, now, the grid circuit impedance has a value of 4,000 ohms, then the total noise resistance is equal to 7,000 ohms, which, according to Equation (200), with an effective band-width of 4,000 c/s corresponds to an equivalent noise voltage of

$$V_N = \sqrt{1600 \times 7000 \times 4000 \times 10^{-23}} = \text{about } 0.67 \ \mu V.$$

If a noise-signal ratio of 1 : 1,000 or better is desired the signal on the grid of this valve must therefore have a voltage of at least 670 μV.

In practice it may sometimes be found easier to have a rule-of-thumb formula to avoid first having to determine the absolute noise voltage to calculate the signal voltage for a given noise-signal ratio that guarantees "noise-free" reception.

Now from measurements made on normal receiving sets it has been established that with a total noise resistance R_{eq} a signal of at least E volts is required to get "noise-free" reception, according to

$$E = 10^{-5} \sqrt{R_{eq}}. \tag{202}$$

This corresponds to a noise-signal ratio of about 1 : 1,200.

When an EF 8 valve is used on a wavelength of 15 metres and with a circuit impedance of the order of 10,000 ohms, i.e. with a total noise resistance of 13,000 ohms, the reception would be free of noise with a signal of

$$E = 10^{-5} \sqrt{13,000} = \text{approx. } 1.14 \ mV.$$

This, therefore, is the signal that should be present on the grid of the EF 8. A signal that is about one-seventieth of that strength is still just intelligible.

The use of a low-noise H.F. valve is in practice only of value for short-wave reception. For other wave bands the circuit impedance preceding the valve is so great (e.g., 100,000 ohms) that the noise of the first stage is caused almost exclusively by the thermal movement in the circuit. If the equivalent noise resistance of 15,000 ohms of a normal R.F. valve were reduced to that of about 3,000 ohms for a valve practically free of noise this would be hardly noticeable. Generally it may be said that the equivalent noise resistance of the first valve in a receiving set must be small compared with the impedance of the input circuit connected to the grid of that valve. For ultra-short-wave reception, where the circuit impedances are usually very small, valves will therefore be sought which have a very low equivalent noise resistance, and it is also for this reason that special ultra-short-wave valves are constructed which for the lowest possible anode direct current possess the highest possible transconductance [1]). Examples of such valves are the pentode **EF 50** with an equivalent noise resistance of about 1,400 ohms and the double pentode **EFF 51**, of which each pentode system has an equivalent noise resistance of 600 ohms.

---

[1]) So that the quotient $\dfrac{\sqrt{I_a}}{g_m}$ or $\dfrac{I_a}{g_m^2}$ is small — vide Equations (198) and (201).

CHAPTER XXV

Short-wave Properties of Amplifying Valves

148. The Damping of the Oscillatory Circuit by Valves and other Circuit Elements Connected in Parallel to it

The selectivity and the amplification of a R.F. or I.F. amplifier are not solely determined by the magnitude of the resistance, the inductance and the capacity of the oscillatory circuits employed, but also by the attenuation resulting from all the elements connected in parallel with those circuits (valves, aerial, etc.). If a resistance is connected in parallel with an oscillatory circuit the resultant impedance will be smaller than the impedance of the circuit alone. As already observed, the impedance Z_0 of an oscillatory circuit at the resonant frequency is determined by the values of r, L and C, according to the formula

$$Z_0 = \frac{L}{rC}. \qquad (203)$$

If, now, the oscillatory circuit is connected in the anode circuit of a valve with an internal resistance R_a, then that internal resistance is in parallel with the circuit and the resultant impedance Z_{res} is found according to the formula

$$\frac{1}{Z_{res}} = \frac{1}{Z_0} + \frac{1}{R_a}. \qquad (204)$$

Usually the alternating voltage between the terminals of the oscillatory circuit is applied to the grid of the next amplifier valve and in order to isolate the anode direct voltage from the grid of that valve a coupling condenser is employed. In that case a d.c. path has to be provided between the grid and the cathode, and a leak resistor usually serves this purpose. In order to minimize the consequences of the negative grid current (see Chapter XXX, Section 184), this leak resistance must not be of too

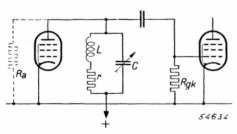

Fig. 305
Damping of an oscillatory circuit connected between two valves by the internal resistance R_a of the preceding valve and the grid-leak resistance R_{gk} of the succeeding valve.

352

high a value, and its effect cannot, therefore, always be ignored [1]).
Fig. 305 shows the method of connecting up an oscillatory circuit
between two valves. The resultant impedance of this circuit with the
parallel connected internal resistance R_a of the preceding valve and the
grid-leak resistance R_{gk} of the following valve can be calculated from
the formula

$$\frac{1}{Z_{res}} = \frac{1}{Z_o} + \frac{1}{R_a} + \frac{1}{R_{gk}}. \tag{205}$$

The resultant impedance is thus smaller than the impedance Z_o of the
circuit alone, and this reduction of the impedance has an effect on the
resonance curve.

In Fig. 306 the resonance curve of the circuit alone is represented by
a broken line, whilst the full line represents the resonance curve of the
circuit together with the additional attenuation due to the resistances
connected in parallel with it. On account of the extra damping, the
peak of the full-line resonance curve is lower than that of the broken-
line resonance curve.

The amplification A of the preceding valve equals the product of the
transconductance g_m and the impedance in the anode circuit Z_{res}. Thus
we have

$$A = g_m Z_{res}. \tag{206}$$

Consequently there exists on the grid of the following valve an alternating voltage $g_m Z_{res}$ times
greater than the voltage at
the grid of the preceding
valve. The greater Z_{res}, the
greater is the amplification.
Therefore the extra damping
influences the amplification.
Besides the internal anode
resistance R_a and the leak
resistance R_{gk}, the anode
capacity of the preceding
valve and the grid capacity
of the following valve are also
in parallel with the oscil-

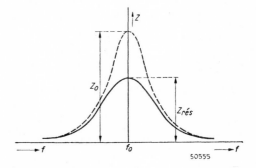

Fig. 306
Broken line: Resonance curve of the oscillatory
circuit without additional damping.
Full line: Resonance curve of the same circuit
with additional damping.

---

[1]) It must also be considered that in the high-frequency range the resistance diminishes
with increasing frequency. Therefore a 1-megohm resistor may have a ten times lower
value at 1 Mc/s.

latory circuit of Fig. 305. Since these are also in parallel with the tuning condenser the tuning capacity C is increased, so that this has to be chosen correspondingly lower to keep the oscillatory circuit exactly in tune. Furthermore the insulation resistances of the anode and of the grid are in parallel with the oscillatory circuit; in the medium and long-wave bands these are generally so great as to have no influence upon the resultant impedance of the circuit. In the case of normal H.F. amplifying pentodes the internal resistance and the maximum allowable grid-leak resistance are relatively high compared with the usual impedance values, so that for the amplification of such valves we have, approximately,

$$A = g_m Z_o \tag{207}$$

(here Z_o is the impedance of the circuit alone).

In the short-wave band, however, the amplification is by no means so great as would be expected from the above equation, because the input and output resistances of the valves for very high frequencies are reduced considerably and as a result cause considerable damping of the oscillatory circuits. When normal receiving valves are used for very short waves these resistances may be so low that the valves no longer produce any worthwhile amplification. Consequently for reception of very-short-wave stations special valves are constructed with high input and output resistances or large transconductance on those wavelengths (e.g., the twin pentode EFF 51, the pentode EF 51 with two cathode-input leads and the so-called acorn valves).

149. Properties of Valves at High and Very High Frequencies [1]

The most important electrical quantities of valves are:

(a) input impedance (between grid and cathode),

(b) output impedance (between anode and cathode),

(c) transconductance (from grid to anode),

(d) feedback impedance (from anode to grid).

These four quantities are quite sufficient even in the short-wave range for calculating the performance of a radio valve in an amplifying stage. The input impedance affects the value of the signal applied to the grid; the transconductance and output impedance, together with

---

[1] See the relevant publications by M. J. O. Strutt quoted at the end of this book, also C. J. Bakker, Some Properties of Receiving Valves on Short Waves, Philips Techn. Review, 1, 1936, p. 171.

the impedance in the anode circuit (single circuit or band-pass filter) influence the magnitude of the output signal and thus the amplification, whilst the feedback impedance is an indication of the influence of the output alternating voltage upon the input alternating voltage and thus the amplification.

The input and output impedances may be regarded as constituting a parallel circuit formed of a resistance—the input or the output resistance, sometimes called the grid or anode resistance of a valve—and a capacity—the input or the output capacity. In the short-wave band the grid and anode resistances are much smaller than in the long and medium-wave bands. The lower grid resistance is to be ascribed to three factors:

1) transit time of the electrons (phase lags due to the transit times of the electrons between the electrodes),

2) inductances of lead-in wires of electrodes and inductive and capacitive couplings between the lead-in wires themselves, and

3) dielectric losses in the insulating materials.

The lower anode resistance is to be ascribed to causes 2) and 3) above concerning the grid resistance.

The considerable feedback from the anode to the control grid in the short-wave band is primarily due to the same causes as produce the lower anode resistance, i.e. inductances of the lead-in wires, mutual inductances and capacities between lead-in wires and electrodes.

(a) Influence of the Transit Time of Electrons upon the Input Resistance

So far it has been assumed that there is no difference in time between the voltage variations on the electrodes and the resultant current variations. Such is the case, however, only when the transit time of the electrons in the valve is negligible compared with the periodic time of the voltage variations. By transit time is to be understood the time an electron takes to move from one point to another in the vacuum of the valve. This transit time will always be of a finite value, though extremely small, and for normal frequencies is of no significance, but when it is of the order of the periodic time of the signal to be amplified it gives rise to an input resistance which causes additional damping of the circuit connected to the control grid.

The transit time causes a phase change of the current to an electrode with respect to the current that would flow if the transit time were infinitely short. Normally the input impedance is purely capacitive, so that between grid and cathode there is only an alternating current which leads by 90° on the voltage. As a result of the transit time this phase lead is reduced. Consequently the alternating current between control grid and cathode acquires a component that is in phase with the voltage, i.e. a wattfull component, which means that an input resistance that causes damping of the grid circuit occurs. According to Bakker and Strutt this input resistance R_{in} due to transit time is approximately of the order:

$$\frac{1}{R_{in}} = f\,g_{mk}\,(\omega\,t_{kg})^2, \tag{208}$$

in which f is a factor approximately equal to $^1/_{10}$, g_{mk} the transconductance of the total cathode current with respect to the grid voltage, ω the angular frequency of the input H.F. voltage, and t_{kg} the transit time of the electrons between cathode and control grid.

From Equation (208) it follows that the input resistance is approximately inversely proportional to the square of the frequency.

(b) Influence of Inductances of the Lead-in Wires and of Inductive and Capacitive Couplings between those Wires upon the Input and Output Resistances

The input and output resistances are considerably influenced by the inductances and capacities of the lead-in wires, the values of which are, it is true, very small but at high frequencies can no longer be ignored. We will first consider the input resistance of a pentode. The H.F. voltage between grid and cathode sets up H.F. currents in the anode, screen-grid and cathode leads due to the various transconductances of the valve. These leads have self-inductances and mutual inductances and the currents flowing through them result in H.F. voltages on the various electrodes. These voltages produce currents through the valve capacities to the control grid, which gives rise to the input resistance.

Let us consider the influence of the inductance of the cathode lead on the input resistance. If an oscillatory circuit is connected to the input of a pentode of which the cathode lead has an inductance L_k

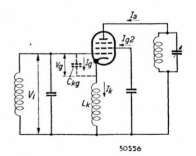

50556

Fig. 307
H.F. pentode with tuned grid and anode circuits and an inductance L_k in the cathode lead.

356

(see Fig. 307), then the alternating voltage V_g between control grid and cathode is approximately:

$$V_g = V_i - j\omega L_k I_k, \qquad (209)$$

in which V_i is the alternating voltage on the grid oscillatory circuit. The alternating current I_g flowing to the control grid as a consequence of the grid-cathode capacity C_{gk} is then:

$$I_g = j\omega C_{gk} V_g = j\omega C_{gk} V_i + \omega^2 L_k C_{gk} I_k. \qquad (210)$$

Now the cathode alternating current I_k as a first approximation equals the product of the transconductance g_{mk} of the cathode current and the voltage V_i across the circuit, thus

$$I_k = g_{mk} V_i. \qquad (211)$$

Consequently we obtain from (210) and (211)

$$I_g = (j\omega C_{gk} + \underline{\omega^2 L_k C_{gk} g_{mk}}) V_i. \qquad (212)$$

Owing to the inductance L_k of the cathode lead the current I_g to the grid has obtained a component which is represented by the underlined term in Equation (212) and which is in phase with the alternating voltage across the circuit. Therefore energy is taken from the circuit. This approximate current component is equal to that which would originate from the connecting of a resistance R_L in parallel with the input circuit, this resistance being given by:

$$R_L = \frac{1}{\omega^2 L_k C_{gk} g_{mk}}. \qquad (213)$$

Here the suffix L indicates the inductive origin of the resistance. According to Equation (213), that resistance component is inversely proportional to the square of the frequency and to the transconductance of the valve.

As regards the output resistance the cause of this is to be explained as follows: owing to the H.F. voltage V_a on the anode and the valve capacities (between anode and metallizing, between anode and suppressor grid and between anode and screen grid of a pentode), currents flow in the electrode leads. These leads have self-inductances and mutual inductances, and the H.F. currents flowing through them set up H.F. voltages between cathode and earth and between grid and earth, hence also between cathode and grid. Owing to the transconductance of the control grid with respect to the anode and the last named H.F. voltage a current flows to the anode. This anode alternating current has a component in phase with the anode alternating voltage and equal to the component which would arise from the connection of the anode circuit in parallel with a resistance R_o, the output resistance, for which, according to Strutt, the following formula holds approximately:

$$\frac{1}{R_o} = \omega^2 g_m C_{ag3} M_{kg3}, \qquad (214)$$

in which g_m is the transconductance from the control grid to the anode, C_{ag3} the capacity between the suppressor grid and anode, and M_{kg3} the mutual inductance of the leads to the cathode and the suppressor grid. According to Equation (214), therefore, the output resistance is likewise inversely proportional to the square of the frequency and to the transconductance.

(c) Influence of Transit Time of the Electrons upon Transconductance

Due to the electron transit time, in the case of extremely high frequencies the transconductance assumes a phase angle, owing to the anode alternating current lagging behind the grid alternating voltage. With an inductance L_k in the cathode lead a phase angle between the circuit voltage V_i and the grid alternating voltage V_g is produced. If we now substitute in Equation (209) the value given by Equation (211) for I_k, we get

$$V_g = (1 - j\omega L_k g_{mk}) V_i, \qquad (215)$$

from which the phase angle due to inductance of the cathode lead between the e.m.f. of the oscillatory circuit V_i and the grid alternating voltage V_g may be found. This phase angle, as follows from Equation (212), is approximately equal to

$$\varphi_L = \omega L_k g_{mk}. \qquad (216)$$

If, now, owing to the electron transit time, the anode alternating current lags behind the grid alternating voltage and thus has a phase angle with respect to V_g which we shall call φ_t, then it is evident that this phase angle increases the phase angle φ_L already resulting from the inductance in the cathode lead and consequently the damping of the input oscillatory circuit. The total phase angle, expressed by φ_A, therefore consists of a component φ_L due to inductance in the cathode lead and a component φ_t due to the transit time of the electrons, thus

$$\varphi_A = \varphi_L + \varphi_t. \qquad (217)$$

The absolute value of transconductance, however, remains unchanged up to very high frequencies (greater than 300 Mc/s).

(d) Feedback Impedance

In Chapter XIV, Section 77a, it has already been explained that the grid-anode capacity gives rise to a back-coupling that may lead to self-oscillation. The lower the impedance of that capacity the greater is the tendency to oscillate. In the ultra-short-wave region, particularly with very high frequencies, we find, instead of the grid-anode capacity, an impedance of a complex nature—the feedback impedance —which may be very much lower than the impedance resulting from the grid-anode capacity. In the ultra-short-wave band, therefore, there may be a tendency towards oscillation, whereas, with the same values of circuit impedance, this is not appreciable on longer waves.

The feedback from the anode to the control grid in the ultra-short-wave range may be explained in essentially the same way as the origin of the output resistance. When there is a H.F. alternating voltage V_a on the anode, currents flow, through the valve capacities, in the leads to the various electrodes. Owing to the self-inductances and mutual inductances of the leads these currents produce H.F. voltages on the electrodes, which in turn set up currents flowing through the valve capacities to the first grid. There is thus apparently an impedance between anode and control grid.

Whereas in the long and medium-wave bands the feedback impedance equals $1/j\omega C_{ag1}$, in the ultra-short-wave band, owing to the influences referred to above, the values of the feedback impedance are entirely different from that given by C_{ag1} of the valve. Instead of the anode-grid capacity C_{ag1} we find a new value, C'_{ag1}, which is approximately equal to

$$C'_{ag1} = C_{ag1} - K\omega^2, \qquad (218)$$

in which, according to Strutt,

$$K = C_o L C_i. \qquad (219)$$

K is consequently a constant of the valve depending on the inductances and capacities of the electrode leads; C_o is the output capacity, C_i the input capacity and L represents the inductance of the electrode leads. The feedback impedance must always be kept high in order to prevent oscillation of the H.F. stage. According to Equation (218), however, the absolute value of C'_{ag1} will be much greater than C_{ag1}, except for angular frequencies in the neighborhood of $\sqrt{\dfrac{C_{ag1}}{K}}$ or below that. The requirements with regard to feedback in the ultra-short-wave range will not usually be so stringent as in the long and medium-wave ranges, because in the former the impedances of the circuits employed are much smaller. Nevertheless it is desirable to keep the feedback small within the largest possible frequency range. A small value of K is favourable for the feedback impedance, for the smaller this factor the greater is the range in which C'_{ag1} is small. Generally the factor K is positive and can be made smaller by selecting a large mutual inductance M_{kg1} between the connections to the cathode and the control grid. It is likewise advisable to increase the mutual inductances M_{g1g3} and M_{g2a}. This can be achieved by choosing the right sequence of connections from the electrodes to the base contacts, in other words by placing alongside each other the connections to screen grid and anode, of cathode and control grid, and of suppressor grid and control grid, which increases the mutual inductances of the wires leading to these electrodes. This principle has been applied in the Philips H.F. pentode EF 22 with a pressed-glass base.

In this connection it should be noted that the measurement of feedback impedance is rendered very difficult by accidental, uncontrollable self-inductances or by mutual inductances with the chassis of the measuring equipment, so that it is hardly possible to give exact figures for valves.

150. Input and Output Capacities

If the transconductance of a valve is reduced by increasing the negative grid voltage, then the grid capacity is changed owing to the increase in density of the space charge in front of the first grid (Section 78).

This capacity change causes a detuning of the oscillatory circuit introduced in the grid circuit, and produces in the first place a reduction of the oscillatory-circuit impedance at the signal frequency. The amplification diminishes, but in practice this is of no great consequence provided in the case of a non-controlled valve the resonant frequency of the input circuit is the correct one, because the gain-control is then only slightly amplified. Of greater importance, however, is the fact that the tuning of the input circuit may become closer to the signal frequency of a nearby transmitter, so that there is more risk of interference. In the ultra-short-wave range one must expect the greatest detuning, because in that region a small capacity change causes a large frequency change.

The detuning of the input circuit due to change of the capacity of the input grid also involves, however, a displacement of the H.F. signal from the middle to the side of the resonance curve, thereby causing unequal amplification of the two side-bands. With unequal side-bands

there arises distortion in the detector, so that it is advisable that the
frequency of the signal should lie as accurately as possible in the middle
of the resonance curve, which means that one should aim at the smallest
possible capacity change.

Gain-control by varying the negative grid potential does not affect
the anode capacity.

Measurements have proved that the input and output capacities are
the same in the ultra-short-wave range as in the long and medium-wave
ranges, whilst the change of input capacity in the ultra-short-wave range
is practically equal to that in the long and medium-wave ranges.

In addition to a variation of the input capacity, transconductance
control effected by changing the negative bias also results in a variation
of the input resistance. Fig. 308 shows, for the pentode EF 50, the
input capacity C_{g1} and the input resistance R_{g1} at a wavelength of
6 metres as a function of the negative grid bias V_{g1}, from which it is
to be seen that unless special measures are taken both these values
vary considerably.

The capacity and resistance variations can in certain cases be reduced
by inserting an impedance in the cathode lead. Such an impedance can
be formed by a resistance of about 30 ohms and a parallel-connected
capacity of 50 $\mu\mu$F. Obviously an impedance of this nature in the
cathode lead somewhat reduces
the effective transconductance
and such a compensation works
well only within a narrow wave-
range.

The lower half of Fig. 308 gives
the variation of the input capacity

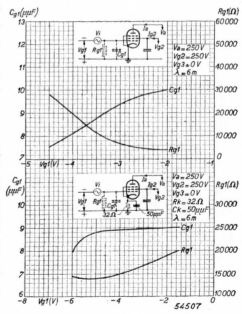

Fig. 308
Valve EF 50

Top: The grid capacity C_{g1} and grid
resistance R_{g1} at a wavelength of 6 m
as a function of the grid potential V_{g1}
when there is no impedance inserted in
the cathode lead.

Bottom: Grid capacity C_{g1} and grid
resistance R_{g1} at a wavelength of 6 m
as a function of the grid potential, with
an impedance in the cathode lead consisting of a 32-ohm resistor connected
in parallel with a 50-$\mu\mu$F capacitor.
For a reduction of transconductance to
$1/_{10}$th of the initial transconductance a
grid-voltage variation from —1.55 V
to —4.5 V is required.

and the input resistance with the grid potential for the pentode EF 50 when an impedance is introduced in the cathode lead.

Another means of counteracting the capacity variation of the grid and the variation of the grid resistance while the transconductance is being changed is to apply the regulating potential simultaneously to the third (suppressor) grid and to the control grid (first grid). Owing to the much smaller transconductance of the third grid the gain-control potential on the third grid must necessarily be very much higher than that on the first grid. As indicated in the diagrams of Fig. 309, the gain-control voltage V_R on the third grid can be attenuated by means of a voltage divider and the attenuated voltage applied to the first grid.

Thus by varying the gain-control voltage V_R it is possible to regulate simultaneously the bias voltages of the third and first grids in a certain ratio. The dividing ratio of the voltage divider must be accurately determined for each case in order to obtain the correct compensation of the capacity variation of the first grid. Fig. 309 gives the results attained with the EF 50 valve, in the upper half without impedance and in the lower half with an impedance (32 ohms in parallel with 50 $\mu\mu$F) in the cathode lead, in both cases with the optimum voltage divider and the transconductance being changed by a factor of 10. This diagram shows that without an impedance in the cathode lead the capacity remains more constant, but that with the stated impedance the value of the input resistance is much more favourable.

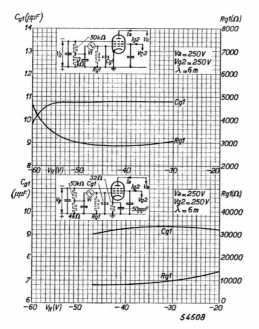

Fig. 309

Valve EF 50

Top: Grid capacity C_{g1} and grid resistance R_{g1} at a wavelength of 6 m as a function of the gain-control potential V_R applied direct to the third grid and via a voltage divider of $50,000 + 3,000$ ohms to the first grid.

Bottom: Grid capacity C_{g1} and grid resistance R_{g1} at a wavelength of 6 m as a function of the gain-control potential V_R applied direct to the third grid and via a voltage divider of $50,000 + 4,000$ ohms to the first grid. Here an impedance of 32 ohms in parallel with 50 $\mu\mu$F is included in the cathode lead.

54508

362

151. Some Measured Values for Philips Valves

In order to give an idea of the valve impedances in the short-wave range the measured values for some Philips valves are tabulated below.

(a) Input Resistance

In the tables for the valves EF 5 and EF 6 the input resistance is given for three wavelengths with a hot cathode.

EF 5 : $V_a = 200$ V, $V_{g2} = 100$ V		
Wavelength (in metres)	Input resistance R_{g1} for $V_{g1} = -3$ V (in ohms)	Input resistance R_{g1} for max. gain control (in ohms)
21	0.18×10^6	1.7×10^6
10.8	0.049×10^6	0.8×10^6
5.0	0.010×10^6	0.17×10^6

EF 6 : $V_a = 200$ V, $V_{g2} = 100$ V	
Wavelength (in metres)	Input resistance R_{g1} for $V_{g1} = -2$ V (in ohms)
21	0.15×10^6
10.8	0.042×10^6
5.0	0.009×10^6

Fig. 310 gives the input resistance R_{g1} of the H.F. pentode EF 22 (pressed-glass base) as a function of the wavelength λ for the normal adjustment at $I_a = 6$ mA (full line) and for maximum control of the transconductance ($I_a = 0$) (dot-dash line).

(b) Output Resistance

The following tables give the output resistances R_o for three wavelengths for the valves EF 5 and EF 6.

EF 5 : $V_a = 200$ V, $V_{g2} = 100$ V		
Wavelength (in metres)	Output resistance R_o for $V_{g1} = -3V$ (in ohms)	Output resistance R_o for max. gain control (in ohms)
23.0	0.27×10^6	0.56×10^6
11.8	0.14×10^6	0.32×10^6
6.15	0.064×10^6	0.19×10^6

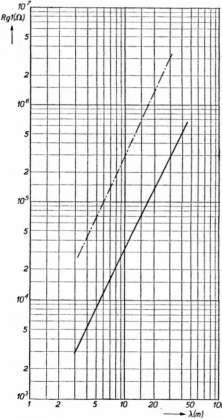

Fig. 310
The input resistance R_{g1} of the pentode EF 22 as a function of the wavelength for $I_a = 6$ mA (full line) and for max. control of the transconductance (dot-dash line), i.e. with $I_a = 0$.

In Fig. 311 the output resistance R_o is plotted for the pentodes EF 22 as a function of the wavelength λ for normal operating point ($I_a = 6$ mA) and for maximum gain control ($I_a = 0$). The full line relates to the normal operating point and the dot-dash line to the maximum control of transconductance.

152. Influence of Short-wave Impedances on the Practical Values of Circuit Impedances

The input and output impedances can be represented by a resistance R and a capacity C imagined as being connected in parallel to the input oscillatory circuit and the output oscillatory circuit of a valve under working conditions.

In order to estimate the actual effect of R and C on the input and output circuits it is of importance first to obtain an idea of the values encountered for the circuit impedance itself. In this connection account must be taken of the fact that this circuit im-

EF 6 : $V_a = 200$ V, $V_{g1} = 100$ V	
Wavelength (in metres)	Output resistance R_o for $V_{g1} = -2$ V (in ohms)
23.0	0.37×10^6
11.8	0.18×10^6
6.15	0.08×10^6

pedance is limited not only by the circuit magnification achieved in practice, but also by the tolerance of the input and output capacities of the valves, and further by the capacity variation due to the control of the negative grid potential of a variable-μ valve. Often when designing receiving sets allowance has to be made for the possibility of interchanging valves without detuning the circuits too much. Such detuning has a relatively greater effect according as the oscillatory circuits have a better quality factor and the tolerances and variations of the valve capacities are greater. For Philips valves the following simple rule can be given: the oscillatory circuits should be so designed that their impedance in case of detuning is as many times a thousand ohms as the number of metres corresponding to the wavelength. Thus, for a wavelength of 12 metres the circuit impedances must not exceed 12,000 ohms. It is quite feasible to make better circuits but when the valves are gain-controlled or replaced the detuning would be too great. The above rule holds for wavelengths of about 5 to 60 m, i.e. for the whole of the ultra-short-wave range. It is to be noted also that a second limitation of the magnification of the circuits is due to the essential band-width for modulation frequencies.

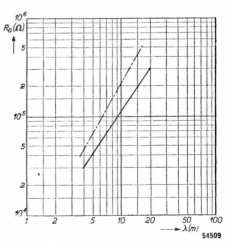

Fig. 311
The output resistance R_o of the pentode EF 22 as a function of the wavelength λ for $I_a = 6$ mA (full line) and for max. control of the transconductance (dot-dash line), i.e. with $I_a = 0$.

As appears from the measured results given here, the impedances of the valves are such that the circuit impedances chosen in accordance with the above rule are not affected by the valve impedances, or hardly so, for wavelengths of 10 m and higher. The normal Philips radio valves can also be very well used for shorter wavelengths, although Philips have designed special valves for that purpose. The feedback impedance, with the circuit impedances encountered, is still high enough not to impair the stability of action of the amplifying stages.

CHAPTER XXVI

Tuning Indicators

153. Object of Tuning Indication

The tuned circuits in the R.F. and I.F. stages of a superheterodyne receiver produce a certain selectivity, and this selectivity can be represented by a curve, namely the resonance curve of the whole set. This is in fact the resultant of the resonance curves of the various circuits. Such a resonance or selectivity curve represents, for instance, the I.F. voltage on the detector diode as a function of the frequency of a R.F. signal with constant amplitude applied to the input of a superheterodyne set when the variable condensers are in a certain position. The most common practice, however, is to plot the ratio of the I.F. signal on the detector with the correct tuning and the I.F. signal for a certain detuning, as a function of the detuning in c/s. Fig. 312 gives a resonance curve of a superheterodyne set. If the set is not accurately tuned to the incoming signal the frequency of its carrier wave will be situated on one of the sides of the resonance curve. This results in an asymmetrical reproduction of the modulation side-bands of the carrier wave. If, for instance, the carrier wave is modulated at a single frequency then it is accompanied by two side-waves and, due to the fact that the frequency of the carrier wave does not correspond to the middle of the resonance curve, one side-wave of the modulation is less amplified than the other. Moreover, in that case the absolute values of the two instantaneous (positive and negative) phase angles

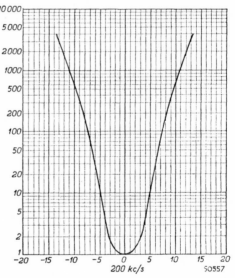

Fig. 312
Resonance curve of a superheterodyne set.
Horizontal axis: deviation of the input-signal frequency from the resonant frequency (200 kc/s) (linear scale).
Vertical axis: ratio of the signal voltage on the detector at resonant frequency and that at a given frequency deviation from the resonant frequency when the input-voltage amplitude has a constant value (logarithmic scale).

of these side-waves with respect to the carrier wave are no longer equal. This unequal amplification and the unequal instantaneous phase angles cause a distortion of the envelope of the carrier wave. The distortion is most marked for the high notes, one side-wave of which may in certain circumstances be almost entirely suppressed.

From the foregoing it will be realised that it is of importance to ensure that the carrier frequency of the desired signal coincides with the centre of the resonance curve. In the case of receivers not provided with automatic volume control this may be attained by tuning to the maximum volume. With sets having automatic volume control it is difficult to tune them so that the converted signal frequency applied to the detector corresponds with the middle of the resonance curve (or in the case of t.r.f. receivers the middle of the resonance curve corresponds with the frequency of the incoming signal), because the attenuation of the I.F. signal by detuning is neutralized more or less by the automatic volume control. The intermediate carrier frequency can then be adjusted only to the middle of the resonance curve by tuning for the smallest possible distortion. It would, therefore, be very useful indeed to have some means of determining the correspondence of the intermediate frequency with the middle of the resonance curve in a visual manner. This can be realized with the aid of the various systems for visible tuning or tuning indication designed for this purpose.

154. The Principle Underlying Tuning Indication

Since the signal voltages usually reach a maximum in the middle of the resonance curve it should be possible to arrange the visible tuning in such a way that the change of the signal voltage across the last oscillatory circuit is indicated as a function of the detuning by means of a voltage indicator. In that case, therefore, one would have to tune in to maximum voltage [1]. As, however, we have to deal with H.F. voltages, a high-frequency voltage indicator is essential. In Chapter XXIII it has already been explained at some length that the direct voltage across the leak resistor of the detector diode is dependent upon the R.F. or I.F. alternating voltage applied, so that the object in view can be attained by making the magnitude of this voltage visible, and it is on this principle that the various systems of tuning indication are based.

---

[1] The voltage will be greatest in the middle of the resonance curve also when the set is provided with automatic volume control. Detuning does not then cause this voltage to drop so quickly, so that it is difficult to judge by ear, by means of the volume, whether in tuning one has reached the middle of the resonance curve; without automatic volume control this is, as a rule, possible. Visible indication can naturally be much more sensitive and it offers a good means of tuning to the maximum voltage in a simple manner.

155. The Various Systems of Tuning Indication

The direct voltage across the leak resistor of a detector diode can be made visible in various ways. Since it is not desirable in practice to connect a measuring instrument directly in parallel or in series with the diode leak resistor, one way is to use the anode direct current of an automatically gain-controlled valve (e.g. an I.F. amplifying valve) for the operation of a small milliammeter. The correct tuning is then obtained by adjusting to the smallest direct current. The gain-controlled valve is thus used at the same time as an amplifier for the tuning indicator. It is necessary that the regulating voltage on the grid of the valve in question should be derived from a non-delayed automatic volume control, as otherwise the tuning indicator would not work with weak signals. Fig. 313 shows the basic circuit of such a tuning indicator. The milliammeter may be of special construction. Usually it is made in the form of a so-called shadow indicator, the moving coil tilting a small plate in a light beam and thus throwing a shadow of variable width upon a frosted glass.

Another method consists in inserting a resistor in the anode lead to a gain-controlled valve and making the direct-voltage variation across that resistor visible by means of a specially designed neon tube (see Fig. 320), the Philips neon tube 4662 for instance. The neon tube and the shadow indicator have, however, been superseded many years ago by the fluorescent-screen tuning indicator, which is much more attractive in appearance and more robust than for example the moving-coil system of the shadow indicator.

The fluorescent-screen indicator works on the principle of a high-vacuum cathode-ray tube and has its own amplifying system by means of which the voltage variations across the leak resistor of a diode can be rendered visible without employing a gain-controlled valve. With this kind of

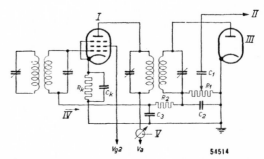

Fig. 313
Principle of a tuning indicator. A milliammeter is introduced in the anode-supply lead of a gain-controlled I.F. valve. The correct tuning is obtained when the meter shows the smallest deflection.

I = I.F. amplifying valve,
II = lead to the A.F. amplifier,
III = detector diode,
IV = A.V.C. supply lead to the grid of the I.F. valve,
V = milliammeter serving as tuning indicator.

369

tube green fluorescent light areas of variable size are shown on a conical fluorescent screen visible from the top, thus giving a very sharply defined indication of the tuning. The inertia-less working of the high-vacuum valve is one of the most important advantages of this type of indicator, which has found application also in measuring instruments and in other ways. This type of tuning indicator is often called magic-eye.

156. The Fluorescent-screen Tuning Indicator

The fluorescent-screen indicator EM 1 consists of the indicator part proper and an amplifying part. The construction of this valve is shown in Fig. 314. The indicator part contains a cathode that is also used for the amplifying part, an anode functioning at the same time as a fluorescent screen, and four deflecting plates.

The anode is conical in shape and is coated on the inside with a fluorescent substance. When the attracted electrons impinge on this substance it glows, and this is visible from the top of the bulb. The valve is mounted in the set in such a way that from the outside only the top of the bulb can be seen, for example through an aperture in the cabinet. Four fluorescent areas in the shape of clover leaves are formed on the screen; they are arranged in the form of a cross. Between the cathode and the anode the four deflecting plates are arranged radially (see Fig. 315), deflecting the electrons projected towards the anode. This deflecting action is increased when the potential difference between the anode and the deflecting plates becomes greater; to the deflecting plates a positive voltage is applied which has a value varying between the anode-supply voltage (e.g. 250 V) and a certain low voltage (about 10 V).

Owing to the deflection from the four plates the shadows thrown on the screen vary in breadth.

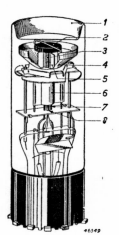

Fig. 314
Construction of the fluorescent-screen tuning indicator EM 1. It contains an amplifying triode (below) and an indicator system which is at the top.
1 = protruding glass rim which produces a darkened space in front of the fluorescent screen;
2 = shielding cup which serves to render invisible the light emanated from the cathode;
3 = space-charge grid of the indicator section;
4 = fluorescent screen;
5 = deflector plates;
6 = anode of the triode section;
7 = grid of the triode section;
8 = cathode.

Fig. 315
a) Fluorescing areas of the screen when the negative voltage on the
   grid of the triode section is low.
b) Fluorescing areas of the screen when the negative voltage on the
   triode section is high.
c) Simplified representation of the arrangement of the electrodes in
   the indicator part. Between the deflector plates D and the anode
   an electrostatic field is formed which has a deflecting action
   (shown by the forces f_1 and f_2) upon the electrons moving towards
   the anode. The line AB indicates the path of an electron from the
   cathode to the anode.

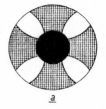
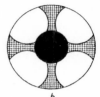
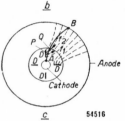

When the voltage difference between the anode
(fluorescent screen) and the deflecting plates is at its
maximum the deflection is the greatest and the
shadows are broadest. Fig. 315a illustrates the lighting
of the fluorescent screen caused by large deflections
and Fig. 315b that caused by small deflections.
The deflecting action is best understood by referring
to Fig. 315c. Here the anode is represented as a cylinder
placed around the cathode, with the four deflecting
plates in between. When these plates D have a lower
voltage than the anode, electric fields are formed
between them and the anode with the lines of force
running roughly as indicated by the broken lines. An electron leaving
the cathode at point A and reaching point P is deflected by the
force f_1. Owing to the deflecting forces in the field the electron
describes a curved path and finally reaches the point B on the
anode. If the difference in potential is great the deflecting field
will be very strong, the deflection great and the shadows behind the
plates will consequently be broad. If, on the other hand, there is no
difference in potential there will be no deflection and the plates will
exercise an attracting force, so that the whole of the anode glows. It
is to be observed that actually the shape of the lines of force is not so
simple as represented in Fig. 315c.
The voltage fluctuations on the deflecting plates are obtained with
the aid of the amplifying triode forming part of this
tuning indicator. The triode part is operated as a

Fig. 316
Essentials of the circuit of the fluorescent-screen tuning indicator
EM 1. The triode section is used as amplifier with resistance
coupling. The voltage variations at the anode a of the triode are
applied directly to the four deflector plates D and in this manner
bring about a change in the breadth of the fluorescent light areas
on the screen S.

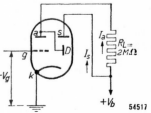

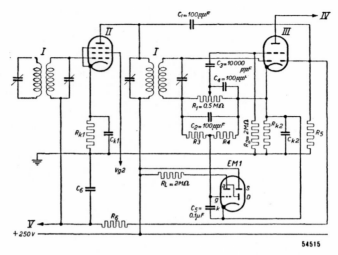

54515

Fig. 317
Connection of the tuning indicator in a superheterodyne receiving set with delayed a.v.c.
The indicator grid is controlled by the direct voltage of the detector diode, so that indication is given also for weak signals. The object of the voltage divider R_3—R_4 is to reduce
the negative direct voltage. The resistances R_3 and R_4 should have the highest possible
values in order to minimize the difference between the a.c. resistance of the diode circuit
and the d.c. resistance. The factor by which the voltage is divided should be so chosen
that the fluorescent screen is fully lighted up by the strongest anticipated signal on the
detector.
   I = I.F. transformers;
  II = I.F. amplifying valve;
 III = detector and A.F. amplifying valve;
  IV = connection to the output valve;
   V = A.V.C. connection.

resistance-coupled direct-voltage amplifier (see Fig. 316). As the negative direct voltage on the control grid increases the anode direct
voltage also increases. Since the deflecting plates of the indicator
part are connected directly to the anode of the triode an increase of
the negative voltage corresponds also to a decrease in the breadth of
the shadow sectors. When during tuning of the set to a signal the
negative direct voltage at the grid reaches its maximum, the fluorescent areas reach their maximum breadth.
Fig. 314 shows the construction of the fluorescent-screen indicator
EM 1. The triode part is mounted at the bottom of the valve and the
indicator part at the top. The two parts have one common cathode
covered at the top with a black cup, which serves as a shield against
the light emitted by the cathode. The method of connection of the
EM 1 in a set with delayed automatic volume control is shown in
Fig. 317.

Fig. 318
Arrangement of the indicator section of the EM 4 and representation of the shadow sectors on the fluorescent screen.
1 = deflecting rod of the sensitive part;
2 = supporting rod for the cathode shield;
3 = deflecting rod of the insensitive part;
4 = shadow sector of the sensitive part;
5 = fluorescent light area;
6 = shield for rendering invisible the light emanated from the
      cathode.

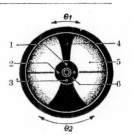

In order to ensure that the electron current to the anode of the indicator part does not become too great, a grid is also provided between the
deflecting plates and the cathode and connected with the latter. The current flowing to the fluorescent screen is limited by the space charge in
front of this grid, so as to avoid deterioration of the fluorescent substance.
A modern form of fluorescent-screen indicator is the EM 4. This has
two sensitivities, making it possible to tune in with the same degree
of accuracy both to stations having a low field strength at the point
of reception and to stations having a high field strength. Outwardly
this valve shows a great similarity with the EM 1. The fluorescent
screen shows only two fluorescent areas instead of four, and the tuning
is regulated according to the breadth of the shadow sectors between
the fluorescent areas. The rate of change in breadth during tuning is
not the same for both shadow sectors, the angle of one sector varying
much quicker than the other; the sensitivity of one section is much
greater than that of the other. By sensitivity is understood here the
angle variation of a shadow sector per volt of direct-voltage variation
on the control grid. These two sensitivities are obtained by replacing
the single triode amplifying system of the EM 1 by two triodes having
a different amplification factor and mounted one above the other
around the cathode (common with the indicator part). These triodes
have also a common grid, which, however, is wound with a different
pitch for each system. The two anodes are electrically separated from
each other, each being connected with one of the two deflecting rods
of the indicator part. (Fig. 318 shows how the deflecting rods are situated in the indicator part.)
These anodes are connected with the positive H.T. voltage of the
receiver (about + 250 V) via series resistors of 1 megohm. The two
triodes are controlled at the same time by the negative direct voltage
on the grid (direct voltage from the detector diode). The triode system
with the large amplification factor produces, for a given grid-voltage
variation, a greater variation of the anode direct voltage, and thus of
the shadow angle, than that produced by the other system.

The consideration underlying the design of this valve was as follows: if with a valve like the EM 1 an indication sufficiently sensitive for weak stations is required then the grid has to be controlled by the direct voltage across the leak resistor of the detector diode without or with very little attenuation. When receiving strong stations, however, the direct voltage on the grid will be so high that the fluorescent areas will cover the whole of the screen long before the middle of the resonance curve of the receiver is reached. If, on the other hand, a good visible tuning of strong stations is preferred, then the direct voltage produced across the leak resistor of the detector diode will have to be so greatly attenuated that there will scarcely be any indication of weak transmitters. Consequently, for a good indication of both strong and weak signals two fluorescent-screen indicators of the type EM 1 would have to be included, one coupled direct to the leak resistor of the detector diode and the other indirectly via a voltage divider.

The EM 4 affords the possibility of replacing these two indicators by one valve having two different sensitivities, whilst moreover the grid can be connected directly to the leak resistor of the detector diode; generally, therefore, there will no longer by any need for a voltage divider.

Finally, mention is to be made of a type of fluorescent-screen indicator having an amplifying system constructed in the form of a pentode, which is used as audio-frequency amplifier with variable amplification. This valve is constructed in a manner similar to that of the EM 1 or EM 4, the amplifying system being at the bottom of the bulb and the indicator system at the top, both mounted around a common cathode. The indicator screen is likewise of conical shape, and there are two deflecting rods between the cathode and the screen. These rods are connected with the screen grid of the pentode part.

The pentode part is so designed as to obtain a variable screen-grid potential, i.e. the screen-grid is fed via a series resistor (see Chapter XXII, Section 131). When the control potential from the automatic volume control is applied to the grid, the screen-grid current drops as the negative grid bias rises, i.e. as the control voltage increases (which is the case as the middle of the resonance curve is approached). This results in a rise of the voltage on the screen grid and at the deflecting electrodes, and as a consequence, as is the case with the EM 1, the deflecting action of the deflecting rods is reduced, so that the shadows on the fluorescent screen become narrower and the light areas broader.

As the screen grid is decoupled by a capacitor, alternating voltages can be conducted at the same time to the control grid without any noticeable effect in the fluorescent light spots.

The pentode part of the **EFM 1** has been so constructed as to be suitable for audio-frequency amplification, and in particular for variable amplification (see also Chapter XXII, Section 132). As already stated, a variable negative direct voltage is conducted to the grid for obtaining the tuning indication. When a strong I.F. signal reaches the detector diode (or the diode for automatic volume control) the grid of the pentode part

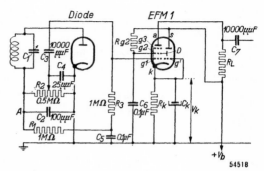

Fig. 319
Essentials of the circuit of the EFM 1 used as gain-controlled A.F. amplifier and fluorescent-screen indicator together with preceding diode detector.

of the **EFM 1** therefore becomes rather strongly negative, thereby reducing the amplification of the valve. This means that the A.F. amplifier takes part in the automatic volume control. Fig. 319 represents the essentials of the circuit of this fluorescent-screen indicator in combination with a detector diode.

157. The Neon Tuning Indicator

As explained in Section 155, the voltage variations caused by the automatic volume control across a resistor in the anode circuit of a gain-controlled valve can be made visible by changing the length of the light column which is formed around the rod-like cathode of a neon tube.

The Philips 4662 has been specially designed for this purpose and consists of a cathode in the form of a long, thin rod, a main anode likewise in the form of a thin rod, but much shorter and connected to the variable-voltage source, and a still shorter auxiliary or striking anode serving to keep the gas in the tube in an ionised state (Fig. 320); this last-mentioned anode avoids having to allow the voltage on the main anode to rise to the striking voltage before the tube begins to operate. Thus after the voltage on the main anode has dropped to a very low value the tube will again begin to work at once as soon as the voltage rises again. The light column is longest when the voltage on the main anode reaches its maximum.

This is, therefore, the case when the negative voltage on the grid of the gain-controlled valve is high, so that the correct tuning is shown by maximum length of the light column around the cathode.

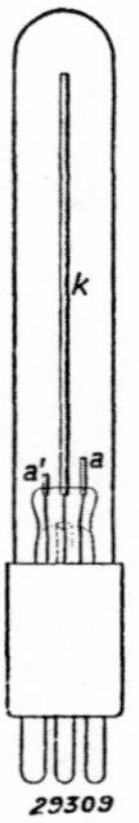

Fig. 320
Construction of the neon indicator tube 4662.
k = cathode
a = main anode
a′ = auxiliary or striking anode

Fig. 320 shows a drawing of this tube, whilst Fig. 321 represents the circuit. The resistance R_1 in the anode circuit of an I.F. valve has to be chosen according to its requirements; in the diagram the value given is for an EF 5 valve, used as I.F. amplifier. The object of the capacitor C_1 is to earth I.F. components at the bottom side of the primary of the I.F. transformer. The resistance R_2 adapts the sensitivity of the indicator to the control-voltage variations. The striking voltage is applied to the auxiliary electrode a′ via the resistance R_3.

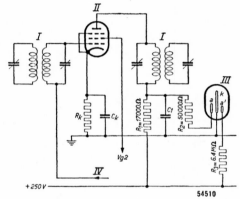

Fig. 321
Neon indicator tube connected up in a superheterodyne receiving set.
  I = I.F. transformers
 II = I.F. amplifying valve
III = neon tuning indicator 4662
IV = A.V.C. connecting lead

CHAPTER XXVII

Negative Feedback

158. Concerning the Reproduction Quality of the Receiving Set

The function of a receiving set is to reproduce through the loud-speaker as faithfully as possible the A.F. modulation of the R.F. signal received. For practical reasons, however, perfectly true reproduction is unattainable, and there will be deviations, large or small, according to the amount expended on assembling the set; the types of valves used, for instance, play a part. Deviations from the true sound are called distortion.

Distortion arising in a receiving set may be classified under two headings, usually named:

(a) attenuation distortion and

(b) non-linear distortion.

By **attenuation distortion** is understood the deviations at the loud-speaker from the relative strength at which the various modulation frequencies are received in the set. Supposing that the modulation of the carrier wave received consists of two frequencies (e.g., 500 and 2,000 c/s) the ratio of the amplitudes of which is 1 : 2, then the ratio of the amplitudes of the 500 and 2,000 c/s notes reproduced by the loudspeaker should likewise be 1 : 2.

Since the reproduction of the loudspeaker, apart from a resonance peak near the lower frequency end of the range, is roughly independent of the frequency if the current passing through the loudspeaker is kept constant, this means that if the modulation depth of the carrier wave received is constant the current through the loudspeaker should be of the same magnitude for each frequency. This, however, does not take into account the directional effect of the loudspeaker.

Owing to this effect certain frequencies are reproduced better in one direction than in another, so that the strength ratio of the various note levels is not the same in every direction. Added to this is another factor, the aural sensitivity for different notes at different strengths. For instance the low notes, compared with a note of 1,000 c/s, when reproduced at low volume are not heard so well as when reproduced at high volume [1]).

---

[1]) See also R. Vermeulen, "Octaves and Decibels" and "The Relationship between Fortissimo and Pianissimo" in Philips Technical Review, *2*, 1937, pp. 47 and 226.

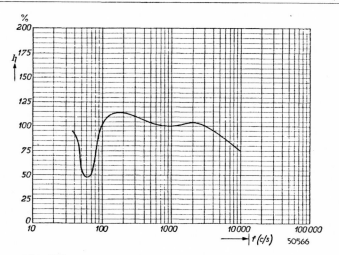

Fig. 322
Frequency characteristic of the A.F. part of a receiving
set (ratio of the loudspeaker current for any frequency
to that at 1,000 c/s as a function of the frequency).

It is no easy matter to take all these factors into account at the same
time, so that as a rule one has to be content by keeping the amplitude
of the alternating current through the loudspeaker as constant as
possible for all frequencies and attenuating if possible the loudspeaker
resonance peaks. The alternating current through the loudspeaker can
be plotted in the form of a curve as a function of the frequency of the
carrier-wave modulation for a constant modulation depth. One then
obtains the so-called **fidelity curve** or **audio-frequency characteristic**
of the set. Generally care is taken to ensure that the audio-frequency
characteristic is as flat as possible for the whole of the aural range.
Only at the resonance peak of the loudspeaker is it desirable that the
current supplied by the set should diminish. Fig. 322 shows such a
frequency characteristic of the A.F. part of a receiving set.

By **non-linear distortion** is understood formation of harmonics of the
A.F. modulation. The principal cause of this formation of harmonics
is the curvature in the characteristics of the valves employed. Harmonics of the A.F. modulation are formed in every stage, and these
are amplified in the following stage and added to those that arise there.
It may also happen that harmonics generated in consecutive stages
compensate each other more or less.

In the R.F., mixing and I.F. stages modulation distortion takes place,
whilst A.F. harmonics are formed in the detector, A.F. preamplifying

and output stages. In order to get an idea of the harmonics arising in sets, the total distortion in the anode circuit of the output stage or in the loudspeaker current can be plotted as a function of the output power, the A.F. modulation of the carrier wave received at the input of the set being assumed to be undistorted.

The curvatures of the characteristics may also result in the formation of combination notes (see Chapter XVII, Section 94a).

In order to obtain undistorted reproduction it is necessary to avoid as far as possible the formation of harmonics and combination notes, or in other words care has to be taken that the relation between the output current and the A.F. modulation of the input voltage is kept linear as far as possible.

Generally speaking the most serious distortion takes place in the A.F. part of the set, in particular in the output stage. By using circuits with **A.F. negative (inverse) feedback** it is possible to prevent to a large extent the formation of harmonics and of combination notes in the A.F. amplifier. At the same time such circuits offer the possibility of improving considerably the frequency characteristic of the A.F. part.

With oscillators (Chapter XX) a part of the output voltage of an amplifier is returned to its input circuit so as to obtain **regeneration.** In order to obtain the desired regeneration the phase of the voltage fed back must be the same as that of the voltage at the input which is the cause of the output voltage. In that case the feedback is called "positive". When the returned output voltage is, however, in antiphase with the input voltage which causes the output voltage, the feedback is termed **"negative"** or **"inverse" feedback.** In the following sections we will deal in considerable detail with the various forms of negative feedback which are used for the reduction of distortion in A.F. amplifiers.

Fig. 323
A.F. amplifier represented by a box with two input terminals a, b and two output terminals c, d. The load resistor (loud-speaker) R_L is connected to the terminals c and d, whilst an alternating voltage V_i is applied between a and b. In this and the next figures the directions in which the currents are taken as positive are indicated by arrows. The directions in which the voltages are taken as positive are indicated by + and — signs on the double-headed arrows.

159. The Principle of Negative Feedback

When an alternating voltage V_i is applied to the input of a L.F. amplifier (see Fig. 323) there is produced in the load resistance R_L an output alternating current I_o, which can be plotted as a function

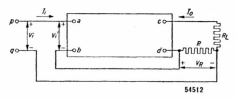

Fig. 324
A.F. amplifier with dynamic transconductance g_{md}, with a load resistor R_L and a resistor R connected in series with R_L. The alternating voltage across R is fed back to the input terminals, thus giving back-coupling. When the back-coupled alternating voltage is in antiphase to the applied alternating voltage V_i one speaks of negative feedback.

of the input voltage V_i, giving a dynamic I_o/V_i characteristic for the amplifier. If the input alternating voltage is sinusoidal then the output alternating current will not, usually, be purely sinusoidal, for in addition to the fundamental frequency (the frequency of V_i) this current contains also harmonics. As a result the dynamic I_o/V_i characteristic of that amplifier is not a straight line.

If a resistance R is connected in series with the load resistance R_L then the current I_o will flow also through that resistance R, across which an alternating voltage V_R occurs; this voltage can be applied to the input of the amplifier.

In Fig. 324 this alternating voltage V_R is applied in series with the input voltage V_i. It is now obvious that the voltage V_R is likewise amplified, so that in the output circuit another current $g_{md}V_R$ is produced (g_{md} representing the slope of the dynamic I_o/V_i characteristic of the amplifier). If the phase of this current $g_{md}V_R$ is opposed to that of the output alternating current I_o then the resultant output alternating current will be smaller than is the case without back-coupling and the feedback is consequently negative. In order to obtain the same output a.c. with negative feedback as without it a greater input signal has to be applied. As V_R contains also the harmonics present in I_o, which in turn when passed through the amplifier give rise to other harmonics in I_o in antiphase to the original I_o harmonics, the resulting harmonics in the output a.c. will be smaller than is the case without negative feedback,, as will be explained later. This feedback of the voltage across the resistance R to the input of the amplifier therefore means a reduction of the transconductance of the amplifier and at the same time reduction of distortion. Since with the circuit represented in Fig. 324 the feedback voltage is proportional to the output alternating current I_o this type of negative feedback is termed simply **Current Feedback.**

160. Feedback Proportional to the Output Alternating Current

(a) Influence upon Amplification, Distortion, Internal Resistance and other Properties

a) Transconductance and Amplification

According to Fig. 324 there is at the input of the amplifier an alternating voltage :

$$V_i = V_i' + V_R = V_i' + I_o R = V_i' + g_{md} V_i' R = V_i' (1 + g_{md} R) \quad (220)$$

Without negative feedback $V_i = V_i'$, so that in order to get the same output current with negative feedback an input signal V_i is required which is $(1 + g_{md}R)$ times as great. With feedback therefore the dynamic transconductance dI_o/dV_i is $(1 + g_{md}R)$ times as small. If the dynamic transconductance of the back-coupled amplifier, that is to say the dynamic transconductance from the terminals p q to the terminals c d, is represented by the term g'_{md}, then we get the equation

$$g'_{md} = \left(\frac{1}{1 + g_{md}R} \right) g_{md} . \quad (221)$$

If the product $g_{md}R$ is very much greater than unity then $\dfrac{1}{1 + g_{md}R}$ can be replaced by $\dfrac{1}{g_{md}R}$, so that the dynamic transconductance then approximates the constant value $\dfrac{g_{md}}{g_{md}R} = \dfrac{1}{R}$. This means that the transconductance becomes independent of the properties of the valves, which also means that when the value of $g_{md}R$ is very great the dynamic I_o/V_i characteristic of the amplifier approaches a straight line. Distortion is then nil.

In order to get the straightest possible dynamic characteristic it is therefore necessary that the negative feedback should be as great as possible. As the dynamic characteristic of an amplifier without negative feedback will usually show a curvature at top and bottom, when the grid swing extends into these curved parts the momentary transconductance g_{md} will be smaller and consequently also the product $g_{md}R$, so that finally this will no longer be much larger than unity and therefore the feedback will be much less effective. Hence, when the grid swing extends into the highly curved parts of the characteristic distortion will still arise in spite of negative feedback.

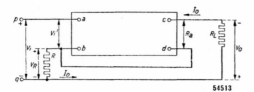

Fig. 325
A.F. amplifier with dynamic transconductance g_{md}, with a load resistor R_L and a resistor R connected in series with R_L. At the input of the amplifier is an alternating voltage V_i' which is the resultant of the two applied alternating voltages V_i and V_R.

β) Reduction of Distortion

From Equation (220) it follows that with negative feedback the input signal must be $(1 + g_{md}R)$ times greater in order to get the same output a.c. as is obtained without feedback. Owing to the feedback with this stronger signal, distortion is reduced by a factor $(1 + g_{md}R)$. This will be further explained [1]) with reference to Fig. 325 where the principle of the circuit is again represented. Suppose that the output a.c. I_o contains a percentage P of the second harmonic, then V_R will likewise contain a percentage P of the second harmonic. Since V_i is sinusoidal, the second harmonic in V_R is also present in V_i'. As V_R is $g_{md}R$ times as large as V_i', the percentage of the second harmonic of V_i' will be $g_{md}R$ times as large as the percentage P of V_R. The percentage P of second harmonic of the output a.c. consists of two components, one originating from the second harmonic of V_i'. The percentage of second harmonic due to this component is equal to $g_{md}RP$. The other component Q is generated in the amplifier by the fundamental wave of V_i' due to the non-linear I_o/V_i characteristic. Q is therefore the distortion due to second harmonic without negative feedback.

These components are in antiphase and partly compensate each other, so that the remaining distortion is less than the distortion arising when the same power is supplied without feedback. Since the percentage of the second harmonic with negative feedback was originally taken as equalling P and the two components referred to are in antiphase, we get the following equation

$$P = Q - g_{md}RP, \qquad (222)$$

or

$$P = \frac{Q}{1 + g_{md}R}. \qquad (223)$$

From this it therefore follows that with equal output power the second-harmonic distortion is $(1 + g_{md}R)$ times as small with negative feedback as without it. If the distortion becomes greater and higher har-

[1]) See also "Inverse Feed-back" by B. D. H. Tellegen in Philips Technical Review 2, 1937, p. 289 et seq. and "Improvements in Radio Receivers" by C. J. van Loon in Philips Technical Review **1**, 1936, p. 264 et seq.

monics arise then the calculation is not so simple. In such a case the harmonics present in V_i' may combine with each other and with the fundamental frequency and give rise to new harmonics, which are added to the harmonics originating from the fundamental frequency. It can, however, be shown that given sufficient feedback all harmonics likewise become smaller.

γ) Increase of Internal Resistance

In addition to a reduction of distortion, current feedback results in an increased internal resistance of the output stage of the amplifier. This is especially of importance when the load resistance (loudspeaker) is dependent upon the frequency.

Let us write the internal anode resistance of the output stage as R_a, and the static transconductance from the input terminals to the output terminals of the amplifier as g_m. Then, since R_a and $(R_L + R)$ are to be regarded as parallel-connected resistances (see Fig. 325), the following applies:

$$I_o = g_m V_i' \frac{R_a}{R_a + (R_L + R)}. \tag{224}$$

Now, according to Fig. 325,

$$V_i' = V_i - I_o R, \tag{225}$$

so that

$$I_o = g_m (V_i - I_o R) \frac{R_a}{R_a + (R_L + R)} \tag{226}$$

or

$$I_o = g_m V_i \frac{R_a}{R_a + R_L + R + g_m R R_a}. \tag{227}$$

Hence

$$I_o R_L = V_o = g_m V_i \frac{R_a R_L}{R_a + R_L + R + g_m R R_a}. \tag{228}$$

The internal resistance with feedback R_a' can now be determined by calculating the short-circuit current I_{ok} and the no-load voltage V_{ol}, since the internal resistance equals the ratio of no-lead voltage to short-circuit current.

The short-circuit current (with $R_L = 0$) is

$$I_{ok} = g_m V_i \frac{R_a}{R_a + R + g_m R R_a}. \tag{229}$$

The no-load voltage (with $R_L = \infty$) is

$$V_{ol} = g_m V_i R_a. \tag{230}$$

383

Therefore the internal resistance R_a' with feedback is equal to

$$R_a' = \frac{V_{ol}}{I_{ok}} = R_a \left(1 + g_m R\right) + R.$$ (231)

In most cases (e.g. when pentodes are used in the output stage) R is relatively small with respect to R_a, so that then

$$R_a' = R_a \left(1 + g_m R\right).$$ (232)

We found that distortion and amplification are reduced by the factor $(1 + g_{md}R)$. When the internal resistance is high compared with the load resistance, the dynamic transconductance g_{md} will be practically equal to the static transconductance. In that case the internal resistance is increased by the same factor as that with which the amplification and distortion are reduced. This increase of the internal resistance due to current feedback results in the output alternating current being less dependent upon the external load, so that not only is the non-linear distortion reduced, but also the attenuation distortion, which may arise for instance from the frequency dependency of the loud-speaker impedance. It is easy to see that the output a.c. becomes less dependent upon the magnitude of the load resistance when it is borne in mind that owing to a possible decrease of the output a.c. due to a higher load resistance the feedback is also reduced, because the voltage $I_o R$ which governs the feedback is lowered. This reduction of the feedback counteracts a drop in the output a.c.

δ) Increase of Input Resistance

In many cases it is also of importance to know how the input resistance of an amplifier is affected by feedback. If we call the input resistance without feedback R_{ab} then, in Fig. 324,

$$R_{ab} = \frac{V_i'}{I_i},$$ (233)

whilst with feedback the input resistance is

$$R_{pq} = \frac{V_i'}{I_i}.$$ (234)

We will now assume that $R \ll R_{ab}$, as is usually the case, so that no account need be taken of the voltage which I_i causes to arise across R.

According to Equation (220)

$$V_1 = V_i' \left(1 + g_{md} R\right).$$ (235)

From Eqs (234) and (235) it follows that

$$R_{pq} = \frac{V_i}{I_i} = \frac{V_i'}{I_i}(1 + g_{md}R) = R_{ab}(1 + g_{md}R).\qquad(236)$$

Therefore owing to current feedback the input resistance increases in the same degree as the amplification and distortion decrease.

ε) Decrease of Input Capacity

As a result of the current feedback the input capacity of an amplifier apparently diminishes. Let us call the apparent capacity between the terminals p and q of Fig. 326 C_{pq} and the capacity between the terminals a and b of the amplifier without negative feedback C_{ab}. The alternating voltage V_i' between the terminals a and b causes a capacitive current I_c to flow through the capacity C_{ab}. This current is equal to

$$I_c = j\omega C_{ab} V_i'.\qquad(237)$$

According to Equation (220) $V_i' = V_i - I_oR$, so that if this value of V_i' is introduced in Equation (237) we obtain

$$I_c = j\omega C_{ab} V_i - j\omega C_{ab} I_oR.\qquad(238)$$

Now $I_o = g_{md} V_i' = g_{md} V_i - g_{md} I_oR$, or

$$I_o = \frac{g_{md} V_i}{1 + g_{md}R}.\qquad(239)$$

Substituting for I_o in Equation (238) the value given by Equation (239), we obtain

$$I_c = j\omega C_{ab} V_i - j\omega C_{ab}\frac{g_{md}V_iR}{1 + g_{md}R} = j\omega\frac{C_{ab}}{1 + g_{md}R}V_i.\qquad(240)$$

The input impedance is then

$$Z_i = \frac{V_i}{I_c} = \frac{1}{j\omega\dfrac{C_{ab}}{1 + g_{md}R}}\cdot\qquad(241)$$

This means that the impedance Z_i between the terminals p and q is formed by an apparent capacity which is equal to $\dfrac{C_{ab}}{1 + g_{md}R}$ and consequently $(1 + g_{md}R)$ times as small as the input

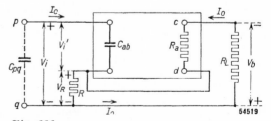

Fig. 326
Amplifier with dynamic transconductance g_{md} and an input capacity C_{ab}. As a result of the negative feedback the apparent capacity C_{pq} between the terminals p and q is the negative-feedback factor $(1 + g_{md}R)$ as small as the capacity C_{ab}.

385

capacity C_{ab} of the amplifier without feedback, $(1 + g_{md}R)$ being the feedback factor.

Thus we have
$$C_{pq} = \frac{C_{ab}}{1 + g_{md}R}.$$ (242)

Here the voltage drop which arises as a result of the current I_e through R is not considered. Generally it is not necessary to take this voltage into account.

ζ) Increase of Constancy of the Amplifier

By constancy, sometimes also called stability, is understood the insensitivity of the gain of an amplifier to various changes occurring within the amplifier. Causes of these changes are for instance the increase or decrease of supply voltages of the valves (e.g. as a result of mains voltage fluctuations), the decrease of the transconductance of the valves due to ageing, modifications in resistances and other components.

All these influences produce in the first place a variation Δg_{md} of the dynamic transconductanc g_{md} of the amplifier without negative feedback. Fluctuations of supply voltages of valves have a direct influence upon their transconductance.

The relative variation of the dynamic transconductance without feedback is $\frac{\Delta g_{md}}{g_{md}}$, and in analogy with this the relative variation of the dynamic transconductance with feedback is $\frac{\Delta g'_{md}}{g'_{md}}$. As will be shown below, the negative feedback results in a reduction of the relative change of the dynamic transconductance with respect to the transconductance change without feedback, which is equal to the feedback factor $\frac{1}{1 + g_{md}R}$. If, for instance, the dynamic transconductance changes as a result of one of the many causes by 10% and the feedback factor is 10, then the relative change $\frac{\Delta g'_{md}}{g'_{md}}$ of the dynamic transconductance g'_{md} of the back-coupled amplifier is 1%. Consequently, as a result of negative feedback, the amplifier is much more stable. According to Equation (221)

$$g'_{md} = \frac{g_{md}}{1 + Rg_{md}}.$$

By differentiating this equation we obtain

$$dg'_{md} = \frac{(1 + Rg_{md})\, dg_{md} - R\, dg_{md}g_{md}}{(1 + Rg_{md})^2} = \frac{dg_{md}}{(1 + Rg_{md})^2} =$$

$$= \frac{dg_{md}}{(1 + Rg_{md})\, g_{md}} \cdot \frac{g_{md}}{1 + Rg_{md}} = \frac{dg_{md}}{g_{md}} \cdot \frac{1}{1 + Rg_{md}} \cdot g'_{md}, \quad (243)$$

or

$$\frac{dg'_{md}}{g'_{md}} = \frac{dg_{md}}{g_{md}} \cdot \frac{1}{1 + Rg_{md}}. \quad (244)$$

From Equation (244) it follows, as postulated above, that the relative change of the dynamic transconductance $\dfrac{dg'_{md}}{g'_{md}}$ of the back-coupled amplifier is reduced by a factor $\dfrac{1}{1 + Rg_{md}}$, the feedback factor, with respect to the relative change $\dfrac{dg_{md}}{g_{md}}$ of the dynamic transconductance of the amplifier without negative feedback. Consequently, with the aid of negative feedback the amplification can be rendered very constant, this being important for calibrated amplifiers, for instance for measuring purposes.

(b) Practical Realization of Current Feedback

There are two methods for reducing the distortion of an amplifier by means of current feedback, namely:

1) by choosing the largest possible dynamic transconductance g_{md},

2) by choosing the highest possible value of the resistance R,

whilst both methods can, of course, also be combined. For the first alternative valves with a large transconductance have to be used and owing to the loss of power occurring in the resistance R this cannot be chosen arbitrarily, so that as a general rule only a combination of the two methods can be considered.

In receiving sets current feedback is often applied only to the output valve. It is obvious that such feedback can be brought about by dispensing with the condenser for by-passing the cathode resistor of the output valve (see Fig. 327). In Fig. 324

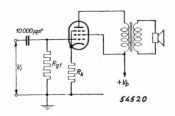

Fig. 327
Circuit of an output pentode with current feedback produced by omitting the cathode condenser.

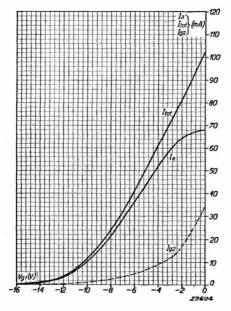

Fig. 328
Anode current I_a, screen-grid current I_{g2} and total current I_{tot} of the pentode EL 3N as functions of the negative grid bias.

the resistor R is then at the same time the cathode resistor R_k which produces the automatic negative grid bias of that valve. In the case of output pentodes, however, simply to leave out the cathode condenser leads to a complication. In addition to the anode alternating current there is also the screen-grid a.c. flowing through the cathode resistor, because in the normal circuit the screen grid is connected through a capacity to the chassis. In a pentode the screen-grid current is proportional to the anode current only over the lower part of the dynamic transfer characteristic and not over the upper part. In the upper part the anode voltage is low, so that the current distribution between anode and screen grid is altered in such a way that a larger part of the electron current leaving the cathode flows to the screen grid than is the case over the lower part of the characteristic (see also Fig. 328).

Since for current feedback a voltage is required that is proportional to the anode a.c., it is desirable to prevent the screen grid a.c. from flowing through the cathode resistor. This can be achieved by feeding the screen grid via a series resistor and by-passing with the aid of a relatively large condenser connected to the cathode (see Fig. 329). With this condenser the resistances R_k and R_{g2} for the anode a.c. are effectively connected in parallel, so that with this circuit the resistance R governing the feedback is equal to

$$R = \frac{R_k R_{g2}}{R_k + R_{g2}}. \qquad (245)$$

Fig. 329
Circuit of an output pentode with feedback by means of a non-bypassed cathode resistor, where the screen-grid alternating current is not flowing through the cathode resistor.

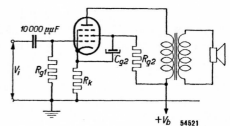

If the resistance R_{g2} is 2,500 ohms then for an EL 3N valve with a cathode resistor of 150 ohms the value of R is

$$R = \frac{150 \times 2,500}{2,650} = 140 \text{ ohms.}$$

The dynamic transconductance can be derived, for instance, from the static transconductance and the internal resistance; it can also be calculated from the sensitivity of the valve (effective grid alternating voltage required for a power output of 50 mW). The sensitivity of valve EL 3N amounts to 0.33 V, so that

$$g_{md} = \sqrt{\frac{50 \times 10^{-3}}{0.33^2 \times R_L}} = \sqrt{\frac{50 \times 10^{-3}}{0.33^2 \times 7,000}} = 8.3 \text{ mA/V.}$$

Therefore

$$\frac{1}{1 + g_{md}R} = \frac{1}{1 + 8.3 \times 140 \times 10^{-3}} = \frac{1}{1 + 1.15} = \frac{1}{2.15}.$$

It is to be remarked, however, that by decoupling the screen grid to the cathode the output power, when driving the stage up to the grid-current starting point, as a result of the negative feedback is less than when using the circuit of Fig. 326 or in the case of a normally decoupled cathode resistor (for valve El 3N about 3.9 W instead of 4.5 W). This may be explained by the following reasoning. With a pentode the distortion at full modulation of the grid consists mainly of the third harmonic, so that with a sinusoidal grid alternating voltage the anode a.c. has the form of a flattened sine curve. Thus in order to compensate this distortion by means of negative feedback it is necessary that the grid alternating voltage has a greater peak value than that which is determined by the sinusoidal form. The maximum output power is consequently more rapidly limited by the grid current than is the case without negative feedback.

To get a greater feedback a higher value would have to be

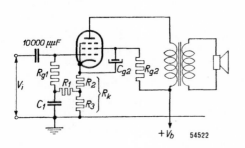

Fig. 330
Circuit of an output pentode with feedback when the value of the non-bypassed cathode resistor is greater than the value required for negative grid bias. In order to obtain the correct negative grid bias the grid-leak resistor is connected to a tapping point on the cathode resistor.

chosen for R_k. As in that case the voltage drop across R_k would be greater than the negative grid bias required for the valve, the excess voltage has to be compensated. This can be done by connecting the grid leak resistor to a point having a positive potential with respect to the chassis, for instance to a tapping-point on the cathode resistor R_k (see Fig. 330). If, for instance, a value of 500 ohms is chosen for R_k, we find for $R_{g2} = 2,500$ ohms:

$$R = \frac{500 \times 2,500}{3,000} = 417 \text{ ohms},$$

so that $1 + g_{md}R = 4.5$.

The maximum power that can be supplied by the EL 3N then amounts, however, to only about 3.3 W instead of the normal 4.5 W, due on the one hand to the fact that the alternating voltage between grid and cathode departs from the sinusoidal wave form (owing to the addition of harmonics), and on the other hand owing to the fact that a part of the output power is lost in the parallel circuit composed of R_k and R_{g2}.
It is to be noted that the load impedance of the output valve is increased by the cathode resistance R_k, so that in determining the load impedance the cathode resistance has to be added to the loudspeaker impedance on the primary side of the output transformer.
The circuit represented in Fig. 327, where only the cathode by-passing condenser is omitted, is of course the simplest. As already remarked, however, this does not yield feedback of the type discussed, on account of the screen-grid a.c. flowing through the cathode resistor in addition to the anode a.c. The feedback voltage V_R of Fig. 325 is therefore determined by the I_{tot} curve of Fig. 328. In this manner only the distortion arising from the curvature of the lower part of the dynamic characteristic is reduced. With a low output power the distortion is consequently appreciably less than that in the circuit without feedback and even still less than that in the circuit of Fig. 329.
Since the distortion attributable to the curvature of the upper part of the characteristic changes but little in the circuit of Fig. 327, the output power supplied when the valve is driven up to the point where grid current begins to flow is approximately equal to that without feedback, so that in that respect the circuit of Fig. 327 is more convenient than that of Fig. 329 where the output power is smaller.
When a pentode is used in the circuit of Fig. 327 it is better to select a slightly smaller impedance than is mentioned in the data for normal power amplification (based on the V_{ao}/I_{ao} ratio), because then there is less curvature of the upper part of the dynamic characteristic. For this

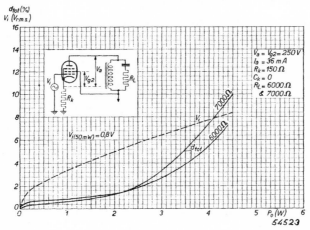

Fig. 331
Distortion with load impedances R_L of 6,000 ohms and
7,000 ohms and with the required grid alternating voltage
V_i for $R_L = 6,000$ ohms, as a function of the power output,
when employing a valve EL3 N with negative feedback
and with the screen-grid alternating current flowing
through the non-bypassed cathode resistor.

reason the distortion with a large power output is much less than that
with the most favourable load impedance given in the data of the
valve. Fig. 331 shows for the EL 3N the measured distortion as a
function of the output power supplied with the normally optimum load
impedance of 7,000 ohms and with the smaller impedance of 6,000
ohms, clearly illustrating the difference.

161. Feedback Proportional to the Output Alternating Voltage

(a) Influence on Amplification, Distortion, Internal Resistance and Other Properties

α) Principle of Voltage Feedback

In Fig. 324 R_L and R form a voltage divider across the output of the
amplifier and a part $\dfrac{R}{R_L + R}$ of the output alternating voltage is
returned in antiphase to the input of the amplifier. This returned
part of the output alternating voltage thus depends upon the value
of the load resistance R_L, the feedback decreasing with increasing
value of R_L, which results in a higher output alternating voltage.
Consequently with increasing value of R_L there is with feedback a smaller reduction of the current through the load resistor

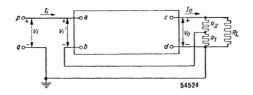

54524

Fig. 332
Essentials of a circuit in which the negative
feedback is proportional to the output alternating voltage.

than without feedback. This is equivalent to higher internal resistance of the output stage of the amplifier [see Section 160 a, Equation (232)].

If the output alternating voltage is to be made as far as possible independent of the value of R_L, it is necessary to introduce between the output terminals c and d a separate voltage divider preferably consisting of resistors of the highest possible values, so as to minimize the loss of output power. The alternating voltage at the tapping-point of the voltage divider is then to be fed back in the right phase to the input of the amplifier. Fig. 332 shows the principle of this method of feedback. The voltage divider between the output terminals is now parallel to the load resistor R_L. In this case the feedback voltage is **proportional to the output alternating voltage,** the returned alternating voltage being determined by the alternating voltage between the terminals c and d of Fig. 332. Therefore this mode of negative feedback is called simply **Voltage Feedback.**

β) Influence upon Amplification and Distortion

If V_i' is the input alternating voltage of the amplifier without negative feedback and V_o the output alternating voltage then the amplification is

$$A = \frac{V_o}{V_i'}. \tag{246}$$

By returning a portion β of the output alternating voltage to the input of the amplifier we obtain:

$$V_i = V_i' + \beta V_o = \frac{V_o}{A} + \beta V_o$$

or

$$V_i = V_o \left(\frac{1}{A} + \beta \right). \tag{247}$$

The amplification A' of the back-coupled amplifier is the ratio of V_o to V_i, thus:

$$A' = \frac{V_o}{V_i} = \frac{1}{1/A + \beta} = \frac{A}{1 + \beta A}. \tag{248}$$

Back-coupling therefore reduces amplification and, similarly to current feedback, it can be shown that distortion is also reduced by a factor $1 + \beta A$. In Fig. 332 the factor β is the attenuation ratio of the voltage divider $R_2 R_1$. If the factor βA is great with respect to unity then the resultant amplification A' approximates the constant value $\frac{1}{\beta}$. This means, therefore, that the output voltage is independent of the load resistance R_L, which corresponds to a very low internal resistance of the output stage. (This is of importance, for instance, for amplifiers used for measurement purposes and for amplifiers with widely varying load.)

γ) Decrease of Internal Resistance

The reduction of internal resistance due to voltage feedback can be calculated in the following way.

Supposing that the amplification factor of an amplifier without feedback (i.e., the amplification with infinitely large load resistance) is equal to μ, then according to Equation (47) (Chapter XII)

$$V_o = \mu V_i' \frac{R_L}{R_a + R_L}. \tag{249}$$

Further, the anode alternating current equals the anode alternating voltage divided by the anode resistance, thus

$$I_o = \frac{V_o}{R_L} = \mu V_i' \frac{1}{R_a + R_L}. \tag{250}$$

According to Equations (247) and (249)

$$V_i = V_i' + \beta V_o = V_i' \left(1 + \beta \mu \frac{R_L}{R_a + R_L} \right) \tag{251}$$

or

$$V_i' = \frac{V_i}{1 + \beta \mu \dfrac{R_L}{R_a + R_L}}. \tag{252}$$

Substituting for V_i' in Equations (249) and (250) the value found in Equation (252) we obtain:

$$V_o = \frac{\mu V_i R_L}{R_a + R_L + \beta \mu R_L} \tag{253}$$

and

$$I_o = \frac{\mu V_i}{R_a + R_L + \beta \mu R_L}. \tag{254}$$

393

The internal resistance with feedback R_a' equals the quotient of no-load voltage V_{ol} and short-circuit current I_{ok}, the latter (for $R_L = O$) being:

$$I_{ok} = \frac{\mu V_i}{R_a}, \tag{255}$$

whilst the no-load voltage (for $R_L = \infty$) is

$$V_{ol} = \frac{\mu V_i}{1 + \mu \beta}. \tag{256}$$

Hence the internal resistance R_a with feedback equals:

$$R_a' = \frac{V_{ol}}{I_{ok}} = \frac{R_a}{1 + \mu \beta} \tag{257}$$

Thus, owing to feedback the internal resistance has been reduced by a factor $(1 + \mu \beta)$.

When R_a is small compared with the external resistance R_L then the amplification without feedback approximates the value of the amplification factor and owing to the voltage feedback the internal resistance diminishes approximately to the same degree as the amplification. Since, however, pentodes are usually employed in the output stage the internal resistance is greater than the external resistance, so that the amplification without feedback is appreciably less than the amplification factor μ and the internal resistance diminishes much more than the amplification. In this manner the same internal resistance can be obtained with a pentode in the output stage as with a triode, while retaining, however, the higher efficiency of the output pentode.

As the voltage divider $R_2 R_1$ needs a negligible current, with this sort of feedback there is not the drawback of an appreciable loss of power. Owing to the low internal resistance, when a loudspeaker is connected to the output terminals of the amplifier, the resonance peaks of the loudspeaker are damped and therefore less noticeable. As the impedance rises at each resonance of the loudspeaker and the output voltage is kept constant by the feedback, the low internal resistance means a drop in the current I_o through the loudspeaker and consequently reduction of the sound volume. The loudspeaker impedance for high frequencies generally increases, so that, since the output voltage is more or less constant, the current passing through the loudspeaker will drop as the frequency becomes higher, with the sound energy likewise diminishing. As a consequence the high notes are less strongly reproduced.

By using circuit elements whose impedance is dependent upon frequency it is possible to reduce the degree of back-coupling in a certain frequency range, thereby causing less diminution of amplification, so that the notes within that frequency range are reproduced in relatively greater strength.

δ) Increase of Input Resistance

As is the case with current feedback, so also with voltage feedback there is an increase of input resistance. With the aid of Equations (247), (233) and (234) we find that

$$R_{pq} = \frac{V_i}{I_i} = \frac{V_i' + \beta V_o}{I_i} = \frac{V_i' + \beta A V_i'}{I_i} = R_{ab} (1 + \beta A). \quad (258)$$

Thus it is seen that owing to voltage feedback the internal resistance increases in the same degree as the amplification and distortion decrease.

ε) Decrease of Input Capacity

As has been calculated in Section 160(aε) for current feedback, it can be determined that the apparent input capacity C_{pq} is $(1 + \beta A)$ times as small as the real capacity C_{ab} between the terminals a and b of the amplifier, $(1 + \beta A)$ being the negative-feedback factor and C_{ab} being the input capacity without feedback. Consequently we may write down:

$$C_{pq} = \frac{C_{ab}}{1 + \beta A}. \quad (259)$$

ζ) Increase of Constancy of the Amplifier

Just as it has been proved for current feedback, we find for voltage feedback that

$$\frac{dA'}{A'} = \frac{dA}{A} \cdot \frac{1}{1 + \beta A}, \quad (260)$$

where $\dfrac{dA'}{A'}$ is the relative change of amplification with feedback and $\dfrac{dA}{A}$ the relative change without feedback. The relative change of the amplification with feedback is consequently the negative-feedback factor $(1 + \beta A)$ times as small as the relative change without feedback.

(b) **Practical Realization of Voltage Feedback**

First of all it is, of course, possible to apply voltage feedback exclusively to the output valve of the amplifier or receiving set. Fig. 333 gives a diagram showing how this is done. The grid of the output valve is coupled to the anode of the preceding valve by an RC element, while the two anodes are connected together by a resistance R_{ag}. The voltage divider parallel to the external resistance is formed by the resistances R_{ag} and R_g, with R_g composed of the anode coupling resistance R_1 of the preceding valve connected in parallel with the grid-leak resistance R_2 of the output valve and the internal resistance of the preceding amplifying valve.

As we shall see, contrary to Equation (258), the input resistance of an output pentode back-coupled in this manner is much lower than that without negative feedback. This can be explained with reference to Fig. 334 [1]). In the first place the input alternating voltage V_i causes a current I_g to flow through the resistance R_g. Secondly the output alternating voltage V_o sets up a current I_{ag} through the resistance R_{ag} which flows back to the cathode. Thus the grid voltage produces a current of $I_g + I_{ag}$ between grid and cathode, so that the input resistance equals $\dfrac{V_i}{I_g + I_{ag}}$. For determining the magnitude of I_{ag} we can use the following reasoning: V_o equals V_i times the amplification A of the valve. When the voltage drops at the point a the voltage at the point c rises A times as much, so that the potential between a and c is $(A + 1)$ times as great as V_i. Therefore the current I_{ag} equals $\dfrac{V_i\,(A + 1)}{R_{ag}}$ or $\dfrac{V_i}{\dfrac{1}{A + 1} \cdot R_{ag}}$. This corresponds to a resistance between grid and cathode connected parallel to R_g which is $(A + 1)$ times smaller than R_{ag}. According to the choice of the resistance R_{ag} the input resistance may therefore be very much smaller

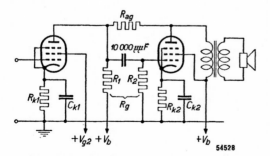

Fig. 333
Principle of voltage feedback applied only to the output valve.

[1]) The difference between the negative feedback according to Fig. 332 and that according to Fig. 334 is that in the first case the back-coupled voltage is in series and in the second case in parallel with the input voltage of the amplifier.

396

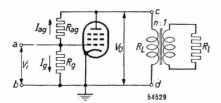

54529

than R_g. If, for instance, the amplification factor of the output valve is 57 (see below), $R_{ag} = 1.64$ megohms and $R_g = 0.1$ megohm, then we find for the input resistance

$$\frac{0.1 \times \dfrac{1.64}{58}}{0.1 + \dfrac{1.64}{58}} \text{ megohm} = 22,000 \text{ ohms.}$$

As a consequence the a.c. resistance in the anode circuit of the pre-amplifying valve is also much smaller than the internal anode resistance R_a, to which is to be ascribed the reduction of amplification due to feedback. When a pentode is used as pre-amplifying valve only the amplification is reduced; the dynamic transfer characteristic remains essentially the same as the static characteristic. When a triode is used for pre-amplifying, the distortion, however, is greater due to the smaller external resistance. This results from Fig. 335. Taking for instance for R_1 a value of 200,000 ohms and for the grid bias —4 volts, then the working point in this figure is P. Now if the anode alternating voltage for a determined grid alternating voltage fluctuates along the resistance line for 200,000 ohms between the points a and b, then Fig. 335 shows that the sections aP and bP are almost equal, so that practically no distortion takes place. If for the

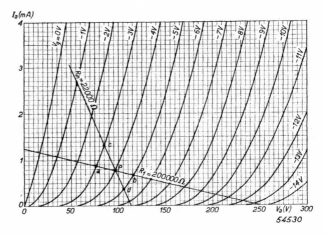

54530

Fig. 335
Plate characteristics of an A.F. amplifying triode operated with an anode coupling resistor of 200,000 ohms and a negative grid bias of —4 V. Owing to the low input resistance of the succeeding back-coupled output valve the load line passes through cd, — not through ab.

397

working point P the a.c. resistance is smaller than 200,000 ohms then the anode voltage fluctuates, for example, along the line cd, and from Fig. 335 it is seen that the sections cP and dP are no longer equal, so that by reason of this fact alone greater distortion occurs.

Owing to the feedback there will be less amplification in the pre-amplifying valve and consequently a higher grid alternating voltage is required for controlling the output stage. With a triode, therefore, the distortion becomes disproportionately large, whilst with a pentode the amplification of which is originally much higher a relatively low grid alternating voltage suffices after back-coupling. For this reason triodes are less suitable as pre-amplifying valves with the described method of back-coupling. Fig. 336 shows the required input alternating voltage and the distortion as a function of the output power supplied by a pentode EL3N with voltage feedback.

The amplification of the EL3N can be calculated from the quotient of the anode alternating voltage for 50 mW power output and the corresponding grid alternating voltage required. With a load resistance of 7,000 ohms the output alternating voltage is 18.7 V, and as the grid alternating voltage required for a power output of 50 mW equals 0.33 V, the amplification amounts to $A = \dfrac{1.87}{0.33} = 57$. Therefore the factor by which the amplification is reduced owing to feedback is

$$1 + \beta A = 1 + \frac{R_g}{R_{ag} + R_g} A = 1 + \frac{0.1}{1.74} 57 = 4.3.$$

Voltage feedback, however, offers still further possibilities. Part of the alternating voltage across the speech coil of the loudspeaker can be returned to the input of the A.F. amplifier and, moreover, the frequency characteristic of the amplifier can be in-

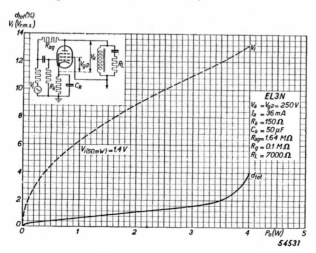

EL3N
$V_a = V_{g2} = 250\,V$
$I_a = 36\,mA$
$R_k = 150\,\Omega$
$C_k = 50\,\mu F$
$R_{ag} = 1.64\,M\Omega$
$R_g = 0.1\,M\Omega$
$R_L = 7000\,\Omega$

$V_{i(50mW)} = 1.4\,V$

d_{tot}

$P_o(W)$

54531

Fig. 336
Distortion and required input alternating voltage as functions of the power output of a valve EL3N with A.F. voltage feedback.

fluenced by elements in the feedback circuit the impedance of which depends on the frequency. By means of the feedback from the output to the input of an amplifier any distortion (for example, that arising in the output transformer) occurring in the various stages can be reduced. Such feedback can operate over several valves.

162. Influence of Negative Feedback upon Hum, Noise, etc.

Feedback results not only in a reduction of the distortion arising in an amplifier but also reduces the effect of interferences, such as hum and noise, that occur in the amplifier itself, because an interfering voltage present somewhere in the amplifier is counteracted by a voltage at the input. The suppression of interference may be calculated as follows, assuming that the amplifier causes no distortion. The output alternating voltage, without feedback, consists of two parts, the amplified input voltage and the interference S. We therefore have (see also Fig. 332):

$$V_o = AV_i' + S. \tag{261}$$

With feedback the input alternating voltage equals

$$V_i = V_i' + \beta V_o = V_i' + \beta AV_i' + \beta S \tag{262}$$

or

$$V_i' = \frac{V_i - \beta S}{1 + \beta A}. \tag{263}$$

From the two Equations (261) and (263) it follows that

$$V_o = \frac{A}{1 + \beta A} V_i + S \frac{1}{1 + \beta A}. \tag{264}$$

Hence it is to be deduced that, with a certain value of the output alternating voltage V_o obtained with feedback by making $V_i (1 + \beta A)$ times as great as it was without feedback, the interference is reduced by a factor $(1 + \beta A)$, i.e., by the same amount as the reduction of distortion and amplification.

163. Correction of the Frequency Characteristic

Fig. 337 shows the principle used for correcting the frequency characteristic. In parallel with the resistor

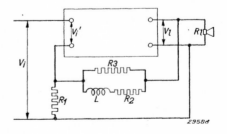

Fig. 337
A.F. negative feedback proportional to the output alternating voltage with a frequency-discriminating element in the feedback circuit used to improve the frequency characteristic of the amplifier.

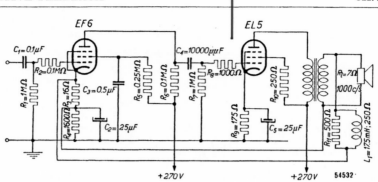

Fig. 338
Circuit of the A.F. part of a receiving set operating with voltage feedback.

R_3 of a voltage divider across the speech coil is an inductance L connected in series with a resistor R_2. At low frequencies L has a negligible impedance, so that R_2 is then effectively directly parallel with R_3. At high frequencies, on the other hand, the impedance of the inductance is much greater, so that the part of the alternating voltage on the loudspeaker which is fed back is much smaller. Consequently for the higher frequencies the amplification is greater and these frequencies are reproduced to a greater extent.

Fig. 338 gives a circuit employing valves EF 6 (pre-amplifying pentode) and EL 5 (18-W power pentode). Part of the voltage across the loudspeaker speech coil is fed back to the cathode of the EF 6. Between the cathode and the self-bias resistor a non-bypassed resistor of 16 ohms is connected. This resistor forms part of the voltage divider connected in parallel with the speech coil. The other part consists of R_{11} with L_1 connected in parallel to it. The resistor R_{11} has a value of 500 ohms, the coil L_1 an inductance of 17.5 mH and a d.c. resistance of 250 ohms. At 50 c/s the impedance of the inductance is only 5.5 ohms, so that practically only the parallel connection of 250 ohms and 500 ohms then remains, i.e. 167 ohms. Across the cathode resistor there is then, therefore, a fraction $\beta = \dfrac{16}{167 + 16} = 0.0875$ of the voltage across the loudspeaker speech coil.

The amplification between the grid of the EL 5 and the loudspeaker speech coil can be calculated in the following way: assuming that the losses in the output transformer amount to 20% and the impedance of the loudspeaker coil is 7 ohms, 50 mW in the anode circuit corresponds to 40 mW in the speech coil, i.e. to an alternating voltage of $\sqrt{0.04 \times 7} = 0.53$ V. The corresponding value of the required grid alternating voltage of the **EL 5** amounts to 0.5 V, as indicated from the data published for the **EL 5**, so that the amplification is approximately 1. The amplification of the **EF 6**, according to the data published for that valve, is 110, so that the total amplification from the grid of the **EF 6** to the loudspeaker coil is about 110. As $\beta = 0.0875$, the amplification and the distortion are thus reduced in the proportion

$$\frac{1}{1 + \beta A} = \frac{1}{1 + 0.0875 \times 110} = \frac{1}{10.6}, \text{ i.e. about } \frac{1}{11}.$$

At low frequencies the feedback is thus 11-fold. At 5,000 c/s L_1 has an impedance of 550 ohms, so that the resultant impedance of this coil in series with 250 ohms and parallel to 500 ohms amounts to about 325 ohms. The ratio β is then $\dfrac{16}{16 + 325} = 0.047$, when disregarding the phase shift, and the reduction in the amplification is $\dfrac{1}{1 + 4.7} = \dfrac{1}{5.7}$. Thus a note having a frequency of 5,000 c/s is amplified about twice as much as a note of 50 c/s; in this manner the less powerful reproduction of the high notes due to increasing loudspeaker impedance with rising frequency is corrected. The feedback is dependent upon the impedance of the loudspeaker coil used, so that a given feedback circuit should be used only with the loudspeaker for which it is designed. Fig. 339 gives the frequency characteristic of the circuit shown in Fig. 338 and illustrates the reduction of the loudspeaker current at the resonance peak of the loudspeaker (at 80 c/s). An I.F. coil with the indicated inductance and resistance ($R/L = 14,000$) can very well be used for the coil in the feedback circuit.

Reduction of feedback, however, is desirable not only for high frequencies but also for low frequencies. In most receiving sets there is not only an electrical attenuation of the low notes (due, for instance,

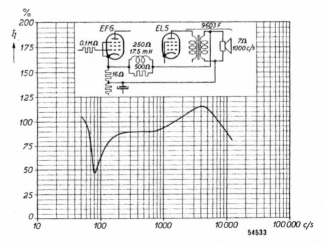

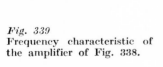

Fig. 339
Frequency characteristic of the amplifier of Fig. 338.

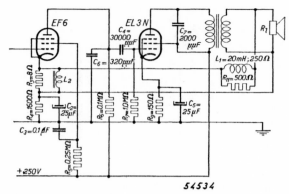

Fig. 340
Circuit of the A.F. part of a receiving set with voltage
feedback. The frequency characteristic is corrected
for the high notes by the choke L_1 and for the low notes
by the choke L_2.

to the no-load impedance of the loudspeaker transformer), but also an
acoustic attenuation, because of the loudspeaker being mounted in a
comparatively small cabinet instead of on a large baffle-board. By
reducing the feedback for the low notes and thus increasing the A.F.
amplification, better reproduction of the low notes can be obtained. This
compensation can be achieved by connecting up a second choke coil
parallel to R_3 (see Fig. 340), but as we are concerned here with low frequencies a coil with a closed iron core will usually have to be employed.
However, as two coils, one of which with an iron core, is a rather
expensive solution, another possibility is indicated where the compensation is obtained both for the high notes and for the low notes by
means of condensers. An example of such a circuit is given in Fig. 341,
where the loudspeaker voltage is applied to a voltage divider consisting
of the resistors R_1, R_2, R_3 and R_4. The voltage across R_3 and R_4 is
applied to the cathode of the A.F. pre-amplifying valve, these resistances being so chosen that together they give the exact value for the
cathode resistor of the EF 6. Parallel to the resistors R_2 and R_4 are the
condensers C_2 and C_4, having such a value that the resistor R_2 is
practically short-circuited by the condenser C_2 in the middle of the
frequency range, while R_4 is not short-circuited by C_4 until the high
frequencies are reached. By this means the part of the loudspeaker
voltage applied to the cathode is reduced. Consequently the feedback
decreases for the high frequencies. Since at low frequencies R_2 is not
short-circuited by C_2 the feedback is also reduced in the lower fre-

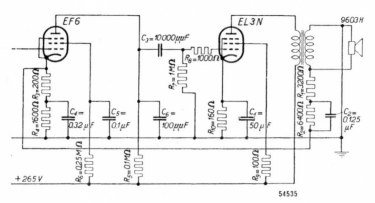

Fig. 341
Circuit of the A.F. part of a receiving set operating with voltage feedback. The frequency characteristic is corrected for the low and the
high notes by the condensers C_2 and C_4. Owing to the non-bypassed resistor of 1,800 ohms in the cathode lead of valve EF 6 there is risk of hum.

quency range. With this arrangement, however, the cathode of the EF 6
is not earthed for alternating voltages. In the cathode lead is a resistance
of 1,800 ohms which is practically non-bypassed at low frequencies, so that
under certain conditions hum voltages may arise across that resistance [1]
(see Chapter **XXVIII**), and therefore this is not ideal. In this respect
the ideal solution is to apply the feedback voltage not over a non-bypassed part of the cathode resistor but via a resistor to the grid of
the EF 6, as shown in Fig. 344 for the EBC 3. The drawback then,
however, is that either the feedback or the amplification is reduced.
The feedback with the arrangement of Fig. 341 is 30-fold at 500 c/s.
Curve a in Fig. 342 represents the frequency characteristic measured

Fig. 342
Frequency characteristic of the
circuit shown in Fig. 341 (loudspeaker current relative to that
at 800 c/s):
a) without negative feedback;
b) with negative feedback,
without correction of the
frequency characteristic;
c) with negative feedback and
with correction of frequency characteristic.

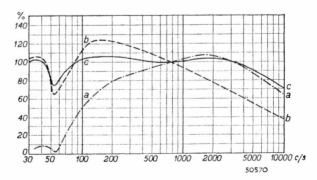

[1] The difference between the arrangements of Figs 338 and 340 lies in the much greater
non-bypassed resistance.

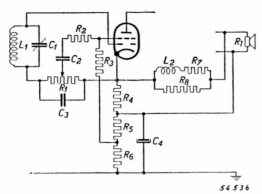

Fig. 343
Circuit of a double-diode-triode as pre-amplifying valve in the A.F. part of a receiver with voltage feedback.

for this circuit without negative feedback. With negative feedback but without correction (i.e. without the condensers C_2 and C_4) the frequency characteristic b was observed. These curves show that by applying negative feedback the reproduction of the low notes is considerably improved, whereas that of the high notes is reduced. If the reproduction of the high notes is improved by introducing C_4 then the low notes are sacrificed. Therefore the combination of C_2 and C_4 is chosen in such a way that both the high notes and the low notes are uniformly reproduced (curve c).

Compared with correction by means of coils (Fig. 340), the circuit of Fig. 341 affords the great advantage that the reproduction of the low and the high notes can be very easily modified by varying the value of C_2 or C_4 respectively.

164. Multiple Valves in the A.F. Pre-amplifying Stage

When a valve like the EBC 3, in which the triode amplifying part is combined with diodes for detection and automatic volume control, has to be used in circuits with negative feedback this leads to complications. Since in this case the diode and the first amplifying valve have one common cathode, it is necessary to arrange for the feedback voltage to be applied to the amplifying section of the valve without applying this voltage to the diode at the same time. Fig. 343 indicates how this can be done.

The diode is connected in the normal manner, with the feedback voltage across R_4 not being applied to the diode circuit. As, however, this voltage is applied, via the grid-leak resistor R_3, between the grid and the cathode of the EBC 3 it is necessary to have a large resistance interposed between the grid and cathode, as otherwise the feedback voltage would still reach the diode. For this reason R_2 is connected between the grid and the volume control.

The tap between R_5 and R_6 is provided because of the fact that the

404

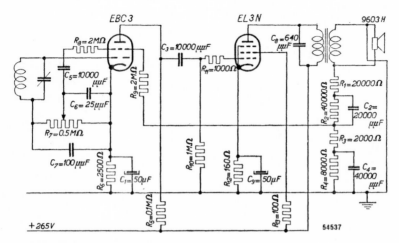

Fig. 344
Circuit of the A.F. part of a receiver operating with voltage feedback employing a double-diode-triode as pre-amplifier. The frequency characteristic is corrected by condensers C_2 and C_4 for low and high notes. As there is no non-bypassed resistor in the cathode lead there is no risk of hum.

delay voltage for the automatic volume control is not equal to the required negative grid bias of the EBC 3. The following values may, for instance, be employed:

$$R_2 = R_3 = 1.6 \text{ megohms},$$

$$R_5 + R_6 = 7{,}000 \text{ ohms}.$$

R_2 and R_3 act as a voltage divider both for the feedback voltage and for the A.F. signal coming from the diode, in the ratio of 2 : 1. This voltage reduction must therefore be taken into account when determining the factor β of the negative feedback, because it is thereby reduced.

If a fairly large negative feedback is applied to the EBC 3 and the EL 3N for instance, the amplification following the diode becomes so low that there is a possibility of considerable distortion in the I.F. valve.

A practical form of circuit for the EBC 3 in combination with the EL 3N is shown in Fig. 344, in which the frequency characteristic is corrected according to the principle of Fig. 341. The feedback voltage in this case is not applied between cathode and chassis but conducted via a resistance of 2 megohms to the grid. With this arrangement a 4-fold

405

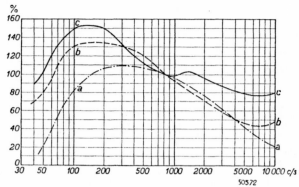

Fig. 345
Frequency characteristic of the circuit shown in
Fig. 344 (loudspeaker current relative to that at
800 c/s):
a) without negative feedback;
b) with negative feedback, without frequency-
   characteristic correction;
c) with negative feedback and with frequency-
   characteristic correction.

feedback is obtained.
Fig. 345 gives the fre-
quency characteristic
of this arrangement
without negative feed-
back (a), with negative
feedback (b), and with
correction (c). It will
be seen that the low
notes are reproduced
in somewhat greater
strength, so that, in
spite of a small baffle-
board being used for
the loudspeaker, high-
quality reproduction
is obtained. The re-
production of the low
notes can, however, be reduced by increasing C_2.
The circuit arrangements described here can of course be applied also
to other output valves.

165. Phase Conditions and Stability in Feedback Circuits

In a single-stage amplifier with resistance coupling the alternating
voltage at the anode is displaced 180° in phase with respect to the grid
alternating voltage. In a two-stage amplifier, as often employed in
receiving sets, the phase of the alternating voltage at the anode of the
last valve should be the same as that of the alternating voltage on the
grid of the first valve, because it is displaced twice 180° in the two
stages, but owing to the coupling elements between the valves, which
do not consist only of pure resistances, there will be some phase shift.
Consequently the output alternating voltage will have a phase angle
with respect to the phase it should have if there were no phase shifting
in the different stages of the amplifier. The feedback circuit, too, will
as a rule cause a phase shift, so that the feedback voltage at the input
of the amplifier has again a phase angle with respect to the phase it
would have were it not for the phase shift in the feedback circuit. The
phase shifts in the amplifier and in the feedback circuit are of course
dependent upon the frequency. Under certain conditions, owing to
these phase shifts, the amplifier may show in a certain frequency range
regeneration instead of degeneration. If this regeneration is of sufficient

magnitude, then, as will be shown below, oscillation will occur in the amplifier. In the construction of an A.F. amplifier with negative feedback, therefore, one has to make sure that no such regeneration can occur at any frequency, for the reason, too, that even if oscillation does not occur regeneration causes an extra emphasis of the voltages in a particular frequency range, whilst the voltages in the degenerating range are attenuated, thus spoiling the reproduction characteristic of the amplifier (increased attenuation distortion). Further, the regeneration causes increased distortion and increased effect of interference, instead of reducing this.

As a result of the phase displacement in the amplifier and in the feedback circuit the terms A and β in Equation (248) (see Section 161a β) are not real ordinary numbers but quantities which contain a phase angle and which in vector analysis are represented by complex numbers. When β A is real and positive the amplification according to Equation (248) is reduced, and when β A is real, negative and in absolute value less than unity then the amplification is increased (the numerator of Equation (248), $1 + \beta$ A, is then smaller than unity and A' therefore greater than A). In that case we therefore have to deal with regeneration (the same as the back-coupling in oscillators). When β A equals —1 the numerator of Equation (248) is zero and consequently A' is infinitely great. The amplifier is then unstable. In such a case it is possible, without any input signal, that an undamped oscillation is set up, i.e., the amplifier oscillates.

When β A is real and negative and has an absolute value greater than unity one would suppose that the amplifier is always unstable, but this need not invariably be the case (cf. H. Nyquist, Regeneration Theory, Bell System Techn. Journal 11, 1932, p. 126; B. D. H. Tellegen, Philips Techn. Review, 2, 1937 pp. 289 et seq..

For these reasons, when applying negative feedback it is necessary to examine carefully the phase displacement occurring at all frequencies; the magnitude of negative feedback that can be applied in practice is limited.

CHAPTER XXVIII

Hum Arising from the Mains

166. Introduction

With receiving sets and amplifiers that are fed from a.c. mains it is often noticed that the music or speech reproduced by the loudspeaker is accompanied by hum. This hum is apparently caused by the alternating voltage of the mains, which affects the amplifying stages of a set. Magnetic induction from the mains transformer can take place in several parts of the set, for example in the output transformer, loudspeaker, R.F. or I.F. coils. Further, hum may be caused by electrostatic induction in the wiring, particularly in leads to the valve grids, due to inadequate smoothing of the anode, screen-grid and control-grid voltages and to the a.c. feeding the valve filaments. Magnetic induction from the mains transformer may also cause hum in the valves, because of the transit of the electrons and consequently the anode current being influenced by the magnetic field. Under normal conditions, when there is no magnetic field, electrons pass across to the anode, but under the influence of such a field they may shoot past the anode.

It is hardly possible to make provision in the valve for eliminating or neutralizing the effect of external magnetic fields (this would require, for instance, a 3 to 5 mm thick, hermetically sealed, iron screen), so that in constructing a set care has to be taken that the valves are not situated in a strong stray field of the mains transformer.

In many cases hum can be avoided by judicious arrangement of the component parts. Hum arising in the valves themselves due to the indirect heating of the cathodes by a.c. can be partly avoided by a suitable valve construction, whilst it can be further suppressed to a great extent by the use of suitable circuits.

167. The Components of the Mains Alternating Voltage

Apparently hum arises in a valve when an alternating voltage originating from the mains is applied, for some reason or other, between the screen grid, the anode, the control grid or the cathode and the chassis (earth). When a set is energized by 50-cycle mains the various electrodes are subjected to alternating voltages having a frequency of 50 and sometimes of 100 c/s, the latter being, for instance, the result of frequency doubling in the rectifier and accompanied by a series of harmonics, since the ripple is distorted.

The alternating voltage from most mains is not purely sinusoidal but somewhat distorted (the power-station generators can supply only an approximately sinusoidal alternating voltage; moreover, distortion arises also in transformers and other equipment), so that the mains alternating voltage contains not only the fundamental frequency of 50 c/s but also harmonics of that frequency. Consequently when investigating the sources of hum we have to consider mains voltages having a fundamental frequency of 50 c/s and harmonics. Experience shows that it is best to start with a purely sinusoidal alternating voltage of 50 c/s with a 3% harmonic of 500 c/s, which will, therefore, be used as the basis of the following considerations.

168. Influence of Aural Sensitivity varying with Frequency

First and foremost in hum investigations it is of importance to know what hum voltage is permissible at the various electrodes. As the ear is not equally sensitive to all frequencies the hum voltage permissible for an electrode also depends upon the frequency. For a frequency of 50 c/s the sensitivity of the ear is extremely low, whereas for 500 c/s it is very much higher. The ratio of the just audible volume of sound to the just audible sound volume at 1,000 c/s expressed in dB [1]) is shown in Fig. 346 as a function of the frequency (threshold of hearing at different frequencies with respect to the threshold of hearing at 1,000 c/s). According to this curve a relatively high hum voltage is permissible at low frequencies (50 c/s). To be able to express hum by a single figure Philips use in their measurements a filter having a frequency characteristic corresponding to the sensitivity of the ear for low volumes of sound.

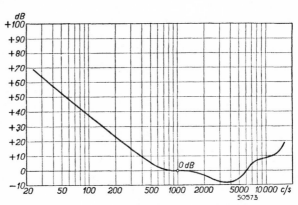

Fig. 346
Curve representing the ratio in dB of the sound at the threshold of audibility to the sound volume at the threshold of audibility at 1,000 c/s, as a function of the frequency. The threshold of audibility of a note of 1,000 c/s corresponds to a sound intensity of 10^{-16} W/cm².

---

[1]) For the definition of dB, see Appendix, page 508.

169. Definition of Hum Voltage

By hum voltage is to be understood here the root-mean-square (r.m.s.) value of the 500-cycle voltage at the electrode in question which gives the same impression of sound volume as the disturbing hum note. The ear being an unreliable instrument for comparison, in the testing of valves the hum voltage is measured after a filter which imitates the frequency-discrimination of the sensitivity of the ear. As regards hum arising in the valves a distinction has to be made between **hum-modulation** and **direct hum.** The definition of hum-modulation has already been given in Chapter XV. This arises in the R.F., mixing and I.F. stages, owing to the carrier wave being modulated with the hum voltage as a result of the curvature of the valve characteristics.

By **direct hum** is understood the hum reaching the loudspeaker as a consequence of direct A.F. amplification.

In the case of R.F., mixing and I.F. valves it is the hum-modulation that is of importance, whereas in detector diodes, A.F. pre-amplifying and output valves the direct hum predominates.

170. Permissible Hum Voltage

Since of all the electrodes the grid is usually the most sensitive to hum, the permissible hum voltage generally relates to the voltage between control grid and cathode. Experience has proved that with a frequency of 500 c/s and a loudspeaker of normal sensitivity a ratio of about 1/2,000 between the hum voltage and the alternating voltage producing a power of 50 mW in the loudspeaker (the voltage standardized for sensitivity measurements—see Chapter XXXIII) is just permissible. Consequently, when using a loudspeaker of average sensitivity the hum voltage—in the case of direct hum—permissible on an electrode corresponds to the voltage applied to that electrode which gives across the primary of the output transformer at most 1/2,000 of the voltage required for the standardized output power of 50 mW.

For hum-modulation, on the other hand, experience has shown that a maximum modulation depth of the carrier wave of $0.25^0/_{00}$ as a result of the 500-cycle component of the mains alternating voltage is permissible.

For R.F., mixing and I.F. valves, therefore, a certain maximum permissible ratio between hum and sound-volume level is fixed, whilst in the case of direct hum an absolute value has been taken for the maximum level of the hum. This is due to the fact that A.F. pre-amplifying valves and output valves are usually connected after the volume control, so that for these valves the setting of the volume control does not affect the hum level. Provision has to be made, therefore, to ensure that there is no audible hum when the volume control is set for low volume.

In the case of hum produced by R.F., mixing, I.F. and detector valves, however, the hum level is in fact affected by the setting of the volume control, together with the desired A.F. signal. Once a certain ratio exists between hum level and signal strength, by turning back the volume control the hum and A.F. signal are both attenuated and consequently the hum does not increase relative to the A.F. signal.

(a) Determining the Permissible Hum Voltage of Output Valves

For output valves the maximum permissible hum voltage can be derived from the data of these valves. As a rule the sensitivity is quoted, i.e. the control-grid alternating voltage required for a power output of 50 mW. In the case of the high transconductance output valves, such as the EL 3N, the EBL 1, the EBL 21, etc. the sensitivity is about 0.3 V (r.m.s.), so that the permissible hum voltage equals

$$\frac{0.3}{2,000} \text{ V} = 150 \ \mu\text{V}.$$

(b) Determining the Permissible Hum Voltage of A.F. Pre-amplifying Valves

The hum voltage permissible on the grid of an A.F. pre-amplifying valve is determined by the amplification between that grid and the input of the final stage and by the sensitivity of the output stage. This hum voltage equals the quotient of the hum voltage permissible on the grid of the output valve and the amplification between that grid and the grid of the A.F. pre-amplifying valve. If this amplification is very large (as is the case, for instance, with public-address amplifying installations using a microphone pre-amplifying stage) the permissible hum voltage thus calculated will usually be very much lower than the noise voltage caused by the Brownian thermal electron movement in the grid circuit and by the hail effect in the valve. With a grid impedance of 0.5 megohm the noise voltage due to Brownian thermal movement amounts to about 10 μV (measured after a filter having the same frequency discrimination as the sensitivity of the ear), without taking any account of the valve noise. It is pointless to demand a hum voltage much smaller than the existent noise voltage, so that in this case the condition can be specified that the hum voltage may not exceed the noise voltage.

411

(c) Determining the Permissible Hum Voltage of Detector Diodes

When the detector diode is combined with another valve it is usually the hum voltage permissible at the grid of that valve that is taken as a criterion. In the case of a separate diode valve account has to be taken of a large A.F. amplification, and the A.F. amplification attainable with a pentode is taken as a guide. It is to be observed here that it is not necessary to make such stringent demands of a detector diode as of the following A.F. pre-amplifying valve, since as a rule the volume control is connected between the former and the grid of the latter.

Only in a very exceptional case will the volume control be set at its maximum. When the control is turned back the hum voltage is less than when it is at its maximum. As a matter of fact when the volume control is at its maximum other factors play a part, for that position is generally used for the reception of a very weak signal, and due to various causes that signal is accompanied by a good deal of interference (for example, noise or atmospherics), so that under such circumstances no purpose is served by stringent requirements as regards hum.

(d) Determining the Permissible Hum Voltage of R.F. and I.F. Valves

The criterion for the permissible hum voltage on the grid of R.F. or I.F. valves is the depth of modulation of the carrier wave. As explained in Chapter XV, for R.F. and I.F. valves curves are published which give, as a function of the transconductance, the hum voltage on the grid which gives a carrier-wave modulation depth of 1%. For determining the permissible hum voltage, however, it is assumed that for 500 c/s a modulation depth of $0.25^0/_{00}$ is just tolerable. Therefore the permissible hum voltage is equal to 1/40 of the minimum value of the hum modulation curve.

(e) Determining the Permissible Hum Voltage of Mixing Valves

As regards the modulator or mixing part of frequency-changing valves, the same applies as stated for R.F. or I.F. pentodes. In the oscillator part (e.g., the triode part in triode-heptodes and triode-hexodes) the oscillator voltage can also be modulated with a hum voltage. This would result in hum-modulation of the I.F. carrier wave.

Thus in the case of the ECH 21 with normal operation of the mixing part ($V_a = 250$ V, $V_{g2,4} = 100$ V, $V_{g1} = -2$ V) and of the oscillator part ($V_a = 100$ V, $R_g = 50,000$ ohms, $I_g = 210$ μA) the I.F. carrier wave is modulated to a depth of 1% by an A.F. voltage of 0.8 V at the grid of the triode. Consequently a hum voltage of 20 mV would just be permissible on that grid for a maximum modulation depth of $0.25^0/_{00}$. With lower oscillator voltages, however, the modulation depth increases; with an oscillator current of 75 μA through a leak resistor of 50,000 ohms the I.F. carrier wave is modulated to a depth of 1% already at 0.18 V, so that it is not desirable to permit a hum voltage higher than 1 mV on the grid of the oscillator valve. Added to this is the fact that a hum voltage on the oscillator grid also causes frequency modulation of the I.F. carrier wave at the hum frequency. If the frequency of the I.F. carrier wave is displaced from the centre of the resonance curve of the I.F. band-pass filter in the anode circuit of the heptode part of the valve considered above the frequency modulation will also cause amplitude modulation, which is much more troublesome than the direct hum-modulation. In this manner, with an oscillator d.c. of 210 μA of the ECH 21 a hum voltage of 80 mV with a normal band-pass filter causes a hum-modulation depth of 1% (the same occurs when I_g = 75 μA and the hum voltage is 18 mV).

171. The Various Causes of Hum in the Valve

It was stated in the foregoing that one of the principal causes of hum in the valve lies in the a.c feeding the filament of indirectly-heated cathodes. Therefore the magnitude of the alternating voltage between the terminals of the filament is partly an indication of the hum. The valve filaments of radio sets may be connected in parallel (a.c. sets) or in series (d.c./a.c. sets). In the latter case the hum is usually stronger than that with parallel connection, because for the majority of the valves of a receiver the alternating voltage between filament and cathode is higher in series feeding by alternating current. As most of the modern Philips valves can be used both for parallel and for series filament connection, in the examples given below the less favourable condition of series connection will be considered.

These considerations apply also for the more recent d.c./a.c. valves, such as the U series with pressed-glass base.

(a) Anode Hum

By anode hum is to be understood the hum voltage that exists, from one or more causes, between anode and cathode or, what generally amounts to the same thing, between anode and chassis (earth). This does not include, however, the hum voltage that may exist at the anode owing to the control of the anode current by a hum voltage present between one of the other electrodes (grids) and the cathode and which sets up a hum voltage across the anode resistor, if any.

Fig. 347 gives an example of a simple circuit of a valve without any impedance in the grid and cathode leads. An amplifier for measuring the hum voltage is connected to the load resistor R_L. There are various causes of hum, which together produce the **anode hum**.

Causes of anode hum are:

1) The capacity between filament and anode.

2) Electrons emitted by the cathode which first pass through the various grids in the normal way but shoot just past the edge of the anode; during the half-cycle of the filament alternating voltage when the filament is positive these electrons may strike the filament, whereas during the other half-cycle when the filament is negative they reach the anode after making a detour. This sets up a hum voltage on the anode, which disappears if the filament voltage is maintained considerably negative, for instance by means of a battery. This cause of hum should be avoided by a suitable construction of the valve.

3) The filament-anode capacity. Generally this capacity can be kept sufficiently small.

Anode hum, if present at all, is found only in the A.F. output and A.F. pre-amplifying stages. (In R.F. and I.F. valves there is no appreciable A.F. impedance in the anode circuit. Moreover, an A.F. voltage on the anode causes only a very slight modulation depth.)

In a circuit as shown in Fig. 347 hum may also arise from other less frequently occurring causes. It is theoretically possible, for instance, for the insulation between filament and anode to be bad, but in practice this does not occur. It is also possible for the filament to control the electron current between cathode and anode, acting more or less as a grid. This phenomenon, too, can easily be avoided in the construction of the valve.

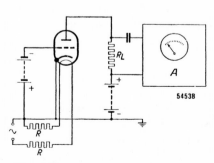

Fig. 347
Essentials of the circuit for measuring the anode hum of a valve.

Further it may happen that a part of the filament not shielded by the cathode emits electrons. During part of the cycle the filament is usually negative with respect to the whole electrode system and the electrons then emitted by the filament may pass over to all the electrodes, hence also to the anode and the grid. When there is a high impedance in the grid circuit even very small currents may be disturbing, on account of the grid circuit being so highly sensitive. In anode hum no account need be taken of this electron emission from the filament, because when it does occur in the first place the grid hum becomes too strong.

Finally mention has to be made of magnetic hum. Generally the filament of indirectly-heated valves is fed by a.c., so that a magnetic alternating field may be produced by the filament. In certain circumstances this alternating field may influence the course of the electrons passing to the anode, thereby giving rise to anode hum. In certain modern valves such an effect is very limited owing to the filament feed and return leads lying alongside each other, and the field of one coil neutralizes that of the other (so-called coiled-coil filament).

(b) Grid Hum

By grid hum is to be understood the hum voltage arising through one or more causes between grid and chassis (earth). For a study of grid hum we may use the diagram of Fig. 348, which differs from Fig. 347 in that it shows a resistor R_g in the grid circuit. Here it is assumed that with a short-circuited grid resistor the anode hum is sufficiently weak. When that short-circuiting is removed the increase in the hum may be termed grid hum.

The main causes of grid hum are:

1) bad insulation between grid and filament;

2) too great a capacity between grid and filament;

3) emission from the filament to the grid.

The minimum value of the insulation resistance between filament and grid and the maximum value of the respective capacity permissible are easily calculated. Let us take the impedance between grid and cathode of the EBC 3 as, say, 0.5 megohm and the amplification in the triode part of that valve as equal to 20. Assuming that in a given case a hum voltage of 10 μV is permissible (if the sensitivity of the output valve is, for instance, 0.4 V and the voltage on the grid of the EBC 3

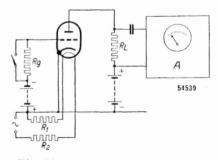

Fig. 348
Circuit for measuring the grid hum of a valve.

is 0.02 V, $\dfrac{1}{2,000} \times 0.02 \times 10^6 = 10\ \mu\text{V}$), we find that the minimum permissible value of the leak resistance between grid and filament is about 9×10^9 ohms ,if the EBC 3 occupies the first place in the heating-current circuit of a d.c./a.c. set (one side of filament connected direct to the chassis, see Chapter XXXII). In this case the alternating voltage between filament and chassis (see Fig 348) at the hum-producing end of the filament is equal to 6.3 V and hence 6.3×10^5 times as great as the admissible hum voltage across the grid-leak resistor of 0.5 megohm (10 μV). The insulation resistance must therefore also be 6.3×10^5 times as great as the grid-leak resistance, i.e. $6.3 \times 10^5 \times 0.5 \times 10^6 \approx 3 \times 10^{11}$ ohms. Taking into account the fact that, owing to the ear being less sensitive to a frequency of 50 c/s, of the 6.3 V heater voltage only the 3% having a frequency of 500 c/s plays a part, we find the minimum permissible value of the insulation resistance to be $0.03 \times 3 \times 10^{11} = 9 \times 10^9$ ohms. Under the same conditions the filament-grid capacity may not exceed 0.034 $\mu\mu$F (for 500 c/s the impedance of 0.034 $\mu\mu$F amounts to about 9×10^9 ohms). For R.F. and I.F. valves the permissible values of insulation resistance and grid-filament capacity can be calculated in a similar manner.

It is difficult to determine the limit for the permissible emission from the filament to the grid. Generally in the case of indirectly-heated valves this emission is very low and the electron current is accordingly already saturated at very low voltages. Due to this saturation, however, the hum voltage set up in this way contains a large number of higher harmonics, which are more disturbing than a voltage consisting of only the fundamental frequency. The degree to which the electron emission current may be disturbing can therefore be judged only by measuring the hum behind a filter imitating the frequency discrimination of the ear. It will be evident that all phenomena related to grid hum depend considerably upon the value of the A.F. grid impedance; consequently this impedance should not be chosen higher than is absolutely necessary.

(c) Cathode Hum

By cathode hum is to be understood the hum voltage existing between cathode and chassis (earth), which is therefore effectively between cathode and control grid when the latter has the potential of the chassis. If between cathode and chassis there is an impedance of a fairly high value owing, for instance, to the cathode resistor being shunted by a condenser of insufficient capacity, or when there is no cathode condenser at all (as sometimes occurs in negative-feedback circuits), such an impedance may give rise to hum. Cases of this are met with in output valves in which the negative feedback is obtained by omitting the cathode condenser, and also in R.F. and I.F. valves in which the cathode resistor is shunted by a cathode condenser which is effective only at the high or intermediate frequencies, the condenser having a high impedance at the frequency of the hum voltage. The cathode impedance at the hum frequency is then formed principally by the cathode resistance. Here it is to be observed that the hum voltage between cathode and chassis does, it is true, control the anode current, but that the resultant anode alternating current is in antiphase to the alternating current passing through the cathode impedance as a result of the cathode hum.

There is, therefore, a certain amount of negative feedback, which results in the effect of the hum voltage being less than one would expect if that back-coupling were overlooked. Consequently a somewhat higher hum voltage is permissible.

The main causes of cathode hum are:

1) bad insulation between cathode and filament;

2) emission from cathode to filament;

3) emission from filament to cathode.

The filament-cathode capacity is generally very small compared with the cathode-chassis capacity and does not give rise to hum. Also when the cathode resistor is not by-passed the impedance of the cathode-filament capacity at hum frequencies is generally very great compared with the cathode-resistance value.

When the impedance of the cathode condenser for hum frequencies is great (e.g., in R.F. valves a condenser of only 10,000 $\mu\mu$F is often employed) the insulation resistance between filament and cathode plays an important part.

The minimum permissible value of the insulation resistance between cathode and filament is easily calculated for any valve. Suppose that for the triode-hexode valve ECH 3, for instance, having regard to hum-modulation, a hum voltage of 225 μV is permissible on the grid. If the filament of the ECH 3 is the third link in the chain of valve filaments then the filament-chassis alternating voltage at the filament end which is most productive of hum is 19 V, given that each of the two preceding valves has a heater voltage of 6.3 V. Bearing in mind the variat￼ aural sensitivity with frequency, it is again assumed that voltage only the 3% of frequency 500 c/s is instrumental i hum arising from faults in the insulation. If the cathode resistance is 215 ohms (the normal value for this valve) then the minimum value for the filament-cathode insulation resistance should be

$$215 \times \frac{0.03 \times 19}{225 \times 10^{-6}} \text{ ohms} = 0.55 \text{ megohm.}$$

Here no account has been taken of the aforementioned negative feedback, which has a factor of $(1 + g_k R_k)$. In the case of the ECH 3 the cathode transconductance g_k is roughly equal to 2.3 mA/V, so that $1 + g_k R_k = 1 + (2.3 \times 10^{-3} \times 215) = 1.5$. Consequently the insulation resistance between filament and cathode must be at least $55 : 1.5 = 37$ megohms.

Thanks to the modern methods applied nowadays in manufacture it is possible to obtain such an insulation resistance value at the usual cathode temperature (it is to be borne in mind that the insulation resistance may vary with the voltage between cathode and filament).

In the emission from the cathode to the filament the electrons may pass to the latter both inside the cathode and round the outside of it. In the latter case two possibilities are to be distinguished:

a) the electrons leave the emitting layer at the bottom end of the cathode (where the filament is led in) and pass directly to the filament, thereby setting up a hum voltage across the cathode resistor;

b) the electrons pass through the various valve grids, just touch the anode during one half-cycle and strike the filament during the other half-cycle, thereby producing a hum voltage directly on the anode, as already explained.

Emission around the outside of the cathode is easily suppressed by employing a screen, whilst the emission inside the cathode can be kept low by carefully avoiding any contamination of the insulating material. Emission within the cathode and that from the filament to the cathode are generally very low and already saturated at a low voltage, so that, as mentioned in the case of grid hum, strong harmonics of the mains frequency occur.

172. Hum due to the A.C. Supply of Directly-heated Valves

Finally there remains to be mentioned the hum occurring as a result of the a.c. supply of directly-heated valves. It is only in exceptional cases that directly-heated valves are fed with alternating current and then the secondary winding of the heater-current transformer is generally provided with an earthed centre-tap, thereby reducing the alternating voltage between the filament ends and earth to one half of the heater voltage. Thus it is possible, in the case of direct heating, that owing to the low heat capacity of the filament the electron emission varies at twice the mains frequency, thereby giving rise to an alternating current of that frequency in the anode circuit. Further it is possible that the filament may not emit electrons uniformly over its entire length, so that the hum is not weakest when the centre-tap of the heater-current winding is earthed. Even if the emission is uniform over the greater part of the length of the filament the electron current may be influenced owing to the voltage drop in the filament, thus giving rise to hum. Moreover, this results in a variable space charge in front of the control grid, causing fluctuations in the control-grid capacity and, consequently, in the tuning of the circuit connected with it.

For these reasons it is generally preferred not to apply direct heating for valves that have to be fed with alternating current. In the case of A.F. pre-amplifying valves the difficulties connected with hum are so great as to make indirect heating essential, and the same is true of detector diodes, and R.F. and I.F. valves. As regards output valves it is possible with care to reach a fairly low hum level, and in the past directly-heated power valves were in fact on the market. At the present day, however, these are almost entirely replaced by indirectly-heated valves, for the additional reason that with the equi-potential cathode a much greater transconductance is attainable, so that one is quite prepared to overlook the somewhat higher heating power required for the indirectly-heated cathode.

CHAPTER XXIX

Microphonic Effect

173. Introduction

As will have been understood from Chapters VI and VII, a radio valve consists of a large number of parts which are not absolutely rigid or are not tightly connected with each other. Small variations in the position of the control grid or, in the case of directly-heated valves, of the filament may affect the anode current. If a valve in a circuit receives a mechanical shock it may happen that the grid, the cathode or some other parts become slightly displaced and a damped oscillation, though small, is set up (through the damped vibration of the respective parts). In such an event the geometry of the valve is somewhat disturbed, thereby affecting the electrical values, such as current and amplification. The anode current then shows oscillations similar to the mechanical vibrations set up in the valve when it is knocked; these oscillations are amplified in the following valves and reproduced in the loudspeaker as one or more damped notes.

Generally the cause of the mechanical vibration of a valve is an acoustic vibration radiated by the loudspeaker. The valve is then comparable to a condenser microphone and this phenomenon is called **microphonic effect.**

Microphonic effect may be due to any one of a large number of causes and it may manifest itself in different ways. It is possible, for instance, that the loudspeaker does not produce a note but rather a rustling or crackling noise.

The electrical vibration caused by acoustic vibrations being conducted from the loudspeaker to a valve sensitive to microphonic effect may, however, be amplified to such an extent and reach the grid of the output valve in such a phase as to set up a sustained oscillation. Even if the original alternating voltage at the valve grid is omitted, the noise from the loudspeaker will still continue, so that by feedback a continuous note is produced.

174. Causes of Microphonic Noise

Microphonic noise may, in principle, arise from two causes, viz.:

1) through loose connections between the parts in a valve which have to make contact with each other;

2) through mutual displacement of parts which influence the anode current.

As regards the first cause, there should not as a rule be any loose connections at all in a valve. There is, however, one danger point, in the cathode of an indirectly-heated valve, for the coiled filament, which is surrounded with a layer of insulating material, is not rigidly fixed in the cathode tube and if the valve should suffer a mechanical shock it is possible that the filament will be moved; as a result disturbances in the form of crackling noises may arise in spite of the insulation (see Section 179).

As remarked in the introduction, a valve that is sensitive to microphonic effect will produce a damped note when it is knocked. In the operation of a receiving set (turning the switches, etc.) or when the set is subjected to shocks, this is apt to lead to unpleasant disturbances.

In the case of microphonic effect due to the reaction of the loudspeaker upon valves sensitive to that effect, three possibilities are to be distinguished:

a) the reaction of the loudspeaker upon an A.F. valve (pre-amplifying stage or output stage);

b) the reaction of the loudspeaker upon a R.F. or I.F. valve;

c) the reaction of the loudspeaker upon a frequency-changing valve.

175. A.F. Microphonic Effect

If the reaction of the loudspeaker upon an A.F. valve is great enough, owing to microphonic effect in the loudspeaker an undamped A.F. note may occur, which will maintain itself if the grid alternating voltage on the valve that caused the original note in the loudspeaker should disappear. In that case, therefore, there is self-excitation, which leads to a disturbing note; this can be stopped by reducing or eliminating the reaction (removal of the loudspeaker or reduction of the amplification following the valve).

The electrode system of a valve consists of very fine rods, wires and plates, a large number of which have natural vibrations which lie in the A.F. range (50—10,000 c/s), and as a rule the damping of these vibrations is extremely small. If a valve vibrates at the natural frequency of such a component part microphonic effect can be produced, assuming that that part happens to influence the anode current. This effect can be determined with the aid of the measuring arrangement shown in

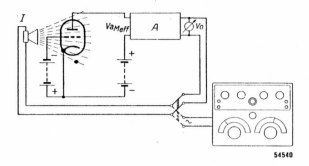

Fig. 349. With this arrangement the amplification following the valve under examination is increased to such an extent and the loudspeaker approached so close to the valve that any external disturbance or any irregularity in grid voltage will cause self-excitation. The loudspeaker is then switched from the amplifier to an audio-frequency generator of variable frequency, and one then measures the voltage at the output of the amplifier while varying the frequency of the audio-frequency generator. The result of these measurements can be plotted on a graph having the frequency on the horizontal axis and the output voltage V_o of the amplifier on the vertical axis. Fig. 351 illustrates the effect obtained with a good valve and Fig. 350 that obtained with a faulty valve.

It will be noticed that the output voltage of the amplifier rises suddenly when during scanning of the frequency range with the A.F. generator the frequency of the latter passes the natural frequency of a microphonic resonance. The resonance peaks are usually so sharp that one has to be very careful in varying the frequency of the A.F. generator, otherwise one may pass beyond them unnoticed. Often it is possible to distinguish a fundamental frequency and its harmonics. Other resonances do not show any mutual relation and apparently indicate vibrations of different component parts. It will be found that the resonance peaks are scattered over the entire A.F. range.

Microphonic effect is influenced by:

1) the frequency characteristic of the amplifier following the valve;

2) the phase of the acoustic feedback.

Obviously the resonances of microphonic effect that are amplified to the greatest extent in the frequency range of the A.F. amplifier will manifest themselves most strongly. Further, the phase of the acoustic feedback must be of such a nature that positive feedback occurs. The phase in which the mechanical vibration reaches the valve depends upon the distance between the loudspeaker and the valve. If that

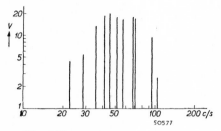

Fig. 350
Vertical axis: Output voltage (in volts) of the amplifier at the various resonance peaks.
Horizontal axis: Frequency of the audio-frequency oscillator in c/s. This diagram refers to a valve having unsatisfactory properties as far as microphonic effect is concerned.

distance equals the wavelength of a sound wave the incident vibration will be in phase with the loudspeaker membrane, whereas if the distance equals half the wavelength the vibrations at the two points will be in antiphase. By moving the loudspeaker in a straight line to the valve it will be found that at a relatively high frequency microphonic effect appears and disappears alternately, this taking place at those points which for sound vibrations in air are situated at intervals of half a wavelength. (This is based on the assumption that the loudspeaker is set up free in the open air.)

If the loudspeaker and the valve are linked by a chassis, i.e. by a solid, since the velocity of propagation through that solid is generally greater than in air, the incident mechanical vibration will still be in phase with the loudspeaker. Should this phase happen to be unfavourable for the feedback it may occur that regeneration takes place when there is free suspension but not when the coupling is constrained, for instance by a connection with the chassis; in the latter case microphonic effect is just avoided, because the altered phase has a predominating influence.

From the foregoing it is evident that feedback can take place in two ways, through the air and through a solid. A familiar means of counteracting feedback via a solid is the mounting of the loudspeaker and the chassis on rubber. The path then consists of: loudspeaker—rubber—cabinet—rubber—chassis. Formerly valves which were highly sensitive to microphonic effect were packed in rubber to counteract acoustic feedback via the air. Furthermore they were placed in resilient holders to counteract feedback via the chassis. In any case a set has to be constructed in such a way that the valve most sensitive to microphonic effect, e.g. the first

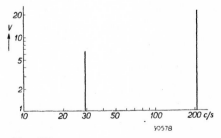

Fig. 351
Vertical axis: as Fig. 350.
Horizontal axis: as Fig. 350.
This diagram refers to a valve having good properties as regards microphonic effect.

A.F. amplifying valve, is placed as far as possible away from the loudspeaker. In the construction of the valves, too, measures should be taken to eliminate microphonic effect as far as possible, which can be done by heavily damping the component parts that are apt to be set in vibration by the loudspeaker. To this end the play between the cathode and the supports for the grids and the anode on one hand and the holes in the mica or ceramic plates on the other hand should be kept as small as possible. There is a limit, however, to the reduction of the play referred to, because if the supports are fixed in an absolutely rigid manner they are liable to bend when the temperature is raised, so that not only will the characteristic be affected but the grid-to-cathode path of the valve may even be short-circuited. Another measure is to use for the assembling of the electrode system parts that are as short and as rigid as possible. In battery valves the filament has to be properly tightened by means of springs, and lateral vibrations should be damped (for instance by a small plate of mica pressed up against the middle of the filament—see Fig. 27). In Philips valves steps have been taken so that in normal circuits, with normal A.F. sensitivity (normal amplification following the valve) and with a not too unfavourable positioning with respect to the loudspeaker, satisfactory results are obtained. In the A.F. part of a set the pre-amplifying valve is the most sensitive to microphonic effect, because it is followed by the greatest amplification. Output valves will not as a rule give rise to any microphonic effect. In the construction of Philips pre-amplifying valves, such as the EBC 3, EF 6, EF 9, EF 22, etc., it is specified that by normal mounting in the set and when a loudspeaker of the usual sensitivity is used the A.F. sensitivity measured on the grid of these valves may not exceed 6 mV (for a loudspeaker power intake of 50 mW).

In the case of public-address and microphone amplifiers the loudspeaker is usually set up at a great distance away from the set and there is no fixed connection between loudspeaker and chassis, so that a much greater amplification can be employed following the input valve.

In microphone amplifiers, owing to the excitation of the input valve by the acoustic waves destined for the microphone, the amplification following that valve must be limited, otherwise it would act as a microphone itself. Further, a limit is set for the gain in microphone and public-address amplifiers by the fact that small shocks and vibrations are apt to cause reverberation.

176. Influence of Operating Voltages on A.F. Microphonic Effect

The influence of vibrations of parts of the valve upon the anode current or the anode voltage depends also upon the voltages applied to the valve electrodes. With equal amplification following the grid of a triode, microphonic effect will be more pronounced with a high anode voltage than with a low one.

If microphonic effect occurs in a valve at a certain resonance then there will arise across the coupling impedance or the coupling resistance in the anode circuit an alternating voltage V_{aM} of a magnitude depending on the strength of the acoustic effect. The effective (r.m.s.) value $V_{aM\ eff}$ of the anode alternating voltage occurring at the given resonance owing to a certain acoustic influence can be plotted as a function of the anode direct voltage, producing a curve similar to that of Fig. 352 (relating to a triode with resistance coupling). From this curve it appears that microphonic effect becomes greater as the anode voltage rises, so that it is advisable to choose the lowest possible anode voltage (for triodes) and a correspondingly larger amplification following the valve. In the case of screen-grid valves the screen-grid voltage is of importance and should be kept low. Since for triodes with transformer coupling usually a higher anode direct voltage is applied than with resistance coupling, given the same amplification following the grid the input valve of an amplifier with resistance coupling will be less sensitive to microphonic effect than an input valve with transformer coupling. Finally, it is as a rule advisable to use valves having a large amplification factor.

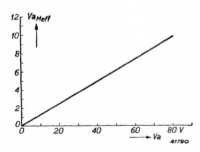

Fig. 352
Anode alternating voltage $V_{aM\ eff}$ occurring when a triode with resistance coupling is subject to a determined acoustic effect, as a function of the direct voltage on the anode.

177. H.F. Microphonic Effect

If a mechanical vibration acts upon a valve that is controlled by a R.F. or I.F. oscillation it is possible that this R.F. or I.F. oscillation will be modulated by the mechanical vibrations of parts in the valve. When this modulated R.F. or I.F. oscillation is further amplified and rectified in the detector the A.F. oscillation that caused the valve parts to vibrate will be reproduced in the loudspeaker. If the valve is subject to the acoustic vibrations of the loudspeaker there is then a possibility of self-oscillation,

425

provided the phase is favourable for it and the feedback is adequate. This phenomenon is called **H.F. microphonic effect.** The modulation of the H.F. output voltage of the valve in question may be due either to a fluctuation in the amplification of the valve in sympathy with the A.F. vibration of the valve parts, or from changes in the capacities. If the amplitude of the transconductance variation equals $m_M g_m$ and that variation takes place at a low angular frequency p ($= 2 \pi f$) the resultant transconductance will be

$$g_m{}' = g_m (1 + m_M \sin pt). \tag{265}$$

If the voltage of the H.F. carrier wave on the grid of the valve equals V_i then at the output of the valve there will be a current

$$V_i g_m{}' = V_i g_m (1 + m_M \sin pt). \tag{266}$$

The result is similar to that which would be obtained if a modulated carrier wave were present on the grid with a voltage of $V_i (1 + m_M \sin pt)$ which is amplified in the valve with a constant transconductance g_m.

This modulated carrier wave is amplified in the normal way by the following stages and rectified by the detector. We then get in the loudspeaker an alternating voltage of a magnitude proportional to m_M and a frequency corresponding to the angular frequency p caused by the vibration of the valve. The factor m_M is proportional to the strength of the mechanical vibration striking the valve and is related further to the construction of the valve. In this connection it is remarkable that the probability of microphonic effect becomes greater as amplification increases, regardless of the stage where that increase takes place, either before or behind the respective valve, in the high or low-frequency stage.

If the H.F. amplification is increased before the valve a correspondingly stronger signal acquires a modulation depth m_M. If, on the other hand, the amplification is increased behind the valve the signal with a modulation depth of m_M is further amplified. In both cases the signal strength at the detector is increased with a modulation depth of m_M and consequently also the signal at the output valve. If the A.F. amplification is increased we again get the same result. In this respect H.F. microphonic effect differs from A.F. microphonic effect, where only the amplification following the grid of the respective valve has any effect.

There is, therefore, as regards H.F. microphonic effect, no definite optimum distribution of the total amplification in a receiving set. The I.F. valve in a superheterodyne set may show the same microphonic effect as the H.F. valve. H.F. mirophonic effect is most pronounced when the aerial signal is strongest and the A.F. volume control is at the maximum position (max. amplification in the set). The amplification in the set depends, however, also upon the manner in which the automatic volume control works.

As stated above, H.F. microphonic effect may also be caused by capacity variations at the input (grid) of the valve. Though small, these capacity variations may affect the resonant frequency of the connected tuned circuit. If the I.F. carrier wave lies on one side of the resonance curve, owing, for instance, to the oscillator frequency not being adjusted to the correct value, then in consequence of the capacity variations the amplification of the I.F. wave changes periodically. This results in an amplitude modulation that can be compared to that caused by changes in the transconductance, which is determined by the product of signal and total amplification. It is to be noted, however, that here other factors also play a part. In the first place the sharpness of the resonance curve of an I.F. circuit has some influence, and secondly, if the I.F. is adjusted exactly to the centre of the resonance curve, the amplification will change twice during each cycle of the capacity variation. Feedback at the original frequency of the mechanical vibration is then impossible. It is typical, therefore, that this kind of microphonic effect occurs only if the high or intermediate frequency does not correspond exactly to the centre of the resonance curve. Naturally the absolute value of the capacity changes is smaller the smaller the grid capacity, and consequently the variation as a percentage of the total capacity is smaller. Further, these changes have least influence when the total capacity of the circuit is large. The variation of frequency has the greatest absolute value at high frequencies (short-wave range). In the short-wave range, however, the resonance of the respective H.F. circuits is much less sharp, so that here again the amplification variation will be small and there is little likelihood of self-oscillation occurring. In the medium- and long-wave ranges no microphonic effect has ever been found to occur in this manner. It is to be added that H.F. valves having a high negative grid bias, for instance as a result of the action of the automatic volume control, are more sensitive to microphonic effect.

178. Microphonic Effect in Mixer Valves

Microphonic effect may occur in mixer valves in the same way as in R.F. or I.F. valves. Owing to the variations in transconductance in the modulator part of a mixer valve the conversion conductance will likewise vary in sympathy with the acoustic feedback, and when conditions are unfavourable the valve may produce continuous oscillations. Also capacity changes of the R.F. grid will have the same effect as in

H.F. valves. Further, as is the case with H.F. valves, the sensitivity of a mixer valve with respect to microphonic effect is not influenced by the manner in which the amplification is distributed before and after the valve, and the probalibity of microphonic effect is greatest when the aerial signal is strong and the set has the highest possible sensitivity.

In addition to the possibilities mentioned in connection with H.F. valves, in the case of mixer valves there is another source of microphonic effect, namely in the oscillator part, for there the grid capacity is apt to be influenced acoustically (the capacity, for instance, of the triode grid in a triode-heptode or a triode-hexode). If the triode grid is in parallel with the oscillatory circuit, as is often the case, the tuning of that circuit is periodically changed by the variations in the grid capacity. This causes the intermediate frequency to be changed periodically, so that the result is comparable to that of grid-capacity change in a R.F. or I.F. valve. In the latter case, however, the resonant frequency of the circuit changes and the frequency of the signal remains constant, whereas in the former case the resonant frequency of the I.F. circuits remains constant and the signal frequency changes. There is a difference in that the capacity change of the oscillator grid will usually have more effect than the capacity change of the grid of an I.F. valve, because the oscillator frequency is generally higher than the frequency of the signal and consequently very much higher than the intermediate frequency. A small capacity change of the oscillator grid will therefore result in a much larger percentage deviation of the intermediate frequency than is the case with the control grid of a R.F. or I.F. valve. This is especially the case in the short-wave range.

Microphonic effect due to variation of the intermediate frequency depends also upon the steepness of the resonance curve of the whole I.F. amplifier, whereas in the case of a variation of the grid capacity of a R.F. or I.F. valve only the steepness of the resonance curve of the respective grid circuit is of importance. This, too, makes the effect with mixer valves more probable than the H.F. microphonic effect due to variation of grid capacity.

With triode-heptodes it is advisable to interpose the oscillator circuit in the H.T. lead to the triode anode and to connect the back-coupling coil to the triode grid. With the normal coupling between circuit coil and back-coupling coil any variation in the control-grid capacity is then stepped down, and consequently the variation in the tuning of the oscillator circuit is also reduced. There is, however, still the detuning due to capacity changes of the triode anode to be taken into account, but usually these changes are smaller than those of the triode grid.

179. Crackling and Scratching Noises

In the beginning of this chapter it was stated that in the case of indirectly-heated cathodes the filament in the cathode tube may give rise to microphonic effect, even though it is insulated from the cathode (see Chapter VII). In that case this effect manifests itself in the form of crackling or scratching noises.

This kind of microphonic effect, however, will occur only in certain circuit arrangements, that is to say when between the cathode and the filament there is an alternating voltage that influences the high, intermediate or low frequencies amplified in the valve or else the oscillator frequency. Such is the case, for example, in circuits with A.F. feedback where part of the cathode resistor or the whole resistor is not shunted by a condenser (see Fig. 338).

The A.F. alternating voltage across that resistor is apt to be modified by interfering frequencies arising from displacement of the filament, due to acoustic feedback or other mechanical shocks. Displacement of the filament causes variations in the filament-cathode capacity and, since the filament is usually earthed direct, also in the cathode-earth (chassis) capacity. These capacity changes will, of course, affect the cathode impedance for the alternating voltage in question, so that this voltage too, will vary.

This kind of microphonic effect may also occur in oscillator circuits where the feedback voltage lies between earth and cathode (Figs 260, 261 and 264). Capacity variations modulate the oscillator frequency.

The steps to be taken to avoid this effect are obvious, i.e. to ensure the best possible fixing of the insulated filament in the cathode tube. It is advisable, however, having regard also to the irregular nature of the filament-cathode insulation, to avoid any circuits where alternating voltages occur between filament and cathode which are likely to influence the reproduction of the loudspeaker in any way. With feedback circuits the unshunted part of the cathode resistor should as a rule be kept small (for instance, in pre-amplifying valves less than 50 ohms; in output valves the whole of the cathode resistance may be left unshunted).

180. Rattling Noises

It sometimes happens that a rattling noise is heard. Although this does not really belong to microphonic effect we will refer to it briefly here. It originates in the valve itself and can be heard when placing the ear close up to the valve. It is caused by an acoustic vibration from the loudspeaker when this is close to the valve. If the loudspeaker is caused to reproduce all frequencies in succession, then, owing to resonance, certain component parts in the valve will vibrate mechanically, and it is these mechanical vibrations that are heard as rattling noises (if these vibrations affect also the path of electrons then we have the A.F. or H.F. microphonic effect described in preceding sections).

181. Final Observations Regarding Quality of the Valves

In the preceding sections we have dealt with the various causes of microphonic effect. One might be inclined to conclude that microphonic effect is a frequently occurring phenomenon making it necessary to take special precautions in the construction of radio sets or other electronic apparatus. In the construction of Philips valves, however, every possible precaution against microphonic effect is always taken. It will be found in practice that these valves show no tendency to this effect, except, of course, where a circuit is exceptionally sensitive and the component parts are unfavourably mounted on the chassis. Directly-heated battery valves are naturally the most sensitive to microphonic effect, and a too large A.F. amplification is to be avoided (usually it is advisable to have no more than one A.F. stage preceding the output stage), whilst there is hardly any likelihood of H.F. microphonic effect leading to difficulties. A normally constructed set should not show any microphonic effect if the output stage is fully driven by the grid-control voltage with a small modulation depth (e.g. 10%) of the received aerial signal.

CHAPTER XXX

Phenomena Occurring During the Life of a Valve

The useful life of a valve is just as much limited as that of an incandescent lamp. In the course of time material inside the valves becomes used up and as a result they deteriorate; for instance, electron emission from the cathode diminishes, at first very slowly indeed or even imperceptibly, but after the valve has been in use for some time this emission decreases more rapidly until it becomes advisable to replace the valve.

In the following sections it will be explained how ageing and other actions take place in valves. It is to be noted that the working of a valve depends not only on the electron emission but also upon other factors, such as residual gases, the contact potential between grid and cathode, the negative grid current, the grid emission and secondary-emission phenomena (this last factor will be dealt with in the next chapter), all of which have a disturbing influence upon the normal working of a valve.

182. Deterioration of Emission

First we will deal with the effect of diminishing electron emission upon the action of a valve, taking as a basis the transconductance value.

As already observed in Chapter IV, in a diode the anode current plotted as a function of the anode voltage takes the shape of the curve 1 drawn in Fig. 353. The same curve holds for a valve having more than two electrodes where the current to the anode and the grids is measured as a function of the effective potential in the control-grid plane (the grid-control voltage). At a sufficiently high voltage V (say 100 volts) the curve makes a bend and then follows a more or less horizontal direction, the saturation current then having been reached.

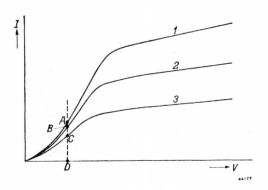

Fig. 353
Relation between the current to the anode and the grids and the effective potential of the control grid. Curve 1 applies to the beginning of the life of a valve, curve 2 to a certain, short working time, and curve 3 to the end of the practical, serviceable life.

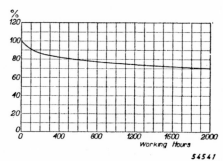

Fig. 354
Average curve of the transconductance (vertical axis) expressed as a percentage of the initial transconductance, as a function of the number of working hours (horizontal axis), derived from measurements taken with a large number of valves.

Normally the saturation current diminishes during the life of a valve, owing to a falling off of the emission faculty of the cathode. After a relatively short working time a valve originally having the curve 1 will be found to give a curve similar to curve 2 in Fig. 353. Now in modern valves the saturation current is generally much higher than the cathode current (anode current plus current to the various positive grids) flowing under normal working conditions of the valve. If we assume that the necessary cathode current equals DA (Fig. 353) it appears that the reduction of the saturation current in the first period of a valve's life has no great influence upon the cathode current and transconductance, but when after a longer period of time the emission has declined to such an extent that the characteristic appears like curve 3, the cathode current will have dropped to DC and consequently transconductance will be appreciably reduced.

In order to form an idea of the change of transconductance with time a mean curve has been plotted for a large number of indirectly-heated valves (Fig. 354). Here the transconductance is expressed as a percentage of the initial value for a brand-new valve and plotted as a function of the number of working hours. It will be seen that in the first two hundred hours or so there is a fairly steep drop. This period coincides with the continuation of the formation process of the cathode started during the "burning out" of the valve (see Chapter IV, Section 21a). Fig. 354 shows further that during 2,000 hours the decrease in transconductance is about 30%, which is relatively small, but of course as this is only an average much larger percentages may also occur. After a still longer time than 2,000 hours a more rapid deterioration in transconductance is to be expected as a result of the afore-mentioned steeper drop in emission, and the usefulness of the valve then decreases rapidly.

Traces of residual gases are apt to have a considerably adverse effect upon the action of a valve during its life (for instance due to poisoning of the cathode). By way of illustration Fig. 355 gives an instance of a valve which showed traces of gas in the beginning. It is seen that after 50 working hours its transconductance had dropped 23 %, whilst after 100 hours it was only 12 % less than the initial value, owing to the gas having meanwhile become partially bound. In the succeeding period the direct influence of the gas had disappeared and the curve shows the normal change such as that for valves that show no trace of gas at the beginning. This curve demonstrates how essential it is to degas the metal parts thoroughly and to evacuate a valve carefully.

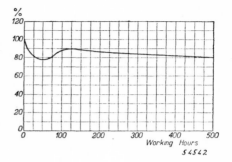

Fig. 355

Transconductance, as a percentage of the initial transconductance (vertical axis), as a function of the number of working hours (horizontal axis) of a valve which showed traces of gas at the beginning.

183. Variation of the Control-grid-to-cathode Contact Potential

It is possible that the anode current and the positive grid current, measured for a fixed negative grid voltage, in course of time increase or decrease without any appreciable change in the transconductance measured for a fixed anode current. On closer investigation it appears that the anode-current-versus-grid-voltage curves $[I_a = f(V_g)]$ and the positive grid-current-versus-grid-voltage curves $[I_g = f(V_g)]$ are displaced in the direction of the positive or negative grid voltage. This displacement of the I_a/V_g and the I_g/V_g curves is to be ascribed to the variation of the contact potential (see also Chapter IV, Sections 18 and 21d).

A variation of the contact potential may occur during the life of a valve and thus affect its working. The work function of the grid, of the cathode, or of both may assume other values, thus changing the contact potential. For instance traces of barium and barium oxide may occur on the grid or be precipitated on it during the "burning-out" process of the valve, covering the grid with a layer having a different work function, so that the potential changes with respect to the cathode. As a consequence the electrostatic field between the surfaces of the grid and the cathode changes (that is to say, the layer of precipitated barium or barium oxide takes the place of the surface of the pure metal of the grid).

A variation of the contact potential may result in too high an anode current in a valve that was originally properly adjusted, with a fixed bias. This will mainly be the case with valves having a very high transconductance. This is another reason why it is always advisable to employ automatic negative grid bias (by means of a cathode resistor). Increasing anode current then causes an increase of the grid bias, which in turn partly neutralizes the increase in the anode current. Furthermore, for the same reason it is better to use a series resistance instead of a fixed voltage for feeding the screen grid. As the grid voltage falls so the screen-grid current increases, just as the anode current rises. When a series resistance is used the screen-grid voltage is lower, due to the increased screen-grid current, and as a consequence the increase in screen-grid current and in the anode current is less than is the case with a fixed screen-grid voltage.

184. Negative Grid Current

During the life of a valve, owing to various causes a grid current may begin to flow in a direction opposite to that of the electron current emitted by the cathode. We will use for this current the term **negative grid current.**

A negative grid current may be brought about by:

(a) an ion current due to residual gases or to gases set free from parts of the valve during its life;

(b) the grid emission;

(c) the insulation current.

(a) Ion Current

It is possible that a little gas may in course of time be set free in the valve and become ionized by the electrons passing from the cathode to the anode. This results in the negative grid attracting the positive ions, thus giving rise to a current in the grid circuit to the cathode, that is in the opposite direction to that of the normal electron current to the grid.

Fig. 356 shows the variation of the grid current as a function of the grid voltage of a valve with traces of gas. It is seen that the ion current

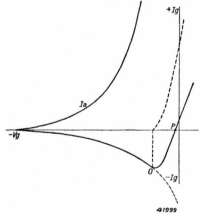

Fig. 356

Grid current I_g as a function of negative grid bias. The broken-line curve on the right below the horizontal line represents the negative grid current resulting from traces of gas. The broken-line curve at top right shows the variation of the positive grid current due to the electron current to the grid. The full-line curve is the variation of the resulting grid current.

434

disappears as soon as the anode current becomes equal to zero, with a large negative grid voltage; there is then no longer ionization of the gas.

When the grid voltage is low electrons also flow to the grid. In Fig. 356 the "positive" electron current $+ I_g$ (see the broken-line curve at the top right) begins to flow when the voltage corresponds to the point O. When a "positive" electron current is also flowing the grid current is equal to the positive value of that electron current plus the negative value of the ion current. In Fig. 356 this sum is zero for a voltage corresponding to the point P.

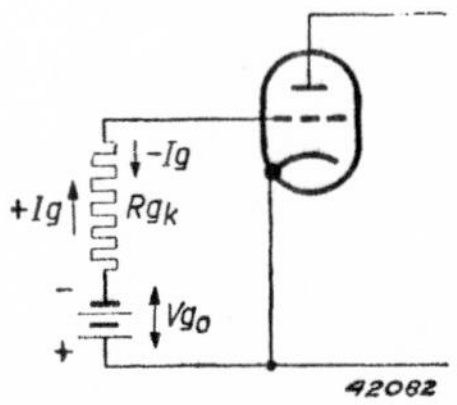

Fig. 357
Leak resistor R_{gk} in the circuit between grid and cathode (or in series with the negative-grid-bias battery). The grid current I_g has to flow through this resistor.

If there is a leak resistor in the grid circuit (see Fig. 357) then the grid current flowing through the leak resistor from the grid to the cathode causes a voltage drop across that resistor, and the original bias V_{go} of the valve changes accordingly. For the resulting bias V_g we then have the equation:

$$V_g = V_{go} - I_g R_{gk}. \qquad (267)$$

In the diagram showing the grid current as a function of the grid voltage this equation can be represented by a straight line (see Fig. 358). When $I_g = 0$ then $V_g = V_{go}$, which means that the line in Fig. 358 must intersect the V_g-axis at $V_g = V_{go}$. When $V_g = 0$ then $I_g = V_{go}/R_{gk}$, and this gives us the point where the resistance line should intersect the I_g-axis, which point, if V_{go} has a negative value, must be at a negative value of I_g. The point B (Fig. 358) where the grid-current curve $I_g = f(V_g)$ intersects the resistance line indicates the negative bias and the corresponding negative grid current of the valve.

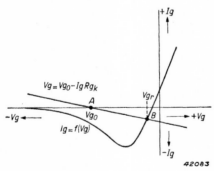

Fig. 358
Grid-current characteristic $I_g = f(V_g)$ of a valve with resistance line $V_g = V_{go} - I_g R_{gk}$, which is determined by the leak resistance in the grid-cathode circuit. The point of intersection B indicates the negative grid bias to which the valve adjusts itself.

From Fig. 358 it is evident that if the resistance value is high and the resistance line is therefore nearly horizontal, there is a large grid-voltage shift due to the negative grid current, and if there is

435

a small leak resistance (i.e. a steep resistance line) that shift will be only a small one. Furthermore, it follows that the grid-voltage shift is in a positive direction, thereby reducing the applied negative bias. Where the leak resistance is of a high value this causes a considerable increase of the anode current and a rise in the temperature of the valve, thereby setting free still more gas and increasing the negative grid current, and so on. In unfavourable cases the valve temperature may even rise so high as to cause damage to the valve, for instance cracking of the glass.

If automatic negative grid bias is applied by means of a cathode resistor then the grid voltage shift will be smaller, because an increase in anode current, due to reduced grid bias, produces a larger voltage drop in the cathode resistor, and thus partly neutralizes the decrease of the negative bias. The anode-current increase will then be very much smaller than in the case of a fixed bias. For this reason it is to be recommended, and in some cases is absolutely necessary, to use automatic grid bias when high values of leak resistance are used.

Ionization of traces of gas may also cause some damage to the cathode, for the ions may collide against the cathode and damage the emitting layer.

(b) Grid Emission

During the life of a valve a negative grid current may arise if the barium oxide and the metallic barium of the cathode evaporate and precipitate on the grid. In that event, if the heat radiated from the cathode is sufficient to raise the grid temperature high enough (sometimes to 300 or 350° C), the grid will be capable of emitting electrons, and one then speaks of **grid emission.** This, too, may cause a displacement of the working point in a positive direction, with the same results as in the case of ionization of traces of gas. In order to avoid grid emission, the temperature of the control grid is kept as low as possible; the grid supports, for instance, are made of a good heat-conducting material (such as copper wire plated with nickel),

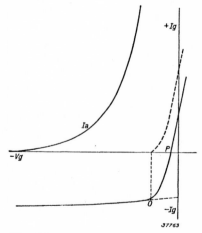

Fig. 359
Negative grid current $-I_g$ caused by grid emission, as a function of the negative grid bias. At a large negative grid bias the grid emission exhibits a saturation effect.

whilst the ends of the supports are often provided with cooling vanes, welded on and given a black colour to promote heat radiation. In Fig. 359 the variation of the grid-emission current is plotted as a function of the grid voltage, which clearly shows the difference between this current and the ion current of Fig. 356; with a large negative grid bias the latter becomes equal to zero, whilst the grid emission current shows a kind of saturation effect.

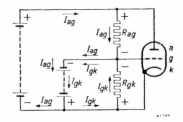

Fig. 360
Diagrammatic representation of the flow of the insulation current between anode and grid and between cathode and grid.

(c) Insulation Current

A negative grid current may also occur due to inadequate insulation between the electrodes, for instance when the barium, barium oxide or evaporated metal from the electrodes is precipitated on the insulating plates, the pinch or the glass beads. In view of this possibility, the insulating parts of the valve are often given a special coating to roughen the surface and thus lengthen the creep.

Fig. 360 shows the direction of the currents through the circuits resulting from an insulation current between anode and grid and between cathode and grid. Curve 1 in Fig. 361 gives the variation of the insulation current between anode and grid combined with the positive grid current, as a function of the grid voltage, when there is inadequate insulation between anode and grid. Curve 2 shows the variation of the insulation current between cathode and grid combined with the positive grid current. Deterioration of the insulation in the valve may lead not only to a variation of the grid potential but also to disturbing crackling noises.

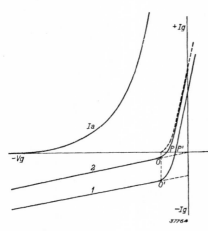

Fig. 361
Negative grid current due to bad insulation combined with the positive grid current, as a function of the grid bias.
Curve 1: Grid current due to inadequate insulation between anode and grid.
Curve 2: Grid current due to inadequate insulation between cathode and grid.

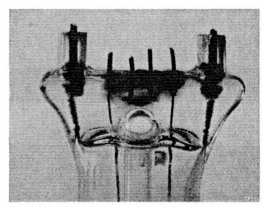

Fig. 362
Electrolysis of the glass (lead trees) of the pinch caused by overloading of the valve.

185. Back-emission.

The evaporation of barium oxide and barium may sometimes cause trouble also in vacuum rectifier valves with oxide cathode. These substances precipitate on the anode and cause an emission of electrons from the anode to the cathode during the time that the anode is negative with respect to the cathode, that is when the current passage should be entirely blocked. One refers to this effect as back-emission. It is even possible that this emission will damage the cathode, owing to its being bombarded by the electrons at a high velocity. In order to avoid this phenomenon it is necessary to provide for good heat radiation, so that the temperature of the anode remains as low as possible; this can be attained, for instance, by blackening the anode and providing cooling vanes.

186. Electrolysis in the Pinch

Electrolysis may take place in the glass of the pinch, especially when there is overloading of valves that already tend to become hot while normally working, such as power amplifying and rectifying valves. This electrolysis is characterized by the formation of so-called "lead trees", lead being drawn out of the glass and collecting at the places where the electrodes pass through the pinch. Fig. 362 shows a pinch with lead trees (on extreme left and extreme right). This leads to defects in the insulation and the valve is rendered useless. Electrolysis can be avoided if the bulbs are properly shaped and properly dimensioned to ensure adequate cooling, so that the glass in the pinch does not get too hot.

CHAPTER XXXI

Possible Disturbances Due to Secondary Emission from Insulating Parts in the Valve

The electrons emerging from the cathode of a valve do not all strike the positive electrodes. Some of them pass over to the insulating parts of the valve, whilst others leave the electrode system and, in certain circumstances, collide with the glass wall of the bulb. As a result the working of the valve may be disturbed, owing to secondary emission from those insulating parts or from the glass of the bulb. Although with properly constructed valves such disturbances are limited, it is well to consider the possibility of this phenomenon, and therefore in the following pages these disturbances due to secondary emission will be described.

187. The Charging of Insulators or Insulated Parts

As a result of the secondary emission from insulating parts made of glass or some other material, due to electrons striking up against them, under the influence of the anode voltage such parts are apt to charge themselves to a high positive potential if the secondary-emission factor is greater than unity. The same applies to insulated metal parts mounted in the valve which are not earthed. The electrostatic fields set up by the charged parts may influence the movement of the electrons in the electrode system; furthermore, strong fluctuations of the potential of insulators may give rise to interfering voltages between control grid and cathode, or the secondary electrons passing to the anode may reduce the internal resistance.

First of all we will consider the charging phenomenon, taking for an example a valve having a screen grid (e.g. the tetrode E 442), and with the control grid having a fixed potential with respect to the cathode (see Fig. 363). First a fairly high positive voltage of, say, 150 V is applied to the screen grid g_2. If the plate characteristic of this valve is plotted it will be found to have a shape like that shown in

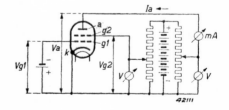

Fig. 363

Diagrammatic representation of the method of measurement for taking the plate characteristic of a screen-grid valve.

439

Fig. 364, from which it appears that between certain limits of anode voltage the anode current is negative. This is to be attributed to the relatively high screen-grid voltage V_{g2}, which is required to comply with the condition that the maximum secondary-emission factor δ of the anode surface is greater than unity.

When $V_a = 0$, I_a is also zero, and when the anode voltage equals V_{a1} the anode current is also zero, which means in this case that the secondary-emission current equals apparently the primary electron current to the anode and that consequently the secondary-emission factor δ is equal to unity. When increasing the anode voltage above V_{a1} the anode current is increasingly negative, until a certain maximum negative anode current is reached. When the anode voltage V_a rises further and nears the value V_{g2}, in consequence of various effects—such as space charge, emergence of secondary electrons in different directions (space spreading), and low potential difference between anode and screen grid—the transition of the secondary electrons to the screen grid will become more difficult; unless they have a sufficiently high velocity they are unable to reach the screen grid and are compelled to return to the anode. When the anode voltage rises still further, even the more rapidly moving secondary electrons are unable to reach the screen grid, so that ultimately the anode current becomes positive again.

As already explained in Chapter XIII, from then on the anode current increases gradually with the anode voltage until a comparatively constant value is reached.

From Fig. 364 it appears that corresponding to an anode current value equal to zero there are three different values of anode voltage, namely $V_a = 0$, $V_a = V_{a1}$ and $V_a = V_{a2}$. If the anode voltage has one of these three values nothing will happen in the event of a break in the anode supply. In that case we have a „floating" anode, the potential of which can only be in one of the three zero-current states described, corresponding to the points A, B or C in Fig. 364. The points A and C refer to stable states, while point B is unstable. If the anode potential is in state B then the smallest deviation in the secondary-emission factor will result in the anode voltage jumping to zero or to V_{a2}.

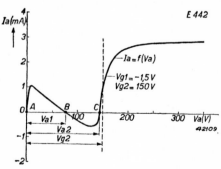

Fig. 364
Plate characteristic of the valve of Fig. 363 at a screen-grid voltage $V_{g2} = 150$ V (control-grid voltage $V_{g1} = -1.5$ V).

A small reduction of the factor δ implies a smaller secondary electron current; consequently the anode receives more negative charge from the primary electron current than is taken away by the secondary electron current, and thus becomes less positive. This leads to a further reduction of δ and, thereby, also of the anode voltage, until the latter finally drops to zero. If, on the other hand, δ increases the reverse process takes place and the anode voltage rises to V_{a2}.

Under the conditions existing at points A or C, however, nothing happens when the factor δ changes. In the case of point C an increase of δ will, it is true, cause the anode voltage to rise slightly, but then the secondary emission diminishes, as is indicated by the rise of the plate characteristic within this anode-voltage range; as a result the anode again becomes less positive and adjusts itself to a new stable value of potential. In the case of point A an increase of δ will likewise lead to a rise of the anode voltage (in this case, however, the higher increase of the primary electron current compared with the secondary electron current restricts the increase in the anode voltage).

If the floating anode potential is in state A (the so-called "low state"), a voltage impulse greater than V_{a1} will bring it into state C (the so-called "high state").

Now, let us imagine the surface of an insulator as taking the place of a floating anode. We will consider, for instance, the inside of the glass bulb of a valve as shown in Fig. 363 with the anode removed and the screen grid g_2—maintained at a high potential (say 150 V)—acting as anode. Usually the secondary-emission factor δ of such a glass surface will be greater than unity even with relatively low voltages (about 70 V). In this case it is possible to have the "high state", which may be brought about, for instance, by a voltage impulse or a low leakage current from the screen grid g_2. The potential of the glass surface then adjusts itself to a value approximately equal to the screen-grid voltage V_{g2}. The glass is thus continuously being struck by the primary electrons shooting through the meshes of the screen grid, while an equally large number of secondary electrons is flowing back to the screen grid. In actual practice one must imagine an anode taking the place of the screen grid g_2, with the glass wall of the bulb or other insulating parts being charged by the electrons shooting past the anode. Sometimes the anode is made of metal gauze and thus the electrons may shoot through the meshes and collide with the glass bulb.

188. Effect of Insulator-charging upon the Working of Valves

(a) H.F. Valves, Switch Effect [1])

With H.F. valves (pentodes) it is important that the internal resistance

[1]) See J. L. H. Jonker, Philips Techn. Review, *3*, pp. 211 et seq.

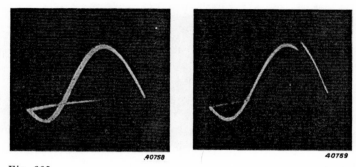

Fig. 365
Right: The irregularity occurring in the anode-voltage-versus-time curve of an output valve due to potential fluctations of insulator surfaces.
Left: Oscillogram of the sinusoidal anode alternating voltage without this effect.

should be as high as possible, the anode current remaining as far as possible independent of anode voltage. If, then, secondary emission takes place in the valve from the wall of the bulb or from insulating parts, in the state C the secondary electron current from the insulator will be equal to the primary electron current flowing to the insulator. Now, as a rule, this primary current fluctuates considerably with the voltage of the insulator or of the glass bulb, which rises and falls in much the same way as the anode voltage. Therefore the secondary electron current flowing to the anode and augmenting the electron current from the cathode will likewise vary with the anode voltage. Consequently, also the resultant electron current to the anode depends considerably on the anode voltage, which means that the internal resistance of the valve is low. If a H.F. tuned circuit is interposed in the anode circuit it will undergo additional damping through the reduction of the internal resistance, and the amplification of the H.F. stage will drop.
We call this phenomenon **switch effect**, because the extra damping disappears when the anode voltage is temporarily switched off, without the heating-current supply being interrupted. An interruption of the anode-voltage supply means that the glass bulb or insulator is brought into the low state, the secondary-emission phenomenon disappears, and the glass bulb or insulator remains in that state until through some cause or other it is charged anew.

(b) Output Valves, Distortion Effect

Another result of the charging of insulators and the inner surface of

442

the glass bulb by electron bombardment, in the case of output valves, is a distortion that manifests itself in the loudspeaker, for instance by a spluttering or scratching noise.

This distortion can be made visible by means of a cathode-ray oscillograph, producing an oscillogram as illustrated in Fig. 365. It occurs mainly at large sound volumes and is to be explained as follows.

When the anode supply of a valve is interrupted, as indicated in Fig. 363, only the two potential states A and C are possible (see Fig. 364). The state C, however, can exist only as long as the screen-grid voltage is high enough. As this voltage drops, the anode-voltage range for negative anode current gradually disappears, so that at relatively low screen-grid voltages (see curve a in Fig. 366 for $V_{g2} = 60$ V) the anode current remains positive for all anode voltages. This, of course, is due to the reduced velocity of the primary electrons at low screen-grid voltages and anode voltages lying below the value of the screen-grid voltage, causing the secondary-emission faculty of the anode to decrease. (When the anode voltage is higher than the screen-grid voltage the secondary electrons from the anode are unable to overcome the field between screen grid and anode, and thus return to the anode.) Let us now imagine, similarly to the case of switch effect, that the anode is replaced by the surface of an insulator (glass bulb, glass bead or mica) situated in the vicinity of an anode in a valve, whilst the screen grid g_2 is replaced by the anode. With anode alternating voltages of great amplitudes, such as occur in output valves with strong signals, it may happen that, when the anode direct voltage is low during one half of the alternating voltage cycle, it is impossible for state C to occur (see curve a of Fig. 366) and the insulators must therefore be at zero potential, whereas during the other half of the alternating-voltage cycle state C is in fact also possible (curve b of Fig. 366). The potential state C may arise, for instance, in consequence of a voltage impulse, which is capacitively conveyed from the anode to the insulator surface.

Thus it is possible for the

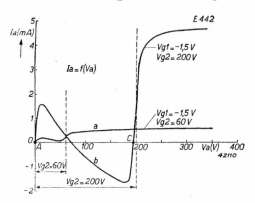

Fig. 366
Plate characteristics of the valve of Fig. 363 for a screen-grid voltage of 60 V (curve a) and for one of 200 V (curve b), the control-grid voltage V_{g1} being in both cases — 1.5 V.

potential of the bulb wall or the surface of one or more insulating parts
to fluctuate periodically with the anode alternating voltage. These potential fluctuations may affect the control grid capacitively and, since they
give rise to many harmonics, may cause the distortion described above.
Generally speaking, the voltage impulse changing the state of the
glass bulb wall from A to C will be applied as a result of a capacitive
voltage divider of the anode alternating voltage formed by the capacities between the surface of the bulb wall and the cathode and between
the bulb wall and the anode. If the capacity between the anode and
the surface of the bulb wall is large compared with the capacity between
that surface and the cathode, then the voltage impulse occurring at the
surface of the bulb wall will be large, and inversely this impulse will be
small if the capacity between anode and bulb wall is small compared with
the capacity between bulb wall and cathode. In the latter case it is possible that the state C cannot be reached and, therefore, distortion does not
arise. It is for this reason that the capacity between the surface of the bulb
wall and the cathode in power valves like the EL 3N, the EL 5 and EL 6
is purposely made large, by applying at the bottom of the bulb, on the
outside, a ring of metallized coating that is connected with the cathode.

189. Means of Avoiding Switch and Distortion Effects

Various means are employed for the prevention of switch effect in H.F.
valves and distortion effect in output valves. In the first place, the
most obvious method is to cover the inside of the glass bulb and/or the
insulating parts in valves with a substance having a secondary-emission
factor less than unity, such as carbon or tungsten oxide.
Further, care has to be taken that the smallest possible number of primary
electrons from the electrode system is able to bombard insulating parts
and the wall of the glass bulb. In modern valves without pinch a cage is
placed around the electrode system for the same reason. This cage is at the
same potential as the cathode and prevents secondary electrons from the
glass bulb reaching the anode. In the case of secondary emission from the
bulb wall the potential variations can also be avoided by metallizing the
bulb on the outside and keeping this metallic coating at a constant low potential (for instance at the cathode potential). In the case of H.F. valves it
is thereby avoided that the potential of the bulb wall reaches the state C,
unless there is a leak due to precipitation between the anode and the bulb
wall. This leak resistance, together with that between the inner wall of the
bulb and the metallization, then forms a voltage divider, which may result
in a high potential of the inner wall and thus give rise to the state C.

CHAPTER XXXII

Feeding of the Valves

190. Adaptation of Valves to the Available Supply Source

Valves in receiving sets can be fed in different ways, according to the power-supply sources available. In former times the valve filaments were fed from an accumulator and the anodes and other electrodes from dry batteries, so-called high-tension and grid-bias batteries. The accumulators had to be regularly re-charged, while the dry batteries had to be replaced by new ones as soon as they became exhausted.

This renewal of dry batteries was rather expensive and inconvenient, so that very soon they came to be replaced by high-tension supply units ("Eliminators") for connecting up to a.c. mains. These H.T. units contained a rectifying valve with a transformer and a smoothing device. Often it was possible to obtain also the grid-bias voltage from this unit by means of voltage dividers. With modern indirectly-heated valves it is possible to feed the filaments with alternating current, and consequently a.c. sets have both a H.T. supply unit and a filament-current transformer. A.c. valves are fed with a low filament voltage in order to avoid difficulties arising from hum, 4 volts being used for the original indirectly-heated Philips valves and 6.3 volts for the later ones. The filaments of indirectly-heated valves are connected in parallel to the secondary side of the filament-current transformer.

Some electricity mains are still supplying direct-current, and with d.c. it is not possible to use a transformer for stepping-down the voltage for heating the cathode, so that the filaments have to be fed direct from the mains. For this reason indirectly-heated valves had to be specially made for d.c. mains; the filaments are made, not for the same voltage, but for the same filament current, because, instead of being connected in parallel they are connected in series and, with a voltage-dropping resistor of appropriate value in series, are fed directly from the mains. In this way valves were made with a filament current of 180 mA and a filament voltage of about 20 V (e.g. the B 2046 and B 2047). The need soon made itself felt, however, for sets that could be connected up to either d.c. or a.c. mains. There were various reasons for this, one of which was that if the owner of a d.c. set moved to a district where the mains were supplying a.c. he naturally wanted to be able to use the same set. Basically there would be nothing against feeding 180-mA valves with a.c., but as they were less suitable for this purpose and, moreover, new methods of manufacture had been introduced, a new set of Philips valves were designed for series feeding of the filaments, both for d.c. and for a.c. mains. The filament current of these a.c./d.c. valves was 200 mA, and the letter C was taken as first letter to indicate the type.

Meanwhile receiving sets came on the market designed for use in cars. As the electrical installations of cars are supplied from an accumulator with a voltage usually of about 6 or 12 V (3 or 5 cells), valves were designed for car sets with filaments connected in parallel which could be connected direct to the battery. The anode voltage is obtained by transforming the low battery voltage into a higher direct voltage of 250 V by means of a vibrator or motor-generator converter. Some of the a.c./d.c. valves of the C series could very well be combined with the valves for 13-V accumulators, as many of the valves of this series were designed for a filament voltage of 13 V with a current of 200 mA. For 6-V accumulators a special series of valves were made for a filament current twice that of the C series in order to produce the same heating power. Thanks to the progress achieved in valve-making technique it was ultimately possible to reduce the required filament-current power of many types of valves to one half of what it used to be, and valves were produced with a filament voltage of 6.3 V and a filament current of 200 mA. As regards adaptability these valves are universal, being suitable both for a.c. sets with valve filaments connected in parallel and for car-radio sets. Moreover, the filaments of these valves can be connected in series and via a ballast resistor connected direct to the mains, so that they are also suitable for d.c./a.c. sets with a filament-current circuit of 200 mA. These are the red "Miniwatt" E type of valves. Some years ago a.c./d.c. valves were also marketed with a filament current of 100 mA and a filament voltage correspondingly higher. These are exclusively intended for d.c./a.c. sets and make it possible to economize in the power consumption, the current in the filament circuit being 0.1 A lower. (With 220-V mains there is a saving of 22 W compared with a set having 200 mA valves.) The valves of this series bear the letter U as type indicator.

The power of the universal cathode (6.3 V, 200 mA) is inadequate for some valve types. A higher heating power means either that the filament voltage or the filament current has to be increased. High-power valves therefore have to be designed for series or parallel filament connection with the corresponding heater voltage and heater-current ratings. This is the case, for instance, with output pentodes, as this kind of valve needs a fairly considerable filament power. The 9-W pentode EL 3N, for example, with a filament voltage of 6.3 V has a filament current of 0.9 A, whilst the corresponding type CL 4 for a.c./d.c. supply has a filament voltage of 33 V and a filament current of 200 mA. The EL 3N, therefore, is suitable only for a.c. sets with a filament-current transformer, whilst the CL 4 can be used only for sets where the filaments are connected in series. Valves for car-radio sets have to be extremely economical in current consumption, to avoid too heavy loading of the accumulator, and for that reason output valves with low filament current and relatively small transconductance are specially made for car-radio sets.

Receiving valves for mains feeding have not altogether ousted battery-fed valves. There are many cases where electricity mains are not available and, moreover, there are portable sets that have to be fed from batteries. Battery valves for such sets must of necessity be sparing in current consumption, as otherwise the filament-current accumulator would have to be recharged too often and the expensive H.T. batteries too frequently replaced.

Originally these valves were made for a filament voltage of 4 V, until it was found that with 2 V more satisfactory results could be obtained, and new battery valves were therefore made for a filament voltage of 2 V, adapted to the accumulator voltage of 2 V. This type of valve was indicated by Philips by the initial letter K (K = 2-V filament voltage).

Ultimately valves were produced for filament-current supply from a dry battery. Until fairly recently dry batteries were hardly ever used for this purpose, because above a certain current dissipation they were very uneconomical in use. Experience has proved that above a current consumption of about 250 mA the accumulator is more economical than the dry battery, whilst with a consumption lower than that figure there is no objection, as regards life of the battery, to using a dry battery for filament-current supply, and there is then an appreciable advantage compared with the accumulator, because the latter has to be repeatedly re-charged and tested for acid content, which needs skilled assistance and involves expense. Furthermore the dry battery is handier for porta-ble sets, there being no risk of acid being spilled. The development of a new technique in filament and valve construction allowed of valves being made with extremely low filament current, while the valve properties, such as transconductance and power output (in the case of output valves), were not inferior to those of valves made for accumulator supply.

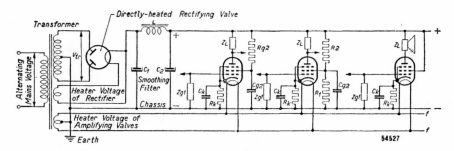

Fig. 367
Method of feeding a receiving set with mains transformer and directly-heated rectifying valve.

For the greater part of its life a dry battery gives a voltage of about 1.4 V. Valves for dry-battery filament-current feeding were therefore designed for a filament voltage of that value, and those of Philips manufacture have as type indicator the letter D. Some of these valves have a filament current of only 25 mA, whilst others take 50 mA or more.

For houses not connected up to electric light mains other solutions are possible, but these are not so often employed owing to their expense. For instance, receiving sets can be energized from a large battery of accumulators charged by a generator driven by the wind or by a petrol engine.

191. Feeding A.C. Sets

The following voltages are required for feeding receiver valves:

a) filament voltage;

b) anode voltage (positive direct voltage with respect to the general negative lead);

c) voltages for other current-carrying electrodes, e.g. screen-grid voltages (positive direct voltages with respect to the general negative lead);

d) control-grid voltages (negative direct voltages with respect to the general negative lead).

Fig. 367 is a diagram showing the principles of feeding an a.c. receiving set.

448

(a) Filament Voltage

As already remarked, the filaments of all a.c. amplifying valves, both the indirectly- and directly-heated (output) ones, are connected in parallel and usually connected to a common filament-current winding of the mains transformer. Consequently the filament voltage of all valves must be equal. Owing to the fact that the required heating power is not the same for all types of valves they have different heating-current ratings. The total filament current is the sum of the currents of the various valves. If valves with different filament voltages are used, for example 6.3 V and 4 V, it is necessary to have a separate filament-current winding for each filament voltage, or else there should be a 4-V tap from the 6.3-V winding. When the directly-heated output triode AD 1 is used in combination with 6.3-V valves it is advisable to provide a separate filament-current winding for the directly-heated valve, because then it is easy to arrange the circuit for the automatic negative grid bias. Usually the centre-tap of the filament-current winding of the amplifying valves is earthed, this being the safest with regard to minimizing hum.

(b) Anode Voltage

As explained in Chapter XIX, with a.c. feeding the a.c. mains voltage is rectified by means of a rectifying valve. The rectified voltage is smoothed and conducted to the anode impedances of the various valves. Both directly- and indirectly-heated valves are suitable for rectification, but generally the former are used. The filament (cathode) of a directly-heated rectifying valve is connected to the positive pole of the rectified alternating voltage, so that it cannot be fed from the filament-current winding for the amplifying valves.

In theory this would be possible with the indirectly-heated rectifying valves; the cathodes of the amplifying valves, however, have a potential about the same as that of the chassis (negative pole of the H.T. voltage). In view of this, as already stated, the filament-current winding is

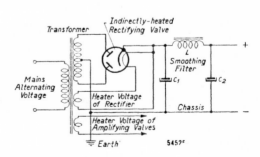

Fig. 368
Method of feeding an a.c. receiver with indirectly-heated rectifying valve.

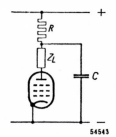

Fig. 369
Extra smoothing of the anode direct voltage of a valve by means of an RC circuit.

earthed, either at a centre-trap or at one of the ends. If, then, the filament of the indirectly-heated recti-fying valve were also connected to that winding, the voltage between filament and cathode of the rectifying valve would be excessive. Therefore the filament current for indirectly-heated rectifying valves has to be drawn from a separate winding, just as is the case with directly-heated rectifying valves, with the winding connected, in this case, to the cathode (see **Fig. 368**).

Since separate filament-current windings are needed for the receiving valves and for the rectifying valve, it is possible to use a rectifying valve with a filament voltage different from that of the amplifying valves, provided the filament-current winding is dimen-sioned accordingly.

Usually the alternating mains voltage is transformed before being rectified. The voltage at the secondary of the transformer is then of such a value that the rectified voltage after being smoothed is about **250-300 V**, under the load formed by the valves in receiving sets. This voltage has to be so chosen as to produce between anode and cathode of the output valve just the right voltage for optimum results; allowance has to be made for the voltage drop in the output transformer and that in the cathode resistor for automatic negative grid bias. Frequently the anode supplies of the amplifying valves preceding the output valve are additionally smoothed by means of special filters. These filters consist of a series resistor with a condenser (Fig. 369) and their primary object is to filter out the ripple, whilst in the second place they avoid coupling between the various receiving valves via the common H.T. supply line.

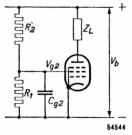

Fig. 370
Feeding a screen grid by means of a voltage divider.

(c) Voltages for the other Current-drawing Electrodes

The other current-drawing electrodes of receiving valves having more than one grid are fed with positive voltages drawn from the high-tension supply; in most cases these electrodes are screen grids. For R.F. and A.F. amplifying valves the screen-grid voltage is generally lower than the anode voltage and the feed voltage has to be reduced accordingly. This can be done by means of voltage dividers (see **Fig. 370**) or with the aid of series resistors (**Fig. 371**). A voltage divider or a series resistor acts at the same time as a smoothing device, since the screen grid should always be adequately earthed by a capacity of low reactance at the frequency of the alternating voltages occurring; otherwise alternating voltages would exist between screen grid and cathode and have an opposing effect upon the anode-current variations due to the control grid.

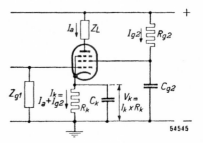

Fig. 371
Feeding a screen grid by means of a series resistor.

Output valves generally require the same voltage for the screen grid as for the anode, in which case the screen grid is connected directly to the anode-supply line. With high-transconductance output pentodes, however, a small resistor has to be introduced in the screen-grid lead, without a by-passing condenser, so as to suppress any H.F. oscillations due to self-oscillation of the valves. (For the same reason a resistor should be connected in series with the control grid of these valves.) Owing to the voltage drop in the output transformer the screen-grid voltage will be higher than the anode voltage.

If the screen grid of a valve is fed via a voltage divider the screen-grid voltage V_{g2} can easily be calculated from the following formula (see also Fig. 370):

$$V_{g2} = V_b \frac{R_1}{R_1 + R_2} - I_{g2} \frac{R_1 R_2}{R_1 + R_2}.$$

For the more recent valves curves are published showing the screen-grid current as a function of the screen-grid voltage for different values of negative grid bias, and these curves can be used for calculating the screen-grid voltage.

(d) Negative Grid Bias

With the indirectly-heated cathode of a.c. valves it is possible to produce the negative grid bias automatically, for by introducing a resistor between the cathode and the nega-

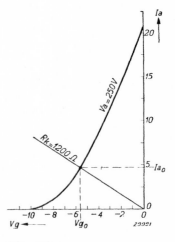

Fig. 372
Method of determining the negative grid bias of a triode for a given value of cathode resistance by means of a resistance line and the transfer characteristic.

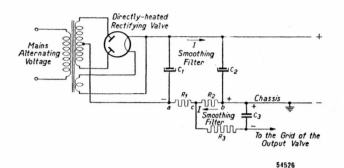

54526

Fig. 373

Tapping the negative grid bias for the output valve from the smoothing resistor in the negative H.T. line. If the cathode current of the output valve is more than 50% of the total current I through $R_1 + R_2$ the grid bias of the output valve is said to be semi-automatic.

tive lead of the supply part the cathode current will cause a voltage drop across that resistor. This cathode current is equal to the sum of the currents to all electrodes (in oscillator valves also the grid current can be included, in so far as it flows through the cathode resistor). As a result of the voltage drop in the cathode resistor the cathode is rendered positive with respect to the negative H.T. lead (chassis). If there is a d.c. path from the grid to the negative H.T. lead the grid then attains a mean potential equal to the potential of that negative lead and is thus negative with respect to the cathode by the same amount as the cathode voltage is positive with respect to the negative lead. In most cases the cathode resistor is shunted by a condenser, so that the cathode is effectively earthed for high, intermediate or low frequencies as the case may be. In order to determine the negative grid bias produced by a certain value of cathode resistor, in the case of triodes the transfer characteristic can be used, by drawing the straight line indicating the relation between current and voltage given by the cathode resistance; the negative grid bias is then found from the point of intersection of the resistance line and the transfer characteristic (see Fig. 372). In the case of multi-grid valves the point of intersection of the resistance line on the curve $I_{tot} = f(V_{g1})$ has to be taken.

The negative grid voltage for output valves is often obtained from the smoothing circuit of the anode-voltage rectifier, in which case the smoothing element (choke or resistor) is inserted in the negative H.T. lead (see Fig. 373). In order to get the correct grid voltage it is sometimes necessary to obtain the negative grid bias from a tapping-point on the choke or resistor (see point c in Fig. 373). By means of a smoothing circuit consisting of a resistor and condenser (R_3C_3) the ripple in the direct voltage between the points b and c is smoothed.

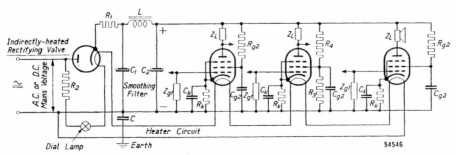

Fig. 374
Method of feeding an a.c./d.c. set. A current-regulating tube (ferro-hydrogen resistor) is recommended in place of the resistor R_2 in series with the chain of filaments. (In this diagram no account has been taken of the right order in which the filaments of the various valves should be connected.)

192. Feeding A.C./D.C. Sets

Fig. 374 shows the essentials of the circuit used for feeding a.c./d.c. sets.

(a) Filament Supply

Since there is no mains transformer, the filaments of the valves are connected in series with a ballast resistor (R_2 in Fig. 374) between the two mains terminals. The filament of the indirectly-heated rectifying valve is likewise included in this circuit. Where dial illuminating lamps are also used they form a link in the filament-current circuit. The total voltage across the filaments is then equal to the sum of the filament voltages of the individual valves, and the difference between that voltage and the mains voltage has to be dropped across the series resistor R_2.

a) Sequence of the Filaments

The chassis is connected direct with one side of the mains and earthed via a condenser C. The chain of filaments begins at the same side of the mains and ends at the resistor R_2. Obviously, the filaments in the chain which are connected at some distance away from the chassis, are carrying a relatively high alternating voltage with respect to the chassis and consequently also with respect to their cathodes. Therefore, not only is it necessary to provide very good insulation between filament and cathode, but there is also a risk of this alternating voltage being fed capacitively to the control grid or affecting the electron paths in the valve in some other way, thereby giving rise to hum, which may be expressed directly as a hum alternating voltage on the grid.

453

In A.F. valves as well as detector valves these hum voltages mix directly with the A.F. voltages and cause a disturbing hum in the loudspeaker. In R.F. or I.F. valves, on the other hand, hum can only arise from hum-modulation. In the case of mixer valves hum-modulation may arise in two ways, viz. due to the curvature of the characteristic and due to the oscillator.

Since the direct hum is more troublesome than hum-modulation the filaments of A.F. and detector valves will be connected nearest to the chassis, followed by the filament of the mixer valve and lastly by the filaments of the R.F. and I.F. valves. As regards the filament of the output valve it has to be considered that the amplification after its grid is very small, so that there is little risk of hum; but this type of valve generally needs a higher filament voltage, so that the valves whose filaments are connected up behind that of the output valve are exposed to a much higher hum voltage. For these reasons the filaments of the output and rectifying valves are taken as the last links in the filament chain (from the chassis onwards).

The rectifying valves of the d.c./a.c. series can withstand peak voltages up to a maximum of 300 V between cathode and filament. Since between these two electrodes in addition to the full, rectified voltage there is also a considerable alternating voltage of mains frequency (the filament-to-chassis voltage), under certain conditions excessively high peak voltages might arise between cathode and filament. It is, therefore, often advisable to connect up the filament of the rectifying valve last but one and that of the output valve last in the chain of filaments.

Generally speaking the sequence recommended for linking the filaments is: chassis—detector valve —A.F. amplifying valve— mixer valve—R.F. amplifying valve—I.F. amplifying valve—output valve —rectifying valve—ballast resistor or current-regulating tube—second mains terminal. If necessary, considering what has been said above, the sequence of the filaments of the output and rectifying valves may have to be changed.

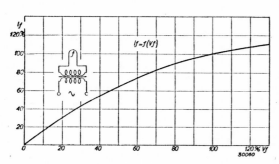

Fig. 375
Curve representing the heating current as a percentage of the nominal value as a function of the heater voltage as a percentage of the nominal value for an indirectly-heated valve.

β) **Filament Overloading Due to Mains-voltage Fluctuations**

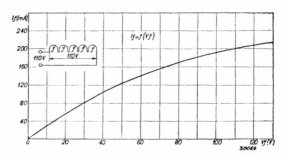

The voltage of nearly all electric lighting mains is liable to fluctuate, but generally not by more than ± 10%, although with some very bad mains fluctuations of 15% and even more may occur. This means that a mains voltage of nominally 220 V may vary between 198 and

Fig. 376
Curve indicating the heater current as a function of the heater voltage (mains voltage) of a heater-current circuit of valves when the filaments are connected in series without a ballast resistor and that circuit is connected directly to the mains.

242 V or perhaps even between 187 and 253 V, but with good mains the fluctuations are much less. Very large fluctuations in mains voltage are liable to overload the filaments of valves and thereby damage them.

When the filaments are connected in series and a ballast resistor is used the effect of mains-voltage fluctuations is greater than when the filaments are connected in parallel and a filament-current transformer is employed. In the latter case the internal resistance of the transformer can be ignored. If the voltage at the filament terminals rises 10% the filament current will increase by much less than 10%, because the resistance of the filament increases as the temperature rises. In Fig. 375 a curve gives the filament current as a function of the filament voltage; this curve shows that with 10% rise in filament voltage the filament current increases only 5%. In the case of valves connected in series the curve of Fig. 376 applies. If, however, a series resistor is introduced in the filament-current circuit its resistance value will not be increased by a rise in the voltage. Therefore the total resistance increase is

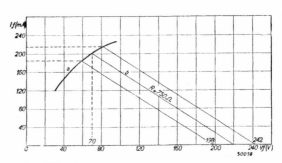

Fig. 377
Graphic determination of the heater-current variation occurring for a mains over-voltage and under-voltage of 10% in the filament chain of valves which is connected in series with a ballast resistor.

455

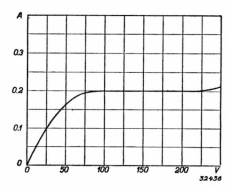

Fig. 378
Current-voltage characteristic of a current-regulating tube designed for a nominal current value of 200 mA.

less than what it would be if the filament-current circuit consisted exclusively of filaments; thus the filament-current increase is greater. Fig. 377 is a graph for determining filament-current variations for an under-voltage and an over-voltage of 10%, from which it is seen that in this case a 10% voltage increase results in a $7\frac{1}{2}$% current rise. The greater the fixed resistance compared with the resistance formed by the filaments connected in series, the greater will be the current variations. It will be obvious that relatively small mains voltage fluctuations may suffice to cause appreciable overloading or underloading of the valves.

By means of current-regulating tubes (also called ballast tubes) it is possible to keep the current in a certain voltage range constant within very narrow limits, owing to the fact that the resistance of such a tube rapidly increases as the current rises.

γ) Current Surges when Switching on Filament Current

If the filaments are cold when the filament current is switched on they have a very low resistance for normal current, so that at that moment there is a current surge of considerable amplitude. This surge does not harm the filaments and the valves are quickly brought up to working temperature. If a ballast resistor is included in the filament-current circuit it limits the amplitude of the current surge when switching on, so that the valves are not brought up to working temperature so quickly. In the case where a current-regulating (ballast) tube is used in the filament-current circuit its resistance will also be low initially and increase gradually as the resistance element warms up, so that here again there will be a strong current surge, which, however, is only of short duration, because a current-regulating tube heats up very quickly. It is not always feasible, however, to use a current-regulating tube instead of a ballast resistor, since its regulating range is limited. If, for instance, the lowest admissible voltage across the tube is 80 V the current can be kept constant up to a voltage of about 200 V, but above that the current increases rapidly until the resistance wire burns out (see Fig. 378, the current-voltage characteristic of a current-regulating tube).

If the voltage drop in the ballast resistor is small compared with that in the heater circuit, then when switching on the circuit a current-regulating tube with a low minimum voltage limit will burn out if it is used instead of a ballast resistor, because. the voltage across the resistor element at the instant of switching on is considerably higher than the upper voltage limit.

If there is a dial-illuminating lamp inserted in the filament-current circuit its filament will be heavily overloaded by the impulse when switching on. Consequently either the lamp used must be specially made for the purpose or the current has to be limited. By employing a current-regulating tube the current impulse is appreciably shortened, so that the risk of the lamp burning out is then less. For current limitation over a longer period current-regulating tubes can be used which include a limiting resistor. Such a resistor has a negative temperature characteristic, that is to say its resistance diminishes as the temperature rises, being very high in the cold state and low in the warm state. When the valves are switched on cold the resistance of the filament-current circuit is mainly that of the limiting resistor, in which the electrical energy is converted into heat. As a result the resistance diminishes, but the time it takes for the limiting resistor to heat up is sufficient for the filament of the current-regulating tube to heat up, so that it can absorb the whole of the over-voltage due to the valves still being cold when the resistance value of the limiting resistor has become low. A dial illuminating lamp in the filament-current circuit will not then be so considerably overloaded and in that case it is not absolutely necessary to use any special lamp made to withstand large current surges.

(b) Anode Voltage

For the anode supply of d.c. sets, of course, no rectifying valve is actually needed, for the H.T. direct voltage can be obtained direct from the mains. It is advisable, however, to have a smoothing circuit because often there is a disturbing a.c. voltage superimposed on the direct mains voltage. Electrolytic condensers cannot be used for this smoothing, because if a wrong mains connection should be made (mistaking the positive for the negative pole) they would be damaged.

Sets for d.c. and a.c. mains must be provided with a rectifier, and for this purpose only a single-phase valve can be considered, which should be indirectly heated because its filament is to be connected in series with the filaments of the receiving valves. An a.c. supply is rectified by the valve and this rectified current can be smoothed in the usual way by two electrolytic condensers and a choke. In the case of d.c.

feeding the rectifying valve acts as a protective device; if, in Fig. 374, the mains terminal which is connected to the anode of the rectifying valve is positive, then current flows through the valve, but if it is negative then the valve blocks the current. Consequently electrolytic condensers can be used, because owing to the protection afforded by the rectifying valve they cannot be erroneously polarized.

In the absence of a mains transformer any mains disturbances can find their way into the set much more easily, and these have to be suppressed by chokes in the supply line. Further, some provision has to be made for proper smoothing of the anode voltage, for only an extremely small voltage drop is permissible in order to get the highest possible H.T. voltage from low-voltage mains.

The smoothing choke should be placed in the positive line. If it were put in the negative line the cathodes of all the valves and the chassis would no longer be connected direct to one side of the mains, and this is apt to give rise to hum.

With d.c. supply the available voltage is generally lower than that with a.c. supply from mains with the same voltage. In the case of an alternating voltage the reservoir condenser of the rectifier is charged to slightly less than the peak value of the voltage, so that the mean value of the voltage on that condenser may be higher than that with a direct voltage.

The rectifying valves for a.c./d.c. sets have a low internal resistance with a view to obtaining the highest possible anode voltage from low-voltage mains. If the internal resistance of the rectifying valve were high then the reservoir condenser of the rectifier would only be charged to a low voltage and consequently the high tension would be low. It is often advisable to connect with high mains voltages a resistor in series with the rectifier (R_1 in Fig. 374). This resistor then serves for limiting the current impulses charging the condenser, as they might damage the rectifier cathode, particularly when large reservoir condensers are used. The required values of resistance in various circumstances are mentioned in the published data relating to rectifying valves. At low mains voltage a limiting resistance is generally not necessary unless the reservoir condenser has an extremely high capacity.

(c) Adaptation to Different Mains Voltages

The mains to which a.c./d.c. sets may be connected differ not only in regard to the nature of the current (alternating or direct) but also in voltage, so that in many cases provision has to be made for adaptation to mains of a different voltage. For a.c. sets with a transformer this can be done by means of taps on the primary winding of the transformer, but this is not possible with a.c./d.c. sets because they have no mains transformer. A simple means of adapting a set to different mains voltages is, for instance in the case of a set built for 110 V, to introduce a resistor in the mains flex for high voltages, but this involves considerable loss of power (converted into heat), and as in certain countries high-voltage mains are more common than lower ones this method is only employed in those countries where low mains voltages are prevalent.

Changing the mains voltage has two results:

(1) the voltage across the filament chain is modified, and

(2) the value of the H.T. voltage is altered.

In consequence of (1) it is necessary to provide for a different value of ballast resistance in the filament chain. If a current-regulating tube is placed in this circuit the heating current will sometimes be maintained to the right value by that tube when the set is connected to mains of higher voltage, but generally the change in voltage is so great and the voltage between the terminals of the filament chain is so high that the regulator tube has to be changed.

As a result of (2) the electrode voltages of the valves change. In a set intended for various mains voltages the circuits of the valves of the C series and of those of the E series, which are suitable for a.c./d.c. operation, are generally established in such a way that at a high anode voltage series resistors or voltage dividers reduce the screen-grid voltages to the level required (for H.F. pentodes usually 100 V). In the case of a low anode voltage these series resistors are either short-circuited or reduced, so that the same screen-grid voltages are again obtained. There are also output valves constructed for a low screen-grid voltage, such as the CL 2 and CL 6, which at a low anode voltage yield a comparatively high power output (about 2 W). For example, with a mains voltage of 110 V the screen-grid and anode voltages will be 90 V, whilst on 220-V mains the anode voltage will be 180 V and the screen-grid voltage will be reduced by a series resistor to about 100 V. When changing over from 220 V to 110 V or vice versa also the ratio of the output transformer has to be altered, by changing either the primary or secondary connections, so that it is necessary to provide this transformer with a tap for two different ratios in order to obtain the most suitable load for the respective anode voltage.

The changing of the current-regulating tube in the filament chain for a different mains voltage has led to this being combined also with a modification of the electrode-current supply circuits of the valves. This can be effected by making different connections in the cap of the current-regulating tubes between non-used contacts. By means of these interconnections, for instance, the common voltage-dropping resistor for the screen grids can be short-circuited and the connection to the primary of the output transformer can be altered.

In the design of a new series of valves for a.c./d.c. sets, viz. the U series, the object was to make it much easier to change over to different mains voltages, without having to alter the screen-grid series resistors, cathode resistors and output transformer. To attain this aim it was necessary, inter alia, to construct an output valve with higher permissible anode dissipation (11 W). With low anode and screen-grid voltages a reasonable power output (about 1 W) is thereby obtained; with high voltage (185 V—the approximate value of the anode direct voltage for 220-V mains) the anode dissipation is increased to the limiting value of 11 W without the values of the screen-grid series resistor and cathode resistor being modified, whilst the anode load impedance can remain the same. In the case of H.F. pentodes, too, the same values of screen-grid and cathode resistors can be retained. The transconductance with low anode voltage is then about 10% lower than that with high anode voltage. Though the properties of a set without a special change-over device for the electrode feed will be slightly inferior on low-voltage mains than those of a set having such a device, the performance will still be very satisfactory.

On low-voltage a.c. mains it is also possible to employ voltage doubling (see Chapter XIX, Section 111) to obtain a high anode voltage for the receiving valves (valve CY 2 for voltage doubling).

193. Power Supply of Car Sets

For car-radio receivers the only power source available is a battery of accumulators with a nominal voltage of 6 or 12 V [1]), to which the filaments of receiving valves of corresponding working voltage may be connected in parallel. The H.T. voltage has to be obtained by converting the low direct voltage from the battery through a vibrator into an alternating voltage and stepping-up this low alternating voltage to a higher one by means of a transformer. Subsequently the higher alternating voltage has to be rectified by means of a valve-type rectifier. Often instead of a rectifier the vibrator is provided with

---

[1]) The average voltage of these batteries is about 6.3 or 13 V, and valves have therefore been designed for these heating voltages.

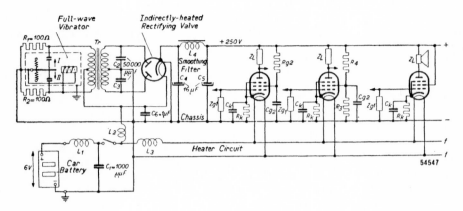

Fig. 379
Method of feeding a car-radio set with vibrator. The vibrator periodically interrupts
the direct current flowing through the two halves of the primary of the transformer Tr.
In the transformer secondary a high a.c. voltage is produced which after rectification
by a full-wave rectifying valve yields a direct voltage of about 250 V.

additional contacts which serve as a commutator and produce the
required direct current at high tension.

Just as is the case with a.c. sets, the H.T. direct voltage has to be
smoothed in the usual way. In Fig 379 a diagram is given showing
the essentials of the feeding circuit of a car-radio set. Between the
battery and the set is a filter to prevent the transmission of engine
interference to the receiver. This filter is followed by a choke L_2 which,
just as the choke L_3, is intended to suppress interference from the
vibrator.

The current impulses through contacts I and II of the vibrator are
attenuated by the resistors R_1 and R_2.

Special vibrators are made to meet the high demands for car-radio sets.
There are various types, some of which work in a high vacuum and
in appearance very much resemble a radio valve. The contacts have
to satisfy special requirements on account of the comparatively strong
currents they have to interrupt. Usually the vibrators are double-
phased, so that when one circuit is opened the other is closed (see
Fig. 379). For the rectification full-wave rectifying valves with indirectly-heated cathode are used. This cathode has a positive voltage
with respect to the chassis and the filament is fed from the car battery,
one side of which is connected to the chassis. Consequently a high
voltage is set up between cathode and filament and the filament of
a rectifying valve for car-radio sets has to be exceptionally well
insulated with respect to the cathode.

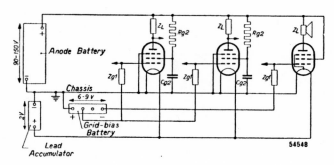

194. Power Supply of Battery Sets

Fig. 380
Method of feeding a battery set.

The circuits used for obtaining power supply from batteries are extremely simple, as shown in Fig. 380. Three batteries are needed, a H.T. battery (often called A-battery), a L.T. battery (often called B-battery) and a grid-bias battery (C-battery). The last mentioned can often be dispensed with by using part of the H.T. battery for the bias.

Usually the negative side of the L.T. accumulator is earthed, and the electrode voltages quoted for Philips battery valves are therefore related to the negative side of the filament.

Much importance is usually attached to having as few connections as possible with the batteries, these being limited to two leads for the H.T. battery and two for the L.T. battery. Consequently the new Philips battery-fed valves have been made either for equal anode and screen-grid voltages or for feeding the screen grids via series resistors, so that it is not necessary to have a connection to a tapping point on the H.T. battery for a lower positive electrode voltage. The screen grid of those valves which are designed for equal anode and screen-grid voltages can be connected direct to the H.T.+lead. Sometimes screen grids have to be decoupled, as shown in Fig. 380, by an RC element.

If it is desired to dispense with the grid-bias battery and thus limit the number of connections to the minimum, the negative grid bias for the output valve can be automatically produced by means of

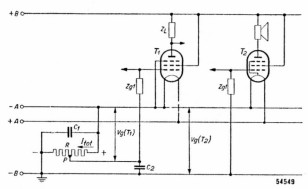

Fig 381
Method of producing automatic negative grid bias in a battery set.

the voltage drop a-
cross a resistor R (see
Fig. 381) interposed
between the negative
pole —A of the L.T.
battery and the ne-
gative pole —B of
the H.T. battery. In
that case the pole
—B is usually earthed
and the resistor R
by-passed by a large
capacitor C_1 effective
at high and low fre-
quencies. Thus the

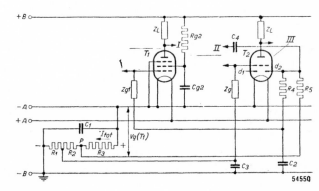

Fig. 382
Delayed A.V.C. with automatic negative grid bias in a
battery set. This method can also be used with a fixed
negative grid bias. I = Output of the I.F. amplifying valve,
II = lead from the I.F. resonant circuit, III = A.F. amplifier.

filament of the valve V_2 is made positive with respect to —B and if
the grid is connected with —B via the grid impedance Z_{g1} the pole
—B is negative with respect to the filament. Since the negative grid
bias for the output valve is usually greater than that for the preceding
valves the grids of the latter should not be connected to —B but
to tapping points on the resistor R to maintain the correct voltage.
When delayed A.V.C. is employed the following has to be borne in
mind. Normally it is not possible to get a delay voltage very much
higher than the negative grid bias of the pre-amplifying valves in the
non-controlled state. Fig. 382 illustrates this for a directly-heated double-diode-triode
used as detector and
A.V.C. rectifier. The
leak resistor R_5 of the
A.V.C. diode d_2 of valve
T_2 is not earthed, as
is done in a.c. sets, but
connected with the tap-
ping point P on the re-
sistor serving to produce
the negative grid bias
of the output valve.
The value of the resistor
R_3 is such that the voltage drop corresponds
exactly to the grid bias

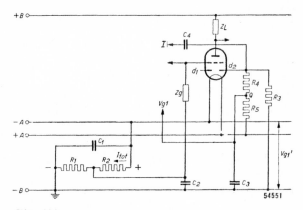

Fig. 383
Method of connections for a directly-heated double-
diode-triode in a receiver requiring a higher delay voltage
for A.V.C. than in the case of Fig. 382. I = lead from
the I.F. resonant circuit.

required for the gain-control valves in the non-controlled state. In the circuit shown in Fig. 382 this voltage is applied to the grid of the I.F. valve T_1 via R_5, the smoothing resistor R_4 and Z_{g1}. Since the diode anode d_2 lies opposite the positive end of the filament, in respect to the part of the filament it uses it already has a negative bias of 1.4 or 2 V when connected to —A. In this case, therefore, the delay voltage equals $2 V + V_{g\ (T1)}$, taking the heater voltage as 2 V. If a larger delay voltage should be necessary special provision will have to be made for it.

In Fig. 383 a circuit is given for providing a larger delay voltage, where the negative grid bias V_{g1}' of the output valve is applied to the diode d_2 via the voltage divider R_3—R_4—R_5, and since R_5 is connected to +A the negative bias on d_2, with respect to the positive end of the filament, is equal to

$$\frac{R_4 + R_5}{R_3 + R_4 + R_5} \times (V_{g1}' + 2)\ \text{V.}$$

The negative voltage of d_2 with respect to —A is reduced by the voltage divider R_4—R_5 to such an extent that the negative voltage at the point Q with respect to —A just corresponds to the negative grid bias required for the automatically controlled pre-amplifying valves in the non-controlled state. Obviously the control voltage for the A.V.C. at d_2 is likewise reduced by the divider R_4—R_5, so that the A.V.C. becomes less effective. If, for instance, $R_3 = 1$ megohm, $R_4 = 0.5$ megohm and $R_5 = 1$ megohm, with an output-valve negative grid bias of 5 V the delay voltage will equal $^3/_5 \times 7$ V $= 4.2$ V. The grid bias of the automatically controlled valves is then equal to 0.8 V before the A.V.C. comes into operation, and only $^2/_3$rds of the control voltage on d_2 is applied to the grids of these valves. If R_5 is connected to a higher positive voltage in the set and V_{g1}' is relatively large a more effective A.V.C. can be obtained.

Some of the pre-amplifying valves of the D series for battery operation are constructed in such a way that they can be used without negative grid bias, the voltage drop across the filament yielding sufficient bias to avoid grid current with weak signals (e.g. the H.F. pentode DF 21 and the octode DK 21). Also the diode-triode DAC 21 of this series can be used without negative bias on the grid of the triode part. The diode uses a part of the negative end of the common filament, so that the part of the filament used for the triode has a positive potential with respect to the negative filament terminal; when to the grid the potential of the negative terminal is applied the grid is sufficiently negative with respect to the filament part used for the triode section to avoid grid current.

195. Power Supply of Battery/A.C./D.C. Sets

In recent years there has been a demand for sets that can be connected up either to batteries, to a.c. mains or to d.c. mains. It is understandable that if one has a portable set and wants to use it at home it is a great advantage to be able to plug it into the mains and thereby save the batteries. Sets have therefore been made to answer these requirements. These employ directly-heated battery valves, the new Philips D series of 1.4-V valves being excellently suited for the purpose. Owing to the hum arising with a.c. supply the filaments cannot be supplied direct by the mains, and therefore with mains feed the rectifying valve employed to obtain the necessary H.T. supply is utilised also to feed the filaments with rectified current. Consequently the best results are obtained when using valves that need only low heating current, so as to reduce the load on the rectifier; the valves of the D series having a very low filament current are therefore particularly suitable for this purpose. By connecting the filaments in series the rectifier has only to supply a filament current corresponding to the highest filament-current value of the valves employed. If valves with a filament voltage of 1.4 V and a filament current of 50 mA are used, such as the octode DK 21, the R.F. pentode DF 22, the double-diode-triode DBC 21, and the power pentode DL 21, the total current that the rectifier has to supply amounts to about 65 mA (heating current 50 mA, anode and screen-grid feed about 15 mA), so that the rectifier valve UY 1N or UY 21 can then be used.

Fig. 384 shows the feeding circuit of such a set as this, where the valves DK 21, DF 22, DBC 21 and DL 21 are used as receiving valves and the UY 1N as rectifier. The filaments of the receiving valves are connected up in the order indicated in the diagram, with the filament of the output valve connected via R_5 to the negative terminal of the filament-feeding circuit, which is also the negative terminal of the anode-feeding circuit. In this way the anode and screen-grid currents of the output valve are conducted direct to the negative pole of the anode-feed circuit (via R_5) and do not pass through the filaments of other valves. The order in which the filaments of the other valves are connected has been chosen in such a way that the voltage drop across the filaments of the mixer valve DK 21, the I.F. valve DF 22 and the A.F. and detector valve DBC 21 can be utilised as delay voltage for the A.V.C. As initial bias for the control grid of the DF 22 the voltage drop across the filament of the mixer valve DK 21 is utilised, whilst to the control grid of the latter no initial bias is applied. With mains supply the bias of the output valve is obtained through

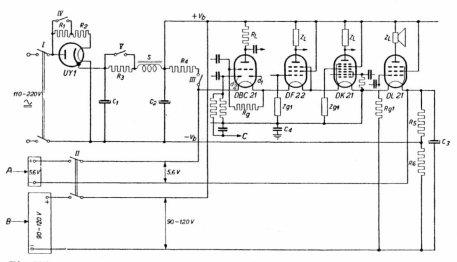

Fig. 384
Method of feeding a receiver designed for battery, a.c. or d.c. mains operation. A = L.T.
battery, B = H.T. battery, C = A.V.C. lead.

the voltage drop in R_5; the filament current and the total electron
current flowing through that resistor produce the desired bias. With
battery operation the filament current does not flow through R_5 and
therefore the negative pole of the H.T. battery is connected via R_6
to the negative line of the H.T. circuit (marked in Fig. 384 by $-V_b$),
in which case the voltage drop in the resistors R_5 and R_6 yields the
negative bias required for the output valve. When the circuit of Fig. 384
is energized by d.c. or a.c. mains the switches I and III are closed
and switch II is opened, the latter cutting out the batteries. The
filament of the rectifying valve UY 1N is then fed from the mains,
and as the heating voltage of this valve is 50 V a resistor is interposed
in the heater-current circuit (R_1 and R_2 in Fig. 384), the value of
which, of course, has to be suitable for the existing mains voltage.
The switch IV short-circuits R_1 for operation on 110-V mains. When
this switch is open the resistors R_1 and R_2 have together the right
value for 220-V mains. The direct current provided by the rectifier
flows through a resistor R_3 and a choke S, the combined d.c. resistance
of which has to be so chosen that the voltage across the second
smoothing condenser C_2 equals that of the H.T. battery used for battery
operation. With the aid of switch V this voltage can be adjusted to
the right value for high and for low mains voltages.
The filaments of the four receiving valves are connected in series and
thus require a total heating voltage of $4 \times 1.4 = 5.6$ V. Since with

mains operation the filament current is supplied by the rectifier, the rectified voltage on C_2 has to be reduced to 5.6 V by a ballast resistor (R_4 in Fig. 384).

With battery operation switches I and III are opened and switch II is closed. The open switch III then ensures that the H.T. battery cannot discharge through R_4 and the filaments. The filaments are then fed in series from a 5.6-V dry battery (of course a 6-V accumulator can be used with a ballast resistor).

Obviously combinations of valves other than that shown in Fig. 384 can be used, with the circuit altered accordingly. It is to be noted, however, that it is not advisable to connect filaments of valves in pairs in parallel, because if one of the valves is taken out this might impair the filament of the remaining valve.

CHAPTER XXXIII

The Sensitivity of a Receiver or an Amplifier

The term "sensitivity" is frequently used in connection with a receiving set or an amplifier, and by this is meant the weakest signal that the circuit can reproduce with a more or less reasonable volume of sound, a more sensitive set being more capable of making weak signals audible than a less sensitive set. For comparative purposes the following definition has been adopted: sensitivity is the r.m.s. value of the signal voltage required to produce an alternating-current output power of 50 mW (this being based on a relatively small sound volume in a living room that can be attained with almost any output valve). For certain output valves it is stated, for instance, that their sensitivity is 0.3 V. This means that for an a.c. output of 50 mW a grid signal of 0.3 V is required. If such a valve is preceded by an A.F. amplifier valve with a gain of 100, the signal required on the grid of that valve will be 100 times smaller for a power output of 50 mW of the output valve. The sensitivity of an A.F. amplifier with these two valves therefore amounts to 3 mV.

It is also of importance to know, with superheterodyne receiving sets having R.F., mixing and I.F. stages, the sensitivity at the various valve grids, for example, the voltage required on the grid of the I.F. valve. Since on that grid there is a modulated I.F. alternating voltage, it is not enough to indicate the value of that voltage, for the modulation depth must also be known. The standard commonly taken for modulation depth is 30%. Therefore when speaking of the sensitivity at the grid of the I.F. valve one understands that it refers to the 30%-modulated I.F. voltage required to obtain 50 mW output power in the anode circuit of the output valve. The sensitivity at the grid of a mixing valve or of the R.F. pre-amplifying valve can be defined in the same way, this always being taken to be the 30%-modulated R.F. signal required to produce an output power of 50 mW of the receiver. The same likewise applies to the detector valve.

Once the amplification of each stage is known it is also possible to calculate—starting from the output valve—the sensitivity at any point in the set. The amplification of the detector diode is easily determined with the aid of the V_{LF}-curve of Fig. 218. Having calculated the A.F. signal required on the leak resistor of the detector diode to get an output of 50 mW, it is easy to read off on that curve the 30%-modulated I.F. signal needed.

In most cases the sensitivity at the input of a receiving set is also of importance, and as the properties of the aerial considerably influence the sensitivity a general standard is taken also for the aerial, the sensitivity being determined for a dummy aerial having a capacity of 200 $\mu\mu$F with respect to earth, an inductance of 20 μH and a resistance of 25 ohms.

Sensitivity is also affected by the frequency of the modulation, because the selectivity of the circuits and the frequency-response characteristic of the A.F. amplifier likewise exercise some influence. For this reason a modulation frequency of 400 c/s is usually taken as basis. As the impedance of the loudspeaker depends upon the frequency, in determining sensitivity the loudspeaker is replaced by a resistance corresponding to the correct load, and the anode of the output valve is fed via a choke having a very high impedance at a frequency of 400 c/s.

A cathode-activation screen for A-technique valves. In the anode and grid circuits of each valve a glowlamp is included which serves for limiting the current and as an indicator of short-circuits. The bottom part contains the power supply; on the right are arranged the maximum-current relay switches for each batch of valves.

APPENDIX

I. UNITS

A. The Various Unit Systems and their Mutual Relationship

(1) The Unit Systems Used so far

Units are distinguished as basic units and derivatives. The following have been taken as basic units:

(a) **The unit of length,** the dimension of which is indicated by the letter L,

(b) **The unit of mass,** the dimension of which is indicated by the letter M, and

(c) **The unit of time,** the dimension of which is indicated by the letter T.

The units of all other quantities can be derived from these basic units by geometrical or physical relationships; speed, for instance, is determined by dividing distance by time, the dimension being LT^{-1}.

The following basic units have been chosen for the absolute or c.g.s. system (centimeter-gramme-second system): the centimetre (cm), the gramme (g) and the second (sec).

All other units, of mechanical, electrical and magnetic quantities, are derivatives. Since many derivatives in the absolute system are either too large or too small for practical use, the so-called practical or technical system of units is used in many cases.

In electrotechnics a distinction is made between electromagnetic and electrostatic units, the former being indicated by e.m.u. and the latter by e.s.u.; both are derivatives. Electromagnetic and electrostatic units bear a certain inter-relationship.

The following table gives the relation between the practical units, the electromagnetic units (e.m.u.) and the electrostatic units (e.s.u.).

TABLE
Relation between Practical, Electromagnetic and Electrostatic Units

Name	Symbol	Practical unit	1 practical unit equals			
			electromagnetic units		electrostatic units	
Charge	Q	1 coulomb	10^{-1} e.m.u.	dim. $Q = L^{1/2}M^{1/2}$	3.10^9 e.s.u.[1])	dim. $Q = L^{3/2}M^{1/2}T^{-1}$
Current Intensity	I	1 ampere	10^{-1} e.m.u.	dim. $I = L^{1/2}M^{1/2}T^{-1}$	3.10^9 e.s.u.[1])	dim. $I = L^{3/2}M^{1/2}T^{-2}$
Potential, E.M.F.	V	1 volt	10^8 e.m.u.	dim. $V = L^{3/2}M^{1/2}T^{-2}$	$1/300$ e.s.u.[1])	dim. $V = L^{1/2}M^{1/2}T^{-1}$
Resistance	R	1 ohm (Ω)	10^9 e.m.u.	dim. $R = L\,T^{-1}$	$1/9.10^{-11}$ e.s.u.	dim. $R = L^{-1}T$
Capacity	C	1 farad	10^{-9} e.m.u.	dim. $C = L^{-1}T^2$	9.10^{11} e.s.u.	dim. $C = L$
Inductance	L	1 henry	10^9 e.m.u.	dim. $L = L$	$1/9.10^{-11}$ e.s.u.	dim. $L = L^{-1}T^2$
Work	A	1 joule $= 10^7$ ergs, dimension of $A = L^2MT^{-2}$				
Power	P	1 watt $= 10^7$ ergs/sec, dimension of $P = L^2MT^{-3}$				

(2) The Rationalized Unit System of Giorgi

The rationalized system of Giorgi uses the following basic units:

(a) **the metre** (m) as unit of length,

(b) **the kilogramme** (kg) as unit of mass, and

(c) **the second** (sec) as unit of time.

For electrotechnics the advantage of the rationalized system of Giorgi is that most equations used become very simple and clear. The units of electric and magnetic field strength are not derived from the attraction and repulsion laws of Coulomb, but direct from inductive effect. Furthermore there are simple relationships between, for instance, mechanical and electrical quantities.

---

[1]) The figure 3.10^{10} cm/sec given as the relation between the electrostatic and electromagnetic units of charge and current is only approximately correct. The exact figure for the velocity of light is: $C = 2.99776.10^{10}$ cm/sec.

(a) Some Derived Mechanical Units

Unit of velocity $= \dfrac{m}{sec}$;

unit of acceleration $= \dfrac{m}{sec^2}$;

unit of force, newton (N) $= \dfrac{kg\ m}{sec^2}$;

energy or work, newton metre (Nm) $= \dfrac{kg\ m^2}{sec^2} = W\ sec = joule$;

power, newton metre per second $= \dfrac{kg\ m^2}{sec^3} = W$.

(b) Electrical Units

As electrical units the conventional volt, ampere, ohm, farad, henry, coulomb, etc. are retained. The units for the electric and the magnetic field are, however, expressed also in volts and amperes. This results in a very clear relationship between electric and magnetic fields on the one hand and electric currents and voltages on the other hand. All electrotechnical units may be expressed in units of length (m), of time (sec), of voltage (V) and of current (A).

(a) Electric Field

When the voltage between the plates of a condenser is increased by 1 volt (direct voltage) the current impulse or charge supplied Q (expressed in coulombs = A sec) is always the same whatever is the rate of voltage increase. The capacity (C) of a condenser gives the magnitude of this current impulse or charge. The unit of capacity is called farad (F). According to the definition given the capacity C equals $\dfrac{Q}{V} \left(\dfrac{coulomb}{volt}\ or\ \dfrac{A\ sec}{V} \right)$. The charging of a condenser requires a work or energy A equal to:

$$\tfrac{1}{2}\ Q\ V = \tfrac{1}{2}\ C\ V^2\ (VA\ sec\ or\ W\ sec).$$

When a condenser is charged an electric field is set up between its plates. It is in this field that the energy required to charge the condenser is stored. The electric flux Ψ of the field existing between the plates is put equal to the charges Q on the positive and negative plates. These charges are due to a potential difference V . . . . (V) between its plates. Thus we have:

$$\Psi = Q \ . . . . . \ (A\ sec),\ and$$
$$\Psi = C\ V \ . . . . \ (A\ sec).$$

In the latter expression C is the capacity of the condenser:

$$C = \dfrac{\Psi}{V} \ . . . . . \left(\dfrac{A\ sec}{V} \right).$$

In a condenser with flat plates having an area S . . . . (m²), the plates being spaced at a distance d . . . . (m), the field is practically homogeneous. In that case at all points between the plates the field strength F is equal to $\dfrac{V}{d}$ . . . . (V/m). The electric induction or displacement D is the flux per unit of area. The unit of electric induction is consequently $\dfrac{A\ sec}{m^2}$. The ratio between the field strength F and the electric induction depends on the dielectric in which the field is present. For vacuum we write

$$D = \varepsilon_0\ F.$$

In this expression ε_0 is the absolute dielectric constant and is equal to:

$$\varepsilon_0 = \dfrac{10^7}{4\ \pi\ c^2} \approx \dfrac{1}{36\ \pi\ 10^9} \ . . . . \left(\dfrac{A\ sec}{Vm} = \dfrac{farad}{metre} \right).$$

For other dielectrics F is furthermore to be multiplied by the relative dielectric constant of the material. This constant has no dimension and consequently we have for these dielectrics the formula:

$$D = \varepsilon\, \varepsilon_0\, F \ldots \left(\frac{A\ sec}{m^2}\right).$$

The capacity of a condenser with flat plates in vacuum is

$$C = \varepsilon_0\, \frac{S}{d} \ldots (F).$$

In this expression S is to be introduced in m² and d in m. The energy contained in the field of a charged condenser per unit volume is for vacuum

$$\tfrac{1}{2}\frac{QV}{Sd} = \tfrac{1}{2}\, D\, F = \tfrac{1}{2}\, \varepsilon_0\, F^2 \ldots (W\ sec).$$

Consequently ε_0 can be measured by means of the charge on the flat plates of a vacuum condenser with a plate area of 1 m² and a plate spacing of 1 m when a potential difference of 1 V exists between the plates and the field is homogeneous.

(β) **Magnetic Field**

When the current intensity in an air coil is increased by 1 A (direct current) the voltage impulse P (V sec) between the coil terminals is always the same independently of the rate of current variation. The inductance (L) of the coil indicates the magnitude of this voltage impulse. According to this definition the unit of inductance, the henry (H), is equal to $\frac{V\ sec}{A}$. The current variation in the coil requires a work or energy equal to

$$\tfrac{1}{2}\, P\, I = \tfrac{1}{2}\, L\, I^2 \ldots (VA\ sec\ or\ W\ sec).$$

The current in the coil generates a magnetic field and the work required to produce the current variation is stored in this field. The current with intensity I (amps) which flows through a coil of a single turn or a straight conductor produces a magnetic flux $\varPhi$ which we put equal to the voltage impulse P in V sec. Thus we have

$$\varPhi = P \ldots \ldots (V\ sec),\ and$$
$$\varPhi = L\, I \ldots (V\ sec).$$

Consequently the inductance equals

$$L = \frac{\varPhi}{I} \ldots \ldots \left(\frac{V\ sec}{A}\right).$$

In an extended solenoid formed by one winding only, for instance by a broad metallic scrip, the area of the cross-section being S .... (m²) and the length l .... (m), the field is practically homogeneous. Outside this solenoid where the field closes the intensity of the latter is negligible. In that case the whole magnetomotive force of the current I enclosed by the lines of force of the field is used for setting up the field at the interior of the winding and the strength H .... (A/m) of the magnetic field is equal to I/l. This magnetic field produces an induction B .... $\left(\frac{V\ sec}{m^2}\right)$ equal to $\frac{\varPhi}{S}$. The ratio of the magnetic field strength H .... (A/m) and the induction B .... $\left(\frac{V\ sec}{m^2}\right)$ depends on the medium in which the field is present. For vacuum we have

$$B = \mu_0\, H \ldots \left(\frac{V\ sec}{m^2}\right),$$

in which expression μ_0 is the absolute permeability and

$$\mu_0 = \frac{4\,\pi}{10^7} \ldots \left(\frac{V\ sec}{Am} = \frac{henry}{metre}\right).$$

For other media H is moreover multiplied by the relative permeability μ of the material which has no dimension:

$$B = \mu\, \mu_0\, H.$$

A straight conductor through which a current flows is surrounded by concentric magnetic lines of force. At the periphery of a circle concentric with the conductor cross-section

473

and having a circumference of 1 m ($r = \dfrac{1}{2\,\pi}$ m) the field strength H is 1 A/m when the current is 1 A.

The inductance of a solenoid of one winding (made of a metallic strip) in vacuum is

$$L = \mu_0 \frac{S}{l} \ldots \text{(H)},$$

where S is again expressed in m² and l in m.

The amount of energy per unit volume of the field is in the case of an air coil:

$$\tfrac{1}{2} \frac{P\,I}{S\,l} = \tfrac{1}{2} \frac{\Phi\,I}{S\,l} = \tfrac{1}{2}\,B\,H = \tfrac{1}{2}\,\mu_0\,H^2 \ldots \text{(W sec)}.$$

Consequently μ_0 can be determined by the flux in a unity coil (1 winding of a metallic strip with breadth of 1 m and cross-section of 1 m²) in vacuum for a current intensity of 1 A when the field within the coil is homogeneous.

(γ) Relationships between Electrical and Magnetic Quantities

In the rationalized Giorgi system the unit of force is the newton (N). As already mentioned under (a), the newton is the force which imparts an acceleration of 1 m/sec² to a mass of 1 kg. In 1938 at Torquay the I.E.C. admitted the newton as the unit of force. The use of the newton greatly simplifies the relationships between electrical quantities and mechanical force. The force exerted on the charge Q .... (A sec = C) in an electric field of intensity F .... (V/m) is equal to

$$f = Q\,F \ldots \text{(N)}.$$

From this it results that 1 newton $= 1\,\dfrac{VA\,sec}{m} = 1\,\dfrac{W\,sec}{m}$.

The attraction between two identical parallel electrodes with an area of S .... (m²) between which a homogeneous field exists is equal to

$$f = \tfrac{1}{2}\,\Psi\,F = \tfrac{1}{2}\,S\,D\,F = \tfrac{1}{2}\,\varepsilon_0\,S\,F^2 \ldots \text{(N)}.$$

The attraction between two charges Q_1 and Q_2 creating symmetrical spherical fields and being at a distance of R .... (m) from each other is equal to

$$f = \frac{1}{4\,\pi\,\varepsilon_0} \times \frac{Q_1\,Q_2}{R^2} \ldots \text{(N)}.$$

The force exerted on a magnetic pole of intensity Φ .... (V sec) placed in a magnetic field of intensity H .... (A/m) is equal to

$$f = \Phi\,H \ldots \text{(N)}.$$

Such a magnetic pole is, for instance, the pole of a very long magnet at the extremity of which the magnetic flux Φ issues or enters.

The attraction between two magnetic poles (Φ_1 and Φ_2) creating symmetrical spherical fields and being at a distance of R .... (m) from each other is equal to

$$f = \frac{1}{4\,\pi\,\mu_0}\,\frac{\Phi_1\,\Phi_2}{R^2} \ldots \text{(N)}.$$

The force exerted on a straight conductor of length l .... (m) in a magnetic field of induction B, when the wire is perpendicular to the lines of force and a current of intensity I flows through it is,

$$f = I\,B\,l = I\,\mu_0\,H\,l \ldots \text{(N)}.$$

The moment produced by the couple of forces exerted by a homogeneous magnetic field of induction B on a coil of n windings the cross-section of which is S .... (m²) and through which flows a current of intensity I, when the longitudinal axis of the coil is perpendicular to the lines of force, is equal to:

$$M = B\,I\,n\,S \ldots \text{(Nm)}.$$

(δ) Some Formulae for Magnetic Circuits

The number of ampere-turns I .... (A) entirely enclosed by the flux in an iron-core magnetic circuit (with air gap) which is required in order to obtain an induction B .... $\left(\dfrac{\text{V sec}}{\text{m}^2}\right)$ is equal to

$$I = H_f 1 + Hd = \frac{B}{\mu_0}\left(\frac{1}{\mu} + d\right) \ldots . (A).$$

In this formula the magnetic circuit has a practically homogeneous field and the iron core with permeability μ has a length l .... (m), whereas the narrow air gap has a length d .... (m).
Very often the field strength H_f required for the iron core can be directly read from the curves which give the induction in gauss and the required field strength in A/cm. In that case the B given in V sec/m² (Giorgi units) must be multiplied by 10^4 in order to obtain it in gauss. The field strength obtained from the curve must be multiplied by 10^2 in order to have H_f expressed in A/m.
The r.m.s. value of the voltage at the terminals of a transformer is:

$$E = n\,\omega\,\frac{\Phi_m}{\sqrt{2}} = 4.44\,n\,\nu\,\Phi_m \ldots . (V).$$

In this expression n is the number of turns of the coil, ω the angular frequency, ν the frequency of the sinusoidal flux in $\dfrac{1}{\text{sec}}$ and Φ_m .... (V sec) the peak value of the flux.

(e) Conversion of the Electrostatic, the Electromagnetic and the Practical Systems into the Rationalized System of Giorgi

In the following table, which gives the factors for converting the units so far used into units of the Giorgi system, the "absolute" volt and ampere are used throughout. All measuring instruments are normally calibrated in "international" volts and amperes. The difference between these values has the following origin. The absolute volt and ampere are deduced exactly from the c.g.s. units by means of the factors in the table.
On the other hand, the **international ampere** is defined as the current intensity which, flowing through a diluted solution of silver nitrate, produces a deposit of 1.118 milligrammes of silver per second.
The **international ohm** is defined by the resistance of a column of quicksilver having a length of 1.063 m, a constant cross-section of 1 mm² and a weight of 14.4521 grammes at the temperature of 0° C.
The **international volt** is the potential difference which is produced, by a current with an intensity of 1 ampere, between the terminals of a conductor free from electromotive force having a resistance of 1 ohm.
The difference between the international and absolute units is very small. It is noticed from the fourth decimal onwards. The table only mentions for the electrostatic system, the electromagnetic system and the practical system the unit or the name of the unit when this is currently used. (Thus the unit of voltage of the electrostatic system is merely indicated by e.s.u. and not by cm, g, sec with exponents.)
In the table the corresponding units of the Giorgi system are expressed in volts and amperes and not in other basic units, since in practice this choice gives a clearer notion of the significance of the complex units.
The value c used in the conversion factors is the approximate numerical value of the velocity of light, $c = 3 . 10^8$ m/sec.
The conversion factors are factors by which the numerical values of quantities expressed in electrostatic, electromagnetic or practical units have to be multiplied in order to find the numerical values in units of the Giorgi system. In this way the conversion factor of the capacity unit in the electrostatic system is $\dfrac{10^5}{c^2}$ or 1.11×10^{-12}. A capacity of 50 cm has consequently the value of $50 \times 1.11 \times 10^{-12} = 55.5 \times 10^{-12}$ farad in the Giorgi system.

TABLE

for converting units of the electrostatic, electromagnetic and practical systems into units of the rationalized Giorgi system

Quantity	Symbol	System	Unit	Conversion factor exact	Conversion factor approx.	Unit in the Giorgi system
Voltage or potential	V	e.s.	e.s.u.	$c/10^6$	300	V
		e.m.	e.m.u.	10^{-8}	—	
Current intensity	I	e.s.	e.s.u.	$1/10c$	3.33×10^{-10}	A
		e.m.	e.m.u.	10	—	
Current density	S	e.s.	e.s.u.	$10^3/c$	3.33×10^{-6}	A/m²
		e.m.	e.m.u.	10^5	—	
		p.	A/cm²	10^4	—	
Power	P	e.s.	erg/sec	10^{-7}	—	VA = W
		e.m.	erg/sec	10^{-7}	—	
Energy or work	A	e.s. ⎫ e.m. ⎬	erg	10^{-7}	—	W sec = = joule = Nm
		p	kWh	3.6×10^{-6}	—	
Resistance	R	e.s.	e.s.u.	$c^2/10^5$	9×10^{11}	V/A = Ω
		e.m.	e.m.u.	10^{-9}	—	
Resistivity	ρ	e.s.	e.s.u.	$c^2/10^7$	9×10^9	Ω m
		e.m.	e.m.u.	10^{-11}	—	
		p.	Ω cm	10^{-2}	—	
		—	$\dfrac{\Omega \text{ mm}^2}{\text{m}}$	10^{-6}	—	
Conductivity	G	e.s.	e.s.u.	$10^5/c^2$	1.11×10^{-12}	A/V = $1/\Omega$ = siemens = mho
		e.m.	e.m.u.	10^9	—	
Specific conductivity, conduction	γ	e.s.	e.s.u.	$10^7/c^2$	1.11×10^{-10}	$\dfrac{1}{\Omega \text{ m}} =$ mho/m = siemens/m
		e.m.	e.m.u.	10^{11}	—	
		p.	$\dfrac{1}{\Omega \text{ cm}}$	10^2	—	
		—	$\dfrac{\text{m}}{\Omega \text{ mm}^2}$	10^6	—	
Capacity	C	e.s.	cm	$10^5/c^2$	1.11×10^{-12}	$\dfrac{\text{A sec}}{\text{V}} =$ F
		e.m.	e.m.u.	10^9	—	
Inductance	L	e.s.	e.s.u.	$c^2/10^5$	9×10^{11}	$\dfrac{\text{V sec}}{\text{A}} =$ H
		e.m.	e.m.u.	10^{-9}	—	
Charge	Q	e.s.	e.s.u.	$1/10c$	3.33×10^{-10}	A sec = coulomb (C)
		e.m.	e.m.u.	10	—	

Quantity	Symbol	System	Unit	Conversion factor		Unit in the Giorgi system
				exact	approx.	
Electric flux	Ψ	e.s.	e.s.u.	$\dfrac{1}{40\pi c}$	2.65×10^{-11}	A sec = coulomb (C)
		e.m.	e.m.u.	$\dfrac{10}{4\pi}$	0.796	
Electric induction	D	e.s.	e.s.u.	$\dfrac{10^3}{4\pi c}$	2.65×10^{-7}	$\dfrac{\text{A sec}}{\text{m}^2} =$
		e.m.	e.m.u.	$\dfrac{10^5}{4\pi}$	7.96×10^3	$\dfrac{\text{coulomb}}{\text{m}^2}$
		p.	$\dfrac{\text{coulomb}}{\text{cm}^2}$	10^4	—	
Electric field strength	F	e.s.	e.s.u.	$c/10^4$	3.10^4	V/m
		e.m.	e.m.u.	10^{-6}	—	
		p.	V/cm	10^2	—	
Absolute dielectric constant	ε_o	p.	F/cm	10^2	—	$\dfrac{\text{A sec}}{\text{Vm}} = \text{F/m}$
Moment of an electric dipole	p or μ	e.s.	e.s.u.	$1/c.10^3$	3.33×10^{-12}	A sec m = coulomb metre
		e.m.	e.m.u.	10^{-1}	—	
		p.	coulomb $\times$ cm	10^{-2}	—	
Magnetic flux	Φ	e.s.	e.s.u.	$c/10^6$	300	V sec = weber
		e.m.	maxwell	10^{-8}	—	
Magnetic induction	B	e.s.	e.s.u.	$c/10^2$	3.10^6	$\dfrac{\text{V sec}}{\text{m}^2}$
		e.m.	gauss	10^{-4}	—	
		p.	$\dfrac{\text{V sec}}{\text{cm}^2}$	10^4	—	
Magnetic field strength	H	e.s.	e.s.u.	$10/4\pi c$	2.65×10^{-9}	A/m
		e.m.	oersted	$\dfrac{10^3}{4\pi}$	79.6	
		p.	A/cm	10^2	—	
Absolute permeability	μ_o	p.	H/cm	10^2	—	$\dfrac{\text{V sec}}{\text{Am}} = \text{H/m}$
Magnetic moment	m	e.s.	e.s.u.	$4\pi\,10^{-10}$	1.257×10^{-9}	V sec m
		e.m.	V sec cm	10^{-2}	—	
Force	f	e.s.	dyne	10^{-5}	—	newton (N)
		e.m.	dyne	10^{-5}	—	
		technical	kg	g	9.81	

B. The Values of some Natural Constants [1])

	Symbol	Value	Unit
Charge of the electron	e	1.6020×10^{-19}	abs. coul.
Specific charge of the electron	e/m_e	1.759×10^{11}	abs. coul/kg
Mass of the electron	m_e	9.1066×10^{-31}	kg
Electronvolt.......................	—	1.6026×10^{-19}	abs. joule
Velocity of the electron at 1 abs. V..	—	5.93×10^5	m/sec
Velocity of light...................	c	2.99776×10^8	m/sec
Mechanical heat equivalent	J	4.1855	abs. joule/cal$_{15°}$
Heat equivalent of the joule.........	—	0.2389	cal$_{15°}$/abs. joule
Boltzmann's constant	k	1.3805×10^{-23}	abs. joule/°K

II. DIRECT-CURRENT CIRCUITS

1) Ohm's law (Fig. 1): $I = \dfrac{V}{R}$; $V = I\,R$; $R = \dfrac{V}{I}$.

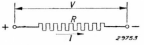

Fig. 1

R expressed in Ohms, V in volts and I in amperes.

2) Power: $P = V\,I$; $P = I^2 R$; $P = \dfrac{V^2}{R}$

P expressed in watts, V in volts, I in amperes and R in ohms

3) Resistances in series: $R = R_1 + R_2 + R_3 + \ldots .$

4) Resistances in parallel: $\dfrac{1}{R} = \dfrac{1}{R_1} + \dfrac{1}{R_2} + \dfrac{1}{R_3} + \ldots .$

Two resistances in parallel (Fig. 2):

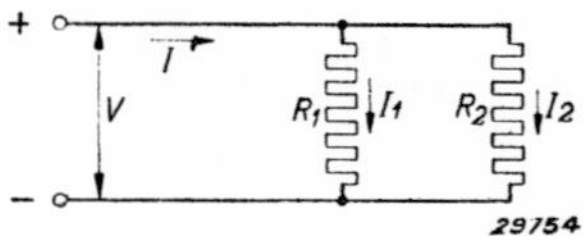

Fig. 2

$$R = \frac{R_1\,R_2}{R_1 + R_2}\,; \qquad I_1 = \frac{R_1}{R_1 + R_2}\,I;$$

$$R_1 = \frac{R\,R_2}{R_2 - R}\,; \qquad I_2 = \frac{R_1}{R_1 + R_2}\,I.$$

$$R_2 = \frac{R\,R_1}{R_1 - R}\,.$$

Approximately:

If R_2 =	1	1.5	2	3	4	5	10	20	$\times$	R_1,
then R =	0.5	0.6	0.67	0.75	0.8	0.84	0.9	0.95	$\times$	R_1.

If R_2 is greater than $10 \times R_1$ a sufficiently accurate approximation is obtained by deducting from the resistance value of R_1 the percentage $\dfrac{R_1}{R_2} \times 100$. If for instance $R_2 = 20\,R_1$, then approximately $R = R_1 - 5\%$.

[1]) See also W. de Groot, Nederlandsch Tijdschrift voor Natuurkunde 9, Dec. 1942, pp. 497—505.

5) Generator with internal resistance (Fig. 3):

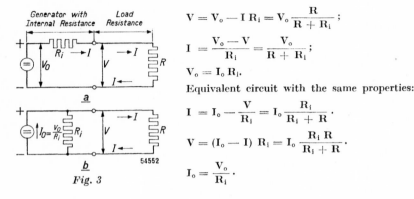

$$V = V_o - I\,R_i = V_o \frac{R}{R + R_i};$$

$$I = \frac{V_o - V}{R_i} = \frac{V_o}{R + R_i};$$

$$V_o = I_o\,R_i.$$

Equivalent circuit with the same properties:

$$I = I_o - \frac{V}{R_i} = I_o \frac{R_i}{R_i + R}.$$

$$V = (I_o - I)\,R_i = I_o \frac{R_i\,R}{R_i + R}.$$

Fig. 3

$$I_o = \frac{V_o}{R_i}.$$

6) Power and efficiency of a generator with internal resistance R_i (Fig. 3):

(a) Power in the load resistance R:

$$P_R = I\,V = V_o\,I - I^2R_i \quad \text{or} \quad P_R = I_o\,V - \frac{V^2}{R_i} \quad \text{or}$$

$$P_R = \frac{V_o\,V - V^2}{R_i} \qquad \text{or} \quad P_R = (I_o\,I - I^2)\,R_i \quad \text{or}$$

$$P_R = \frac{V_o^2\,R}{(R_i + R)^2} \qquad \text{or} \quad P_R = \frac{I_o^2\,R_i^2\,R}{(R_i + R)^2}.$$

(b) Efficiency

$$\eta = \frac{P_R}{P_{tot}} = \frac{R}{R_i + R}.$$

7) Maximum power:

This is reached if $R = R_i$.

The power in the load resistance is then equal to:

$$P_R = \frac{^1/_4\,V_o^2}{R_i} \quad \text{or} \quad P_R = {}^1/_4\,I_o^2\,R_i.$$

The efficiency in this case is

$$\eta = \frac{P_R}{P_{tot}} = 0.50.$$

8) Voltage divider unloaded (Fig. 4):

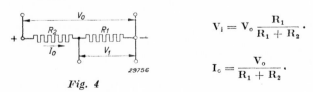

Fig. 4

$$V_i = V_o \frac{R_1}{R_1 + R_2}.$$

$$I_c = \frac{V_o}{R_1 + R_2}.$$

479

9) Voltage divider loaded (Fig. 5a):

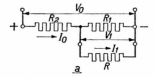

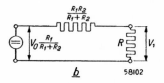

$$\text{Fig. 5}$$

$$V_1 = V_0 \frac{R_1}{R_1 + R_2} - I_1 \frac{R_1 \times R_2}{R_1 + R_2},$$

$$I_1 = \frac{V_0 \dfrac{R_1}{R_1 + R_2} - V_1}{\dfrac{R_1 \times R_2}{R_1 + R_2}}.$$

Equivalent circuit with the same properties (Fig. 5b).

$$I_0 = \frac{V_0}{R_1 + R_2} + I_1 \frac{R_1}{R_1 + R_2},$$

$$I_0 = \frac{V_0 - V_1}{R_2}.$$

10) Voltage divider with two different tapped voltages and two different loads (Fig. 6):

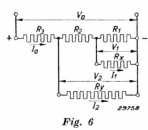

$$V_1 = V_0 \frac{R_1}{R_1 + R_2 + R_3} - I_1 \frac{R_1 (R_2 + R_3)}{R_1 + R_2 + R_3} -$$
$$- I_2 \frac{R_1 \times R_3}{R_1 + R_2 + R_3}.$$

$$V_2 = V_0 \frac{R_1 + R_2}{R_1 + R_2 + R_3} - I_1 \frac{R_1 \times R_3}{R_1 + R_2 + R_3} -$$
$$- I_2 \frac{R_3 (R_1 + R_2)}{R_1 + R_2 + R_3}.$$

Fig. 6

Used for energizing a screen-grid valve.

11) Equivalent of a star and a delta connection (Fig. 7).

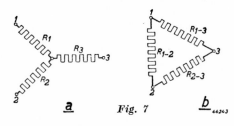

Fig. 7

Fig. 7a is equivalent to Fig. 7b if

$$R_{12} = \frac{R_1 R_2}{R_3} + R_1 + R_2,$$

or

$$R_1 = \frac{R_{12} R_{13}}{R_{12} + R_{13} + R_{23}}.$$

12) Wheatstone bridge connection (Fig. 8)

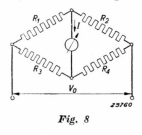

Fig. 8

The state of balance ($I = 0$) is reached if

$$\frac{R_1}{R_2} = \frac{R_3}{R_4}, \text{ or } \frac{R_1}{R_3} = \frac{R_2}{R_4} \text{ or}$$

$$R_1 \cdot R_4 = R_2 \cdot R_3;$$

$$R_1 = R_2 \frac{R_3}{R_4}.$$

III. ALTERNATING-CURRENT CIRCUITS

1) Alternating current and alternating voltage:

$e = E_o \sin(\omega t + \varphi_1)$;
$i = I_o \sin(\omega t + \varphi_2)$;
$\omega = 2\pi f$ = angular frequency in radians per second ($\pi = 3.14\ldots$);
λ = wavelength $\left(\lambda = \dfrac{3.10^8}{f}\right.$, λ in m and f in c/s, $\left.f = \dfrac{3.10^8}{\lambda}\right)$;
φ = phase angle between voltage and current = $\varphi_1 - \varphi_2$;
$f = \dfrac{1}{T}$ = frequency in cycles per second (c/s);
t = time;
T = periodic time;
e = instantaneous value of the voltage;
E_o = amplitude of the voltage;
i = instantaneous value of the current;
I_o = amplitude of the current;

2) Effective r.m.s. values:

$$E = \frac{E_o}{\sqrt{2}}; \quad I = \frac{I_o}{\sqrt{2}}.$$

Mean values (for sinusoidal voltages and currents);

$$\bar{E} = \frac{2}{\pi} E_o = 0.637\, E_o; \quad \bar{I} = \frac{2}{\pi} I_o = 0.637\, I_o.$$

The form factor is the ratio of the effective and mean values:

$$\xi = \frac{I}{\bar{I}} = \frac{E}{\bar{E}} = \frac{\pi}{2\sqrt{2}} = 1.11.$$

3) Impedance:

$$Z = \frac{E}{I}; \quad I = \frac{E}{Z}; \quad E = IZ.$$

4) Power:

$$P = EI\cos\varphi; \quad P = I^2 Z\cos\varphi; \quad P = \frac{E^2}{Z}\cos\varphi.$$

5) A.C. circuit with a resistance (Fig. 9)

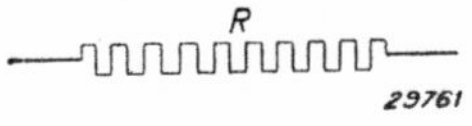

$Z = R; \cos\varphi = 1.$
R expressed in ohms.

Fig. 9

6) A.C. circuit with an inductance (Fig. 10):

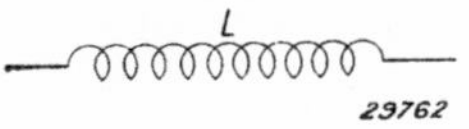

$Z = \omega L, \cos\varphi = 0.$
L expressed in henrys.

Fig. 10

7) A.C. circuit with two inductances connected in series (Fig. 11):

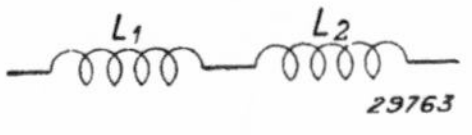

$L = L_1 + L_2.$

Fig. 11

If a mutual inductance M is present between the coils L_1 and L_2:

$$L = L_1 + L_2 \pm 2M.$$

481

8) A.C. circuit with two parallel-connected inductances (Fig. 12):

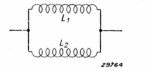

$$L = \frac{L_1 \times L_2}{L_1 + L_2}.$$

Fig. 12

If a mutual inductance M is present between the coils L_1 and L_2:

$$L = \frac{L_1 \times L_2 - M^2}{L_1 + L_2 \pm 2\,M}.$$

9) A.C. circuit with a capacity (Fig. 13):

$$Z = \frac{1}{\omega C}, \quad \cos \varphi = 0.$$

Fig. 13

C expressed in farads.

10) A.C. circuit with two capacities connected in series (Fig. 14):

$$C = \frac{C_1 \times C_2}{C_1 + C_2}.$$

Fig. 14

11) A.C. circuit with two parallel-connected capacities (Fig. 15):

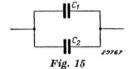

$$C = C_1 + C_2.$$

Fig. 15

12) A.C. circuit with an inductance and a resistance in series (Fig. 16):

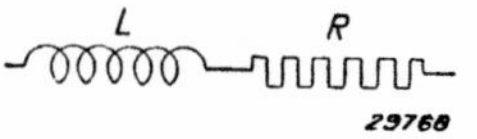

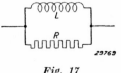

$$Z = \sqrt{R^2 + \omega^2 L^2}.$$

$$\cos \varphi = \frac{R}{Z}.$$

Fig. 16

$$\tan \delta = \frac{R}{\omega L}.$$

$$Q = \frac{\omega L}{R} = \text{quality or magnification of the coil.}$$

$\delta = 90° - \varphi =$ loss angle and may serve for calculating the wattless current and the power; the power $P = E\,I \sin \delta$. In practice it is advantageous to work with $\tan \delta$. If $R \ll \omega L$, $\tan \delta$ can be taken as equal to $\sin \delta$, so that the loss of power in the resistance then equals $P = E\,I \tan \delta$. If R represents the loss resistance of a coil or of a condenser, then $\tan \delta$ is usually practically constant for a variable ω.

13) A.C. current with an inductance and a resistance connected in parallel (Fig. 17):

$$Z = \frac{R\,\omega\,L}{\sqrt{R^2 + \omega^2 L^2}}.$$

$$\cos \varphi = \frac{Z}{R}.$$

Fig. 17

$$\tan \delta = \frac{\omega L}{R}.$$

$$Q = \frac{R}{\omega L}.$$

14) Substitution of a circuit with inductance and resistance in series for one with inductance and resistance in parallel, and vice versa (Fig. 18):

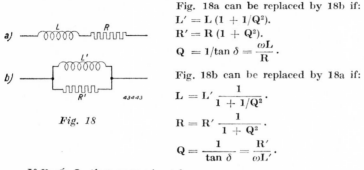

Fig. 18a can be replaced by 18b if:

$$L' = L \left(1 + 1/Q^2\right).$$
$$R' = R \left(1 + Q^2\right).$$
$$Q = 1/\tan\delta = \frac{\omega L}{R}.$$

Fig. 18b can be replaced by 18a if:

$$L = L' \frac{1}{1 + 1/Q^2}.$$
$$R = R' \frac{1}{1 + Q^2}.$$
$$Q = \frac{1}{\tan\delta} = \frac{R'}{\omega L'}.$$

Fig. 18

If $R \ll \omega L$, then approximately:

$$R' = \frac{\omega^2 L^2}{R} \text{ and } L' = L.$$

If $R \gg \omega L$, then approximately:

$$R' = R \text{ and } L' = \frac{R^2}{\omega L}.$$

15) A.C. circuit with a capacity and resistance in series (Fig. 19):

$$Z = \sqrt{R^2 + \left(\frac{1}{\omega C}\right)^2}.$$

Fig. 19

$$\cos\varphi = \frac{R}{Z}.$$
$$\tan\delta = \omega R C.$$

In practice this may represent, e.g., a condenser with a series loss resistance.

16) A.C. circuit with a capacity and resistance connected in parallel (Fig. 20):

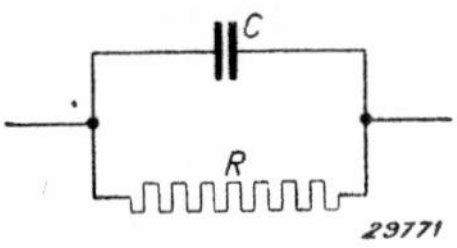

$$Z = \frac{R}{\sqrt{1 + \omega^2 R^2 C^2}}.$$
$$\cos\varphi = \frac{Z}{R}.$$

Fig. 20

$$\tan\delta = \frac{1}{\omega R C}.$$

In practice this may represent, e.g., a condenser with a parallel loss resistance.

17) Replacement of a circuit with capacity and resistance in series by one with capacity and resistance in parallel (Fig. 21):

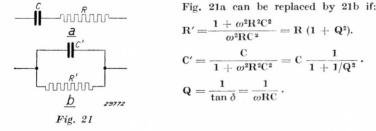

Fig. 21a can be replaced by 21b if:

$$R' = \frac{1 + \omega^2 R^2 C^2}{\omega^2 R C^2} = R \left(1 + Q^2\right).$$
$$C' = \frac{C}{1 + \omega^2 R^2 C^2} = C \frac{1}{1 + 1/Q^2}.$$
$$Q = \frac{1}{\tan\delta} = \frac{1}{\omega R C}.$$

Fig. 21

Fig. 21b can be replaced by 21a if:

$$R = \frac{R'}{1 + \omega^2 R'^2 C'^2} = R' \frac{1}{1 + Q^2}.$$

$$C = \frac{1 + \omega^2 R'^2 C'^2}{\omega^2 R'^2 C'} = C' \left(1 + \frac{1}{Q^2}\right).$$

$$Q = 1/\tan \, \delta = \omega R' C'.$$

If $R \ll \dfrac{1}{\omega C}$ then approximately:

$$R' = \frac{1}{\omega^2 R C^2} \text{ and } C' = C.$$

If $R \gg \dfrac{1}{\omega C}$ then approximately:

$$R' = R \text{ and } C' = \frac{1}{\omega^2 R^2 C}.$$

18) A.C. circuit with a capacity, inductance and resistance in series (Fig. 22):

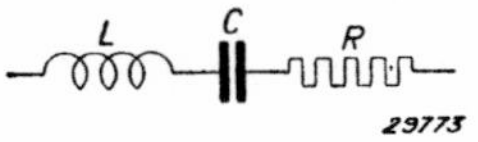

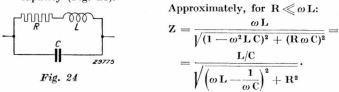

$$Z = \sqrt{\left(\omega L - \frac{1}{\omega C}\right)^2 + R^2}.$$

$$\tan \varphi = \frac{\omega L - 1/\omega C}{R}.$$

Fig. 22

19) A.C. circuit with capacity, inductance and resistance in parallel (Fig. 23):

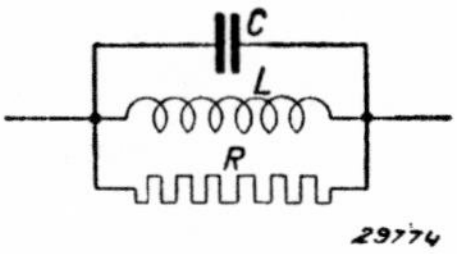

Fig. 23

$$Z = \frac{R}{\sqrt{R^2 \left(\omega C - \frac{1}{\omega L}\right)^2 + 1}} =$$

$$= \frac{L/C}{\sqrt{\left(\omega L - \frac{1}{\omega C}\right)^2 + \left(\frac{L}{C R}\right)^2}} =$$

$$= \frac{\omega L}{\sqrt{(1 - \omega L C)^2 + \left(\frac{\omega L}{R}\right)^2}}.$$

20) A.C. circuit with inductance and resistance in series, connected in parallel to a capacity (Fig. 24):

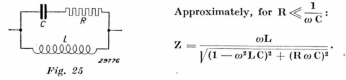

Approximately, for $R \ll \omega L$:

$$Z = \frac{\omega L}{\sqrt{(1 - \omega^2 L C)^2 + (R \omega C)^2}} =$$

$$= \frac{L/C}{\sqrt{\left(\omega L - \frac{1}{\omega C}\right)^2 + R^2}}.$$

Fig. 24

21) A.C. circuit with capacity and resistance in series, connected in parallel to an inductance (Fig. 25):

Approximately, for $R \ll \dfrac{1}{\omega C}$:

$$Z = \frac{\omega L}{\sqrt{(1 - \omega^2 L C)^2 + (R \omega C)^2}}.$$

Fig. 25

IV. OSCILLATORY CIRCUITS

(1) Free Oscillations (Fig. 26)

If in a circuit consisting of L, C and R in series the capacitor C is charged by turning the switch to position 1, and the circuit is then closed by changing the switch to position 2, a damped "free" oscillation can only occur in the circuit (L, C and R) if

$$R^2 < \frac{4\,L}{C}.$$

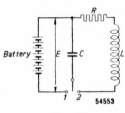

Fig. 26

Under these conditions the angular frequency of the damped oscillation is:

$$\omega = \sqrt{\frac{1}{LC} - \frac{R^2}{4\,L^2}}.$$

The natural logarithm of the constant ratio of two successive amplitudes of the same sign is equal to:

$$\frac{\pi R}{\omega L} = \frac{R}{2\,L}\,T,$$

in which T is the periodic time.

The so-called logarithmic decrement is:

$$\delta = \frac{R}{2\,L}.$$

If E is the voltage to which the condenser is charged and $R^2 \ll L/C$, then the first current amplitude is:

$$I = E\sqrt{\frac{C}{L}}$$

and the succeeding current amplitudes with the same sign as the first are given by:

$$I_n = I_1 \varepsilon^{-n\delta T}.$$

(2) Forced Oscillations

A. Series Connection of C, L and R (Fig. 27)

The impedance of the series circuit of C, L and R is:

$$Z = \sqrt{\left(\omega L - \frac{1}{\omega C}\right)^2 + R^2}.$$

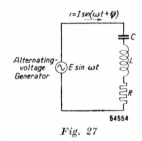

The amplitude of the current in the oscillatory circuit is:

$$I = \frac{E}{Z} = \frac{E}{\sqrt{(\omega L - 1/\omega C)^2 + R^2}}.$$

Fig. 27

The instantaneous value of the current is: $i = I \sin(\omega t + \varphi)$,

in which φ is the phase angle between current and voltage, given by:

$$\tan \varphi = \frac{\omega L - 1/\omega C}{R}.$$

In the case of resonance the impedance of the circuit of Fig. 27 is a minimum and the current maximum.

Resonance occurs when

$$\omega L = \frac{1}{\omega C} \ \text{ or } \ \omega L - \frac{1}{\omega C} = 0 \ \text{ or } \ \omega^2 LC = 1.$$

The angular frequency at resonance ω_0 is thus equal to:

$$\omega_0 = \frac{1}{\sqrt{LC}}.$$

In the case of resonance $\tan \varphi = 0$ and $Z = Z_0 = R$, and the amplitude I_0 of the current then equals:

$$I_0 = \frac{E}{R}.$$

The amplitude of the voltage across the condenser and across the inductance, in the case of resonance, is equal to:

$$E_C = E_L = \frac{I_0}{\omega_0 C} = I_0 \omega_0 L = \frac{E}{\omega_0 CR} = \frac{E \omega_0 L}{R}.$$

Therefore a magnification takes place when

$$\omega_0 CR < 1 \ \text{ or } \frac{R}{\omega_0 L} < 1.$$

When $\omega < \omega_0$ the voltage lags behind the current, and when $\omega > \omega_0$ the current lags.

B. Parallel Connection of C to L and R in Series (Fig. 28)

It is assumed that $R \ll \omega L$.

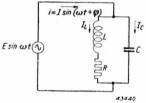

The impedance of the parallel connection of C to L and R in series is then:

$$Z = \frac{L/C}{\sqrt{(\omega L - 1/\omega C)^2 + R^2}} = \frac{\omega L}{\sqrt{(1 - \omega^2 LC)^2 + (R\omega C)^2}}.$$

The amplitude I of the total current flowing through the oscillatory circuit is:

Fig. 28

$$I = \frac{E}{Z} = \frac{E}{\omega L} \sqrt{(1 - \omega^2 LC)^2 + (R\omega C)^2} = E \frac{C}{L} \sqrt{\left(\omega L - \frac{1}{\omega C}\right)^2 + R^2}.$$

The instantaneous value of the current is:

$$i = I \sin (\omega t + \varphi),$$

in which φ is the phase angle between current and voltage, given by:

$$\tan \varphi = \frac{\omega L - \dfrac{1}{\omega C}}{R}.$$

At resonance the impedance of the circuit of Fig. 28 is a maximum and the current minimum. Resonance occurs when approximately

$$\omega L = \frac{1}{\omega C} \ \text{ or } \ \omega L - \frac{1}{\omega C} = 0 \ \text{ or } \ \omega^2 LC = 1 \cdot$$

The angular frequency at resonance therefore equals:

$$\omega_0 = \frac{1}{\sqrt{LC}}$$

In resonance $\tan \varphi = 0$ and $Z = Z_0 = L/CR$. The amplitude I_0 of the current is then:

$$I_0 = \frac{ECR}{L}.$$

The amplitude of the current through the condenser is:

$$I_C = E\omega C$$

and the amplitude of the current through the inductance L is ($R \ll \omega L$):

$$I_L = \frac{E}{\omega L}.$$

C. Parallel Connection of L, C and R[1] (Fig. 29)

The impedance of the parallel circuit of L, C and R[1] is

$$Z = \frac{L/C}{\sqrt{\left(\omega L - \dfrac{1}{\omega C}\right)^2 + \left(\dfrac{L}{CR'}\right)^2}} =$$

$$= \frac{\omega L}{\sqrt{(1 - \omega L C)^2 + \left(\dfrac{\omega L}{R'}\right)^2}}.$$

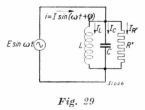

Fig. 29

The amplitude I of the total current through the oscillatory circuit is:

$$I = \frac{E}{Z} = E \frac{C}{L} \sqrt{\left(\omega L - \frac{1}{\omega C}\right)^2 + \left(\frac{L}{CR'}\right)^2} = \frac{E}{\omega L} \sqrt{(1 - \omega^2 LC)^2 + \left(\frac{\omega L}{R'}\right)^2}.$$

The instantaneous value of the current is:

$$i = I \sin (\omega t + \varphi),$$

in which φ is the phase angle between current and voltage, given by:

$$\tan \varphi = \frac{\omega L - \dfrac{1}{\omega C}}{\dfrac{L}{CR'}}.$$

At resonance the impedance of the circuit of Fig. 29 is a maximum and the current minimum. Resonance occurs when:

$$\omega L = \frac{1}{\omega C} \text{ or } \omega L - \frac{1}{\omega C} = 0 \text{ or } \omega^2 LC = 1.$$

The angular frequency at resonance ω_0 is thus equal to:

$$\omega_0 = \frac{1}{\sqrt{LC}}.$$

In the case of resonance $\tan \varphi = 0$ and $Z = Z_0 = R'$. The amplitude I of the current is then equal to that of the current through the resistance, thus:

$$I_0 = \frac{E}{R'} = I_{R'}.$$

The amplitude of the current through the condenser is:

$$I_C = E\omega C,$$

and that of the current through the inductance:

$$I_L = \frac{E}{\omega L}.$$

(3) Selectivity and Loss Factor of Oscillatory Circuits

A. Selectivity of an Oscillatory Circuit (Figs 28 and 29)

The angular frequency of a signal at resonance is:

$$\omega_0 = \frac{1}{\sqrt{LC}} \text{ or } \omega_0{}^2 LC = 1 \text{ (L in henrys and C in farads)}.$$

The frequency of a signal at resonance is:

$$f_0 = \frac{1}{2\pi\sqrt{LC}} \text{ (L in henrys and C in farads)}.$$

The wavelength of a signal of resonant frequency is

$$\lambda_0 = 1885 \times 10^6 \times \sqrt{LC} \text{ } (\lambda_0 \text{ in m, L in henrys and C in farads)}.$$

The deviation $\Delta\omega$ of the resonant frequency ω_0, for an arbitrary frequency ω, is

$$\Delta\omega = \omega - \omega_0.$$

The selectivity for a signal with a frequency differing by an amount $\Delta\omega$ from the resonant frequency ω_0 is defined as the ratio of the amplitude E_0 of a signal of resonant frequency to that of the signal E having a frequency deviating $\Delta\omega$ from the resonant frequency, for equal current amplitudes I.

In the case of Fig. 28 this relation (for small values of $\Delta\omega$) is approximately:

$$\frac{E_0}{E} = \frac{Z_0}{Z} = \sqrt{\left(\frac{2\,\Delta\omega}{R/L}\right)^2 + 1}.$$

In the case of Fig. 29 the relation (again for small values of $\Delta\omega$) is approximately:

$$\frac{E_0}{E} = \frac{Z_0}{Z} = \sqrt{(2\,\Delta\omega\,R'C)^2 + 1}.$$

In the case of Fig. 28 the selectivity is determined by the ratio R/L between the resistance and inductance of the coil, and in the case of Fig. 29 by the product R'C of the parallel resistance and capacity (R/L and 1/R'C are called the loss factor of a circuit).

B. Parallel Connection of a Resistance to an Oscillatory Circuit

The parallel connection of a resistance R' to an oscillatory circuit corresponds to an increase of the factor R/L by ΔR/L:

$$\Delta\,R/L = \frac{1}{R'C}.$$

C. Series Connection of a Resistance to the Inductance or Condenser of an Oscillatory Circuit

If a resistance R is introduced into an oscillatory circuit, in series with the coil or the condenser, this corresponds to the parallel connection of a resistance R' to the circuit:

$$\frac{1}{R'C} = \frac{R}{L}.$$

D. Determining the Magnitude of the Loss Factor R/L or 1/R′C of the Circuit

This factor can be determined by measuring the voltage E_o across the circuit at resonance and the voltage E at a frequency deviating $\Delta\omega$ from the resonant frequency, with equal current amplitudes I:

$$\frac{R}{L} = \frac{1}{R'C} = \frac{2\,\Delta\,\omega}{\sqrt{\left(\dfrac{E_o}{E}\right)^2 - 1}}.$$

$$\text{For } \frac{E_o}{E} = \sqrt{2}, \; R/L = 1/R'C = 2\,\Delta\,\omega.$$

These formulae do not hold for large values of $\Delta\omega$.
The following formulae, however, are of general application.

E. General Formulae for the Selectivity and Reciprocal Quality Factor of Oscillatory Circuits

(a) Definitions:

Relative detuning $\beta = \dfrac{\omega}{\omega_o} - \dfrac{\omega_o}{\omega} \approx \dfrac{2\,\Delta\,\omega}{\omega_o}$.

Reciprocal circuit quality factor $\tan\delta = d = 1/Q = \dfrac{R}{\omega_o L}$ or $\dfrac{1}{R'\omega_o C}$.

(b) Formulae:

Impedance at an arbitrary frequency ω: $Z = \sqrt{\dfrac{L/C}{\beta^2 + d^2}}$.

Impedance at the resonant frequency ω_o: $Z_o = \dfrac{\sqrt{L/C}}{d}$.

Selectivity: $\dfrac{Z_o}{Z} = \sqrt{\left(\dfrac{\beta}{d}\right)^2 + 1}$.

(c) Determining the Reciprocal Quality Factor d (= tan δ = 1/Q):

$$d = \frac{\beta}{\sqrt{\left(\dfrac{E_o}{E}\right)^2 - 1}},$$

in which E_o represents the voltage across the circuit at resonance and E the voltage at a frequency differing by $\Delta\omega$ from the resonant frequency ω_o, at equal current amplitudes I.

If $\dfrac{E_o}{E} = \sqrt{2}$, $d = \tan\delta = \beta$, or $Q = 1/\beta$.

(d) How the Reciprocal Quality Factor is Affected by an Increase of the Series Resistance in the Circuit or of a Resistance Connected Parallel to the Circuit

Any increase ΔR of the series resistance in a circuit or the parallel connection of a resistance R′ to the circuit corresponds to an increase of d by Δ d:

$$\Delta d = \frac{\Delta R}{\omega_o L},$$

or

$$\Delta d = \frac{1}{R'\omega_o C}.$$

V. BAND-PASS FILTERS

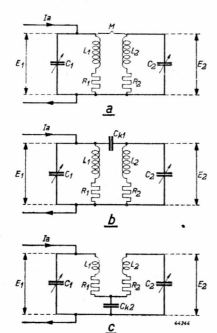

Fig. 30

a: with coupling by a mutual inductance
b: with top-end capacitive coupling
c: with bottom-end capacitive coupling

1) Basic formulae, see Figs 30a, b and c:

$$\omega_0 = \frac{1}{\sqrt{L_1 C_1}} = \frac{1}{\sqrt{L_2 C_2}} \cdot$$

Voltage across the secondary circuit of the band-pass filter:

$$E_2 = I_a \frac{K/d}{1 + K^2/d^2} Z,$$

in which K is a coupling factor. Mean value of the circuit impedance

$$Z = \sqrt{Z_1 \times Z_2},$$

$$Z_1 = \frac{L_1}{R_1 C_1}, \quad Z_2 = \frac{L_2}{R_2 C_2} \cdot$$

Mean value of the reciprocal quality factor:

$$\tan \delta = d = \sqrt{d_1 d_2}, \; d_1 = \frac{R_1}{\omega_0 L_1}, \; d_2 = \frac{R_2}{\omega_0 L_2} \cdot$$

a) Coupling by a mutual inductance (Fig. 30a):

$$K = \frac{M}{\sqrt{L_1 L_2}} \cdot$$

b) Top-end capacitive coupling (Fig. 30b):

$$K = \frac{C_{k1}}{\sqrt{C_1 C_2}} \cdot$$

c) Bottom-end capacitive coupling (Fig. 30c):

$$K = \frac{\sqrt{C_1 C_2}}{C_{k2}} \cdot$$

2) Critical coupling K = d: $\quad E_2 = \frac{1}{2} I_a Z.$

3) Input impedance of the band-pass filter: $\dfrac{E_1}{I_a} = Z_1 \dfrac{1}{1 + K^2/d^2} \cdot$

Input impedance at the critical coupling: $\dfrac{E_1}{I_a} = \frac{1}{2} Z_1 \cdot$

4) Ratio of input and output voltages: $\dfrac{E_1}{E_2} = \dfrac{d}{K} \sqrt{\dfrac{Z_1}{Z_2}} \cdot$

5) Equation of the resonance curve

a = attenuation =

$$= \frac{\text{voltage across secondary circuit at resonant frequency } (= E_{2o})}{\text{voltage across secondary circuit at non-resonant frequency } (= E_2)}$$

$$a = \frac{E_{2o}}{E_2} = \frac{\sqrt{(K^2/d^2 + 1)^2 - 2 \; \beta^2/d^2 \, (K^2/d^2 - 1) + \beta^4/d^4}}{K^2/d^2 + 1} \cdot$$

$$\beta = \frac{\omega}{\omega_0} - \frac{\omega_0}{\omega} \approx \frac{2 \Delta \omega}{\omega_0} \quad \text{(see IV, 3 Ea)} \, .$$

At the cirtical coupling ($K = d$):

$$a = \frac{\sqrt{\beta^4/d^4 + 4}}{2}.$$

These formulae apply only in the vicinity of the resonant frequency ω_0, i.e. as long as $\frac{\omega_0}{\omega}$ does not differ much from unity and d_1 and d_2 do not differ much from each other.

Band-pass Filter Curves, Sheet I

Attenuation a as a function of β/d with K/d as parameter for all kinds of band-pass filters and for oscillatory circuits coupled by valves.

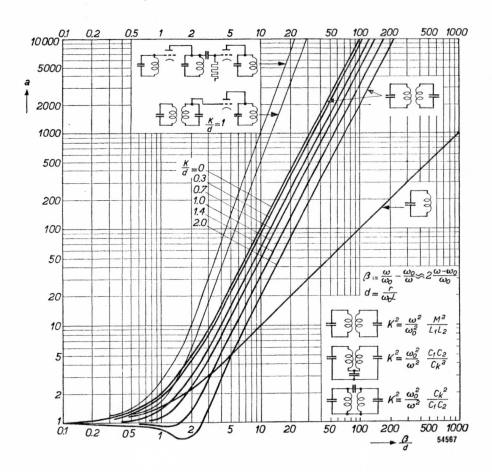

Band-pass Filter Curves, Sheet II

Attenuation a for band-pass filters as a function of β/d with K/d as parameter.

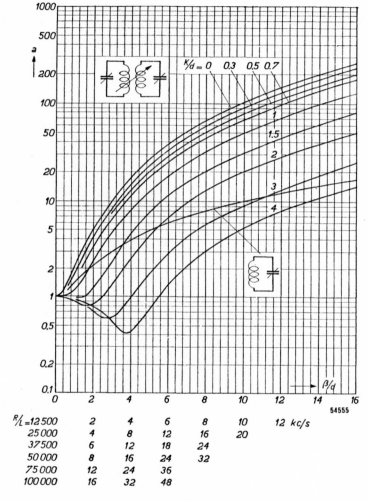

$R/L =$					
12 500	2	4	6	8	10	12 kc/s
25 000	4	8	12	16	20
37 500	6	12	18	24
50 000	8	16	24	32
75 000	12	24	36
100 000	16	32	48

The curves which are given for inductively-coupled circuits apply also to capacitively-coupled circuits. For circuits with R/L = 12,500 the graduation from 0 to 12 kc/s on the horizontal scale applies to the frequency deviation β/d; for circuits with R/L = 25,000 the horizontal scale from 0 to 20 kc/s applies, and so on. Here, too, the condition is that relatively the frequency may not deviate too much from ω_o, i.e. that $\dfrac{\omega_o}{\omega}$ may not differ appreciably from unity. For this reason the scales for β/d for different values of R/L have not been plotted farther than the highest values indicated in the diagram.

VI. RC ELEMENTS IN AMPLIFYING STAGES

(1) Resistance Coupling between two Valves

A resistance coupling between two valves may be represented by the diagram of Fig. 31:

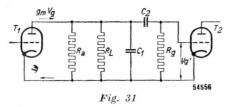

Fig. 31

Here

$g_m V_g$ = current source formed by valve T_1;
R_a = internal resistance of valve T_1;
R_L = anode series resistance of valve T_1;
R_g = grid-leak resistance of valve T_2;
C_1 = parallel condenser for filtering the H.F. and representing the wiring capacity;
C_2 = coupling condenser.

A. Characteristic for High Frequencies

For high frequencies the impedance of the condenser C_2 is small compared with R_g and the circuit of Fig. 31 can be simplified to that of Fig. 32, where

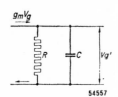

Fig. 32

$$\frac{1}{R} = \frac{1}{R_a} + \frac{1}{R_L} + \frac{1}{R_g} \text{ and } C = C_1.$$

The impedance of R and C in parallel is

$$Z = \frac{R}{\sqrt{1 + \omega^2 R^2 C^2}}.$$

The alternating grid voltage V_g' is

$$V_g' = g_m V_g Z = \frac{g_m V_g R}{\sqrt{1 + \omega^2 R^2 C^2}}.$$

The ratio $a_{(\omega)}$ of V_g' for the frequency ω to V_g' for a very small value of ω ($\omega RC \ll 1$) is:

$$a_{(\omega)} = \frac{1}{\sqrt{1 + \omega^2 R^2 C^2}}.$$

In Fig. 33 $a_{(\omega)}$ is given as a percentage as a function of the frequency for different values of the product RC.

With small values of ω the grid a.c. voltage V_g' approaches $g_m V_g R$.

Fig. 33

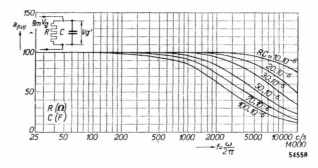

493

B. Characteristic for Low Frequencies

At low frequencies the impedance of the condenser C_1 in Fig. 31 is very large compared with the value of the resistance and its influence can be disregarded. The impedance of C_2 is then no longer infinitely small compared with R_g and voltage division occurs. In this case the circuit of Fig. 31 can be simplified to that of Fig. 34a, where

$$\frac{1}{R'} = \frac{1}{R_a} + \frac{1}{R_L}.$$

In Fig 34 the current source $g_m V_g$ with parallel resistance R' can be replaced by a voltage source $g_m V_g R'$ with resistance R' connected in series, producing the system of Fig. 34b.

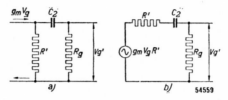

Fig. 34

The alternating grid voltage is

$$V_g' = g_m V_g \frac{R' R_g}{R' + R_g} \frac{1}{\sqrt{1 + \dfrac{1}{\omega^2 (R' + R_g)^2 C_2^2}}}.$$

With large values of ω

$$V_g' = g_m V_g \frac{R' R_g}{R' + R_g} = g_m V_g R.$$

The ratio $a_{(\omega)}$ of V_g' at the frequency ω to V_g' at a very large value of ω $[\omega (R' + R_g) C_2 \gg 1]$ is:

$$a_{(\omega)} = \frac{1}{\sqrt{1 + \dfrac{1}{\omega^2 (R' + R_g)^2 C_2^2}}}.$$

In Fig. 35 $a_{(\omega)}$ is given as a percentage as a function of the frequency for different values of the product $(R' + R_g) C_2$.

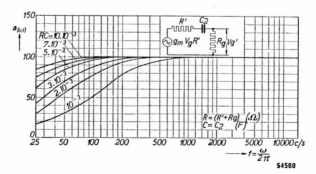

Fig. 35

(2) RC Element for Automatic Negative Grid Bias

Fig. 36 indicates a system for producing automatic negative grid bias of a valve. At an extremely low frequency the impedance of the condenser C is very large compared with R and the effect of C can be ignored (Fig. 36b). If in Fig. 36a the dynamic transconductance of the valve is g_{md} in Fig. 36b this is given by

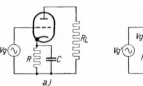

Fig. 36

$$g'_{md} = \frac{g_{md}}{1 + R\,g_{md}}.$$

The RC element in the cathode lead to the valve causes a reduction $a_{(\omega)}$ in the dynamic transconductance g_{md} of the valve at the frequency ω, compared with the dynamic transconductance at a very high frequency ($\omega RC \gg 1$):

$$a_{(\omega)} = \frac{g'_{md}}{g_{md}} = \sqrt{\frac{1 + \omega^2 R^2 C^2}{(1 + Rg_{md})^2 + \omega^2 R^2 C^2}}.$$

This reduction of the dynamic transconductance depends not only on RC but also on Rg_{md}, so that it is not possible to represent the dependency of $a_{(\omega)}$ on the frequency with the aid of a single family of curves.

(3) Decoupling by an RC Filter

The decoupling by an RC filter (see Fig. 37) is determined by the relation between the voltage V_2 across the condenser and the voltage V_1 at the input of the filter.

$$\frac{V_2}{V_1} = \frac{1}{\sqrt{\omega^2 R^2 C^2 + 1}},$$

or as a percentage:

$$\frac{V_2}{V_2} = \frac{100}{\sqrt{\omega^2 R^2 C^2 + 1}}\,\% .$$

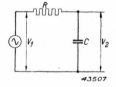

Fig. 37

With the aid of the following simple formula one can calculate quickly and with reasonable accuracy the value of the product RC required for n% decoupling at a given frequency, when $n < 30$:

$$RC = \frac{16 \cdot 10^6}{n \times f},$$

where R = the resistance in ohms,
C = the capacity in μF,
f = the frequency in c/s.

(4) Coupling by a CR Element

When a voltage source V_1 is coupled by means of a CR element (see Fig. 38) the voltage V_2 across the resistor R is given by

$$V_2 = \frac{R}{\sqrt{R^2 + 1/\omega^2 C^2}}\,V_1 = \frac{\omega RC}{\sqrt{\omega^2 R^2 C^2 + 1}}\,V_1.$$

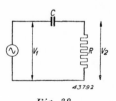

Fig. 38

(5) Decoupling of the Screen Grid

The screen grid of a multigrid valve (pentode or tetrode) is generally fed from a source of higher voltage, the anode supply line for instance, by means of a voltage divider or via a series resistor. The screen-grid voltage is then smoothed by means of a condenser (Fig. 39). If condenser C_{g2} were omitted the anode current would also be modulated by the alternating voltage then present on the screen grid, which alternating voltage is the result of the alternating screen-grid current flowing through the screen-grid series resistor R_{g2}.

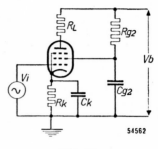

Fig. 39

The modulation of the anode current by the alternating voltage on the screen grid reduces the transconductance of the first grid, because this alternating voltage is opposed in phase to the control-grid voltage V_i. When the capacity of condenser C_{g2} is too low the low-frequency alternating voltage is inadequately smoothed. This results in a reduction of the transconductance for these frequencies and consequently a bad reproduction of low notes.

For a pentode or tetrode the following assumptions may be made with reasonable accuracy:

(a) A variation of the anode voltage does not affect the intensity of the cathode current; it only influences the distribution of this current over screen grid and anode.

(b) A variation of the control-grid voltage does not interfere with the ratio of the screen-grid current and the anode current.

(c) A variation of the screen-grid voltage does not affect the ratio of the screen-grid and anode currents.

(d) The internal anode resistance R_a is very high with respect to the external anode load R_L.

(e) The internal screen-grid resistance R_{a2} ($\dfrac{dV_{g2}}{dI_{g2}}$ at constant values of V_{g1} and V_a) is not small with respect to the external impedance in the screen-grid circuit (the impedance of C_{g2} as far as this is not considerably influenced by R_{g2}).

Under these circumstances the attenuation $a_{(\omega)}$, which is the ratio of the anode alternating current with an angular frequency ω to the anode alternating current with a considerably higher frequency, is equal to

$$a_{(\omega)} = \frac{\omega C_{g2} R_{a2}}{\sqrt{1 + (\omega C_{g2} R_{a2})^2}}.$$

The curves of Fig. 35, giving the attenuation $a_{(\omega)}$ as a function of the frequency $f = \dfrac{\omega}{2\pi}$ with RC as a parameter, are also convenient for determining the response at an angular frequency ω with a screen-grid capacity C_{g2} ($\mu\mu$F) and an internal screen-grid resistance R_{a2} (MΩ).

In the case of an output pentode the power dissipated in the loudspeaker is proportional to the square of the anode alternating current. The attenuation of the loudspeaker power resulting from the impedance $\dfrac{1}{j\omega C_{g2}}$ in the screen-grid circuit compared with the power at a very much higher frequency is consequently

$$a_{(\omega)}{}^2 = \frac{(\omega C_{g2} R_{a2})^2}{1 + (\omega C_{g2} R_{a2})^2}.$$

496

Fig. 40 gives as a percentage the attenuation $a_{(\omega)}^2$ of the power output as a function of the frequency for various values of the product RC (R in ohms and C in farads).

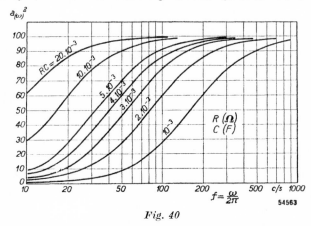

Fig. 40

VII. THE H.F. RESISTANCE OF ROUND WIRE DUE TO SKIN EFFECT

Owing to skin effect the resistance of a wire for currents of high frequency is greater than the D.C. resistance. The H.F. resistance of round wire depends on a quantity x_1. If we call the resistance for high frequencies R and the D.C. resistance R_0 then the ratio R/R_0 is a function of x_1, where x_1 is given by:

in which
$$x_1 = \pi\, r_1 \sqrt{\pi\, \varkappa\, \nu\, 10^{-9}},$$

$r_1 =$ radius of the section of wire in cm,
$\mu =$ permeability,
$\varkappa =$ specific conductivity in 1/ohm cm,
$\nu =$ frequency of the alternating current in c/s.

When x_1 has been determined for a certain section of wire the ratio R/R_0 can be found with the aid of

(a) $R/R_0 = 1 + \dfrac{x_1^4}{3} - \dfrac{4}{45} x_1^8$ for $0 < x_1 < 0.8$;

(b) $R/R_0 = 0.997\, x_1 + 0.277$ for $1.5 < x_1 < 2$;

(c) $R/R_0 = x_1 + 0.25 + \dfrac{3}{64 x_1}$ for $x_1 > 2$.

For values of x_1 lying between 0.8 and 1.5 R/R_0 can be determined with the aid of the following table:

TABLE I

x_1	R/R_0	x_1	R/R_0	x_1	R/R_0
0.80	1.11	1.05	1.31	1.30	1.56
0.85	1.14	1.10	1.35	1.35	1.61
0.90	1.18	1.15	1.40	1.40	1.66
0.95	1.22	1.20	1.45	1.45	1.72
1.0	1.26	1.25	1.50		

Some values are given below for the specific conductivity in 1/ohm cm and the permeability:

TABLE II

Specific Conductivity and Permeability of Some Materials

Material	Specific conductivity in $\dfrac{1}{\text{ohm cm}}$	Permeability μ
Copper	58.2×10^4	1
Brass	14.3×10^4	1
Tin	8.7×10^4	1
Constantan	2.0×10^4	1
Iron	1.0×10^4	1000 [1]
Chromium nickel	0.9×10^4	1

The following table gives the relative increase of the resistance R/R_o due to skin effect for copper wire of various diameters and at certain frequencies.

TABLE III

Increase of Resistance of Copper Wire Due to Skin Effect

Wire diam. in mm	R/R_o at a frequency ν of				
	100 kc/s $(\lambda = 3000$ m$)$	500 kc/s $(\lambda = 600$ m$)$	1000 kc/s $(\lambda = 300$ m$)$	3000 kc/s $(\lambda = 100$ m$)$	10 Mc/s $(\lambda = 30$ m$)$
0.05	1.00	1.00	1.00	1.0	1.2
0.1	1.00	1.00	1.01	1.2	1.5
0.2	1.03	1.04	1.1	1.5	2.6
0.3	1.04	1.06	1.4	2.2	3.8
0.5	1.05	1.4	2.2	3.2	6.1
0.7	1.13	1.9	2.8	4.5	8.8
0.9	1.37	2.6	3.6	6.2	11.1
1.0	1.50	2.8	4.2	6.9	14.3
2.0	2.76	5.4	8.1	9	50

These figures apply only to a straight wire.

The following are the wire gauges in microns $(1\ \mu = 10^{-4}$ cm$)$ of some materials with which, at a frequency of 30 Mc/s, the increase of resistance due to skin effect amounts to 1%.

copper	$20\ \mu$	constantan	$108\ \mu$
brass	40.4μ	chromium nickel	$161\ \mu$

[1]) At high frequencies (10—100 Mc/s) the permeability of iron is about 200.

VIII. CALCULATION OF INDUCTANCES

(1) The Inductance of a Circular Winding of Round Wire (Fig. 41).

The inductance of a circular winding of round wire as shown in Fig. 41 can be calculated from the formula

$$L = 4\pi R \left(\ln \frac{2R}{d} + 0.33\right) 10^{-9} \text{ henry,}$$

in which

R = radius of the winding in cm;

d = diameter of the wire in cm.

Fig. 41

This formula applies to cases where d is much smaller than R. The contribution of the field inside the wire, the so-called internal inductance, is allowed for in the formulae. This is always proportional to the gauge of the wire and amounts to 0.5×10^{-9} henry/cm. At high frequencies, where owing to skin effect the current flows along the surface of the wire, this internal inductance disappears, in which case the term 0.33 in the formula changes to 0.08.

(2) Mutual Inductance between two Equal, Parallel, Circular Windings of Round Wire

The mutual inductance between two equal, parallel, circular windings of round wire can be calculated with the aid of the formula:

$$M = 4\pi R \left(\ln \frac{R}{a} + 0.08\right) 10^{-9} \text{ henry,}$$

in which

R = the radius of the circular windings in cm;

a = centre-to-centre distance between the wires in cm.

(3) The Inductance of a Square Winding (Fig. 42)

The inductance of a square winding as shown in Fig 42 can be calculated with the aid of the formula

$$L = 8s \left(\ln \frac{2s}{d} - 0.524\right) 10^{-9} \text{ henry,}$$

in which

s = length of one of the sides of the square, in cm;

d = diameter of the wire, in cm.

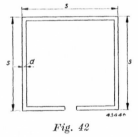

Fig. 42

This formula applies to cases where d is much smaller than s and includes also the internal inductance. Where the latter disappears the term —0.524 changes to —0.774.

(4) The Inductance of Coils (Fig. 43)

The inductance of a cylindrical coil with rectangular winding section (see Fig. 43) can be calculated with the aid of the formula

$$L = n^2 D \, \Phi \, 10^{-9} \text{ henry,}$$

in which

n = number of windings;

D = mean diameter of the windings, in cm.

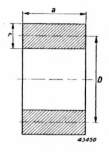

Fig. 43

The value of Φ as a first approximation of the self-inductance can be calculated from the formula given by Prof. E. Löfgren of the Stockholm Technical University

$$\Phi = \frac{9.9}{0.45 + \alpha + \rho + 0.5\, \alpha\rho}$$

in which

D = mean diameter of the windings, in cm;

r = thickness of the coil, in cm;

a = length of the coil, in cm;

α = a/D and ρ = r/D.

This formula has an accuracy of $\pm 2\%$ for $\rho < 0.75$ and $0.25 < \alpha < 3.0$. It is applicable for cylindrical coils of not too great a thickness and with reasonable accuracy to flat coils up to the maximum value 1 of ρ.

The exact value of Φ is found by using the curves of Figs 44, 45 and 46 (taken from J. Hak, Eisenlose Drosselspulen, published by K. F. Koehler, Leipzig, 1938). In these graphs again α = a/D and ρ = r/D. The internal inductance is included in these formulae.

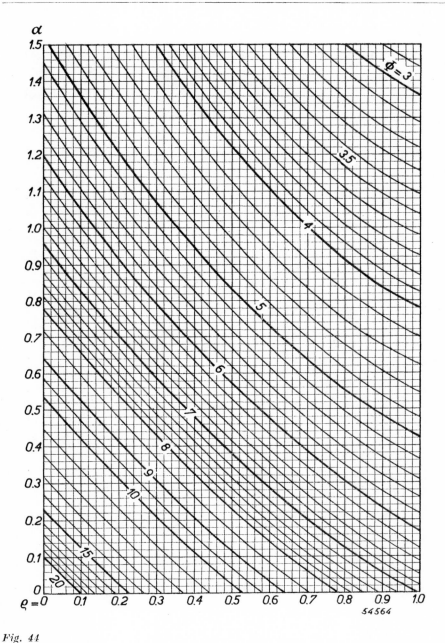

Fig. 44

The value of Φ as a function of $\alpha = a/D$ and $\rho = r/D$ for determining the inductance of a cylindrical coil with rectangular winding section, with the aid of the formula $L = n^2 D \Phi 10^{-9}$ henry, for values of ρ from 0 to 1 and of α from 0 to 1.5.

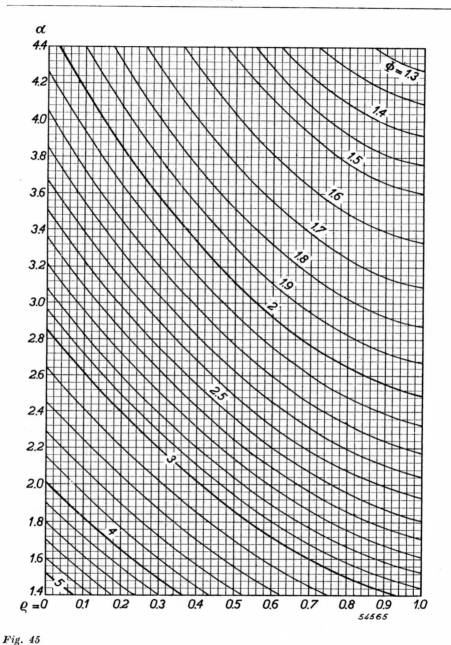

Fig. 45

The value of Φ as a function of $\alpha = a/D$ and $\rho = r/D$ for determining the inductance of a cylindrical coil with rectangular winding section with the aid of the formula $L = n^2 D \, \Phi \, 10^{-9}$ henry, for values of ρ from 0 to 1 and of α from 1.4 to 4.4.

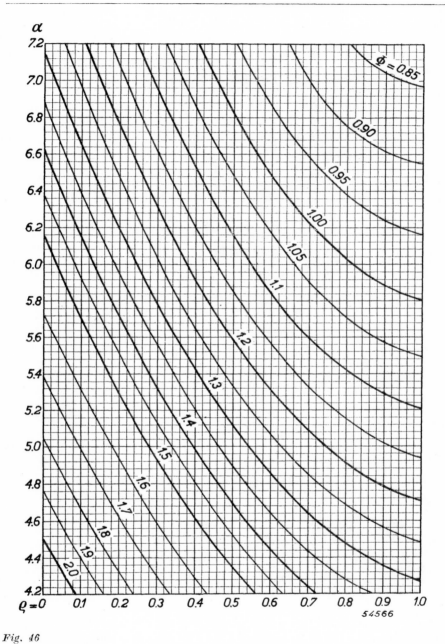

Fig. 46

The value of Φ as a function of $\alpha = a/D$ and $\rho = r/D$ for determining the inductance of a cylindrical coil with rectangular winding section with the aid of the formula $L = n^2 D \Phi 10^{-9}$ henry, for values of ρ from 0 to 1 and of α from 4.2 to 7.2.

IX. PROPERTIES OF LONG LEADS

In the expressions for inductance given below the internal inductances are not included.

(1) Straight, Round Wire Parallel to Earth (Fig. 47)

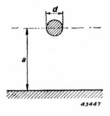

Fig. 47

(a) Capacity per cm length

$$C = \frac{\varepsilon}{1.8 \ln \frac{4a}{d}} \, 10^{-12} \ldots . \, F/cm.$$

(b) Inductance per cm length

$$L = 2\mu \ln \frac{4a}{d} \, 10^{-9} \ldots . \, H/cm.$$

a and d in cm, ε = dielectric constant (for air $\varepsilon = 1$), μ = permeability (for air $\mu = 1$).
These formulae apply where d is much smaller than a.

(2) Two Parallel Wires Suspended in Air (Fig. 48)

(a) Capacity per cm length

$$C = \frac{\varepsilon}{3.6 \ln \frac{2a}{d}} \, 10^{-12} \ldots . \, F/cm.$$

Fig. 48

(b) Inductance per cm length

$$L = 4\mu \ln \frac{2a}{d} \, 10^{-9} \ldots . \, H/cm.$$

(c) Characteristic impedance

$$Z_o = \sqrt{\frac{L}{C}} = 120 \ln \frac{2a}{d} \sqrt{\frac{\mu}{\varepsilon}} \ldots . \, \Omega.$$

a and d in cm, ε = dielectric constant, μ = permeability.
These formula apply where d is much smaller than a.

(3) Concentric Lead or Cable (Fig. 49)

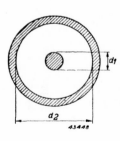

Fig. 49

(a) Capacity per cm length

$$C = \frac{\varepsilon}{1.8 \ln \frac{d_2}{d_1}} \, 10^{-12} \ldots . \, F/cm.$$

(b) Inductance per cm length

$$L = 2\mu \ln \frac{d_2}{d_1} \, 10^{-9} \ldots . \, H/cm.$$

(c) Characteristic impedance

$$Z_o = \sqrt{\frac{L}{C}} = 60 \ln \frac{d_2}{d_1} \sqrt{\frac{\mu}{\varepsilon}} \ldots . \, \Omega.$$

d_2 and d_1 in cm, ε = dielectric constant, μ = permeability.

X. CALCULATION OF CAPACITIES

The Capacity of a Condenser Consisting of two Parallel Flat Plates (Fig. 50)

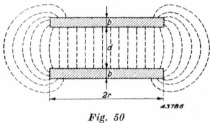

Fig. 50

Disregarding the edge effects, the capacity of two equal, parallel, flat, conducting plates is given by

$$C = \frac{\varepsilon\,S}{4\,\pi\,d}\;cm = \frac{\varepsilon\,S}{3.6\,\pi\,d}\;10^{-12}\;farad,$$

in which

S = surface of the plates in cm^2;

d = distance between the plates in cm;

ε = dielectric constant of the insulating material between the plates.

Taking into account the edge effects, the capacity of two equal, parallel, flat, circular plates is given by

$$C = \frac{\varepsilon r^2}{4d} + \frac{\varepsilon r}{4\pi}\left[\ln\frac{16\,\pi\,r\,(d+b)}{d^2} + \frac{b}{d}\,\ln\frac{d+b}{b} + 1\right]cm,$$

in which

r = radius of the circular plates in cm;

d = distance between the plates in cm;

b = thickness of the plates in cm;

ε = dielectric constant of the insulating material.

XI. TABLES AND GRAPHS

Decibel and Neper

Definitions

A bel is the logarithm, with base 10, of the ratio of two powers, i.e. twice the logarithm of the ratio of the corresponding voltage or current amplitudes when the impedances through which the currents are flowing or across which the voltages are developing are identical.

A decibel (dB) is one-tenth of a bel, so that amplification or attenuation can be expressed in dB as follows:

$$\text{number of dB} = 10 \, \log_{10} \frac{P_2}{P_1} = 20 \, \log_{10} \frac{V_2}{V_1} = 20 \, \log_{10} \frac{I_2}{I_1}. \tag{1}$$

The neper (N) is the logarithm to the base $\varepsilon = 2.718\ldots$ of the ratio of two amplitudes of voltage, current, pressure or velocity, so that amplification or attenuation can be expressed in nepers as follows:

$$\text{number of N} = \log_{\varepsilon} \frac{V_2}{V_1} = \log_{\varepsilon} \frac{I_2}{I_1} \text{ or } N = \ln \frac{V_2}{V_1} = \ln \frac{I_2}{I_1}. \tag{2}$$

Nepers can be converted into decibels and vice versa with the aid of the Formulae (3) and (4)

$$1 \text{ dB} = 0.1151 \text{ N} \tag{3}$$

$$1 \text{ N} = 8.686 \text{ dB}. \tag{4}$$

With the aid of the graph below decibels and nepers can be converted to ratios of power (P_2/P_1) or to ratios of corresponding voltage (V_2/V_1) or current (I_2/I_1).

GRAPH

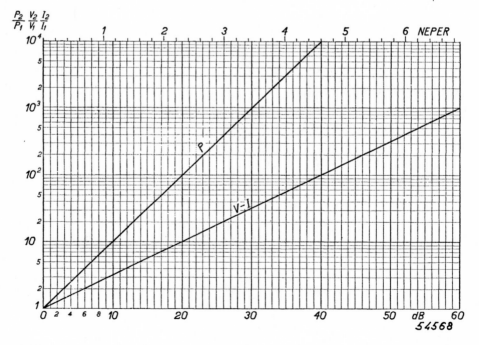

54568

Remarks concerning the Use of the Table of Conversion of Decibels to Ratios of Powers, Voltages or Currents

1) The number of decibels corresponding to a given ratio I_1/I_2 is equal to the number of decibels corresponding to the ratio I_2/I_1, but with the opposite sign. This results from the fact that:

the number of dB $= 20 \log_{10} I_2/I_1 = 20 \log_{10} (I_1/I_2)^{-1} = -20 \log_{10} I_1/I_2$.

The number of decibels corresponding to $I_2/I_1 = 2$ is 6.021 and consequently the number of decibels corresponding to $I_1/I_2 = 0.5$ is equal to -6.021.

2) In order to find ratios of currents or voltages corresponding to a number of decibels higher than 20 it is to be remembered that 20 dB corresponds to a ratio of 10, 40 dB to a ratio of 100, 60 dB to a ratio of 1000, and so on.

In order to obtain the ratios of current corresponding to decibel numbers higher than those quoted in the table, the ratio of current corresponding to a number of decibels higher than 20, 40, 60, 80 etc. must be multiplied by 10, 10^2, 10^3, 10^4, etc. For instance to determine the ratio I_1/I_2 corresponding to 45 dB first the ratio corresponding to 5 dB is looked up in the table, this ratio being 1.778, and then multiplied by 100. In the case of the example given 45 dB corresponds to a ratio of 177.8.

If the number of dB were 35 the ratio corresponding to 15 dB should be looked up (15 dB corresponds to 5.623) and multiplied by 10 (consequently 35 dB corresponds to a ratio of 56.23).

3) Inversely, for each ratio of currents, etc. the value of which is higher than those included in the table the number of dB can be found by dividing this ratio by 10, 100, 1000 or 10,000 and finally adding to the number of dB 20, 40, 60 or 80. In the case of a ratio of 5248 the number of dB corresponding to 5.248 is 14.4. Since we first divided the ratio by 1000, to the number of dB finally found we have to add 60 dB. Thus the number of decibels corresponding to the ratio $I_1/I_2 = 5248$ is $14.4 + 60 = 74.4$.

TABLE for Converting Decibels to Ratios of Current (I),

Decibel	$\dfrac{I_1}{I_2}$ or $\dfrac{V_1}{V_2}$	$\dfrac{P_1}{P_2}$	$\dfrac{I_2}{I_1}$ or $\dfrac{V_2}{V_1}$	$\dfrac{P_2}{P_1}$	Decibel	$\dfrac{I_1}{I_2}$ or $\dfrac{V_1}{V_2}$	$\dfrac{P_1}{P_2}$	$\dfrac{I_2}{I_1}$ or $\dfrac{V_2}{V_1}$	$\dfrac{P_2}{P_1}$
0.1	1.012	1.023	0.9886	0.9772	5.1	1.799	3.236	0.5559	0.3090
0.2	1.023	1.047	0.9772	0.9550	5.2	1.820	3.311	0.5495	0.3020
0.3	1.035	1.072	0.9661	0.9333	5.3	1.841	3.388	0.5433	0.2951
0.4	1.047	1.096	0.9550	0.9120	5.4	1.862	3.467	0.5370	0.2884
0.5	1.059	1.122	0.9441	0.8913	5.5	1.884	3.548	0.5309	0.2818
0.6	1.072	1.148	0.9333	0.8710	5.6	1.906	3.631	0.5248	0.2754
0.7	1.084	1.175	0.9226	0.8511	5.7	1.928	3.715	0.5188	0.2692
0.8	1.096	1.202	0.9120	0.8318	5.8	1.950	3.802	0.5128	0.2630
0.9	1.109	1.230	0.9016	0.8128	5.9	1.972	3.891	0.5070	0.2570
1.0	1.122	1.259	0.8913	0.7943	6.0	1.995	3.981	0.5012	0.2512
1.1	1.315	1.288	0.8811	0.7763	6.1	2.018	4.074	0.4955	0.2455
1.2	1.148	1.318	0.8710	0.7586	6.2	2.042	4.169	0.4898	0.2399
1.3	1.162	1.349	0.8610	0.7413	6.3	2.065	4.266	0.4842	0.2344
1.4	1.175	1.380	0.8511	0.7244	6.4	2.089	4.365	0.4786	0.2291
1.5	1.189	1.412	0.8414	0.7080	6.5	2.114	4.467	0.4732	0.2239
1.6	1.202	1.445	0.8318	0.6918	6.6	2.138	4.571	0.4677	0.2188
1.7	1.216	1.479	0.8222	0.6761	6.7	2.163	4.677	0.4624	0.2138
1.8	1.230	1.514	0.8128	0.6607	6.8	2.188	4.786	0.4571	0.2089
1.9	1.245	1.549	0.8035	0.6457	6.9	2.213	4.898	0.4519	0.2042
2.0	1.259	1.585	0.7943	0.6310	7.0	2.239	5.012	0.4467	0.1995
2.1	1.274	1.622	0.7852	0.6166	7.1	2.265	5.128	0.4416	0.1950
2.2	1.288	1.660	0.7763	0.6026	7.2	2.291	5.248	0.4365	0.1906
2.3	1.303	1.698	0.7674	0.5888	7.3	2.317	5.370	0.4315	0.1862
2.4	1.318	1.738	0.7586	0.5754	7.4	2.344	5.495	0.4266	0.1820
2.5	1.334	1.778	0.7499	0.5623	7.5	2.371	5.623	0.4217	0.1778
2.6	1.349	1.820	0.7413	0.5495	7.6	2.399	5.754	0.4169	0.1738
2.7	1.365	1.862	0.7328	0.5370	7.7	2.427	5.888	0.4121	0.1698
2.8	1.380	1.906	0.7244	0.5248	7.8	2.455	6.026	0.4074	0.1660
2.9	1.396	1.950	0.7161	0.5128	7.9	2.483	6.166	0.4027	0.1622
3.0	1.413	1.995	0.7080	0.5012	8.0	2.512	6.310	0.3981	0.1585
3.1	1.429	2.042	0.6998	0.4898	8.1	2.541	6.457	0.3936	0.1549
3.2	1.445	2.089	0.6918	0.4786	8.2	2.570	6.607	0.3891	0.1514
3.3	1.462	2.138	0.6839	0.4677	8.3	2.600	6.761	0.3846	0.1479
3.4	1.479	2.188	0.6761	0.4571	8.4	2.630	6.918	0.3802	0.1445
3.5	1.496	2.239	0.6683	0.4467	8.5	2.661	7.080	0.3758	0.1412
3.6	1.514	2.291	0.6607	0.4365	8.6	2.692	7.244	0.3715	0.1380
3.7	1.531	2.344	0.6531	0.4266	8.7	2.723	7.413	0.3673	0.1349
3.8	1.549	2.399	0.6457	0.4169	8.8	2.754	7.586	0.3631	0.1318
3.9	1.567	2.455	0.6383	0.4074	8.9	2.786	7.763	0.3589	0.1288
4.0	1.585	2.512	0.6310	0.3981	9.0	2.818	7.943	0.3548	0.1259
4.1	1.603	2.570	0.6237	0.3891	9.1	2.851	8.128	0.3508	0.1230
4.2	1.622	2.630	0.6166	0.3802	9.2	2.884	8.318	0.3467	0.1202
4.3	1.641	2.692	0.6095	0.3715	9.3	2.917	8.511	0.3428	0.1175
4.4	1.660	2.754	0.6026	0.3631	9.4	2.951	8.710	0.3388	0.1148
4.5	1.679	2.818	0.5957	0.3548	9.5	2.985	8.913	0.3350	0.1122
4.6	1.698	2.884	0.5888	0.3467	9.6	3.020	9.120	0.3311	0.1096
4.7	1.718	2.951	0.5821	0.3388	9.7	3.055	9.333	0.3273	0.1072
4.8	1.738	3.020	0.5754	0.3311	9.8	3.090	9.550	0.3236	0.1047
4.9	1.758	3.090	0.5689	0.3236	9.9	3.126	9.772	0.3199	0.1023
5.0	1.778	3.162	0.5623	0.3162	10.0	3.162	10.000	0.3162	0.1000

Voltage (V) or Power (P) and Vice Versa

Deci-bel	$\dfrac{I_1}{I_2}$ or $\dfrac{V_1}{V_2}$	$\dfrac{P_1}{P_2}$	$\dfrac{I_2}{I_1}$ or $\dfrac{V_2}{V_1}$	$\dfrac{P_2}{P_1}$	Deci-bel	$\dfrac{I_1}{I_2}$ or $\dfrac{V_1}{V_2}$	$\dfrac{P_1}{P_2}$	$\dfrac{I_2}{I_1}$ or $\dfrac{V_2}{V_1}$	$\dfrac{P_2}{P_1}$
10.1	3.199	10.23	0.3126	0.09772	15.1	5.689	32.36	0.1758	0.03090
10.2	3.236	10.47	0.3090	0.09550	15.2	5.754	33.11	0.1738	0.03020
10.3	3.273	10.72	0.3055	0.09333	15.3	5.821	33.88	0.1718	0.02951
10.4	3.311	10.96	0.3020	0.09120	15.4	5.888	34.67	0.1698	0.02884
10.5	3.350	11.22	0.2985	0.08913	15.5	5.957	35.48	0.1697	0.02818
10.6	3.388	11.48	0.2951	0.08710	15.6	6.026	36.31	0.1660	0.02754
10.7	3.428	11.75	0.2917	0.08511	15.7	6.095	37.15	0.1641	0.02692
10.8	3.467	12.02	0.2884	0.08318	15.8	6.166	38.02	0.1622	0.02630
10.9	3.508	12.30	0.2851	0.08128	15.9	6.237	38.90	0.1603	0.02570
11.0	3.548	12.59	0.2818	0.07943	16.0	6.310	39.81	0.1585	0.02512
11.1	3.589	12.88	0.2786	0.07763	16.1	6.383	40.74	0.1567	0.02455
11.2	3.631	13.18	0.2754	0.07586	16.2	6.457	41.69	0.1549	0.02399
11.3	3.673	13.49	0.2723	0.07413	16.3	6.531	42.66	0.1531	0.02344
11.4	3.715	13.80	0.2692	0.07244	16.4	6.607	43.65	0.1514	0.02291
11.5	3.758	14.13	0.2661	0.07080	16.5	6.683	44.67	0.1496	0.02239
11.6	3.802	14.45	0.2630	0.06918	16.6	6.761	45.71	0.1479	0.02188
11.7	3.846	14.79	0.2600	0.06761	16.7	6.839	46.77	0.1462	0.02138
11.8	3.891	15.14	0.2570	0.06607	16.8	6.918	47.86	0.1445	0.02089
11.9	3.936	15.49	0.2541	0.06457	16.9	6.998	48.98	0.1429	0.02042
12.0	3.981	15.85	0.2512	0.06310	17.0	7.080	50.12	0.1412	0.01995
12.1	4.027	16.22	0.2483	0.06166	17.1	7.161	51.29	0.1396	0.01950
12.2	4.074	16.60	0.2455	0.06026	17.2	7.244	52.48	0.1380	0.01906
12.3	4.121	16.98	0.2427	0.05888	17.3	7.328	53.70	0.1365	0.01862
12.4	4.169	17.38	0.2399	0.05754	17.4	7.413	54.95	0.1349	0.01820
12.5	4.217	17.78	0.2371	0.05623	17.5	7.499	56.23	0.1334	0.01778
12.6	4.266	18.20	0.2344	0.05495	17.6	7.586	57.54	0.1318	0.01738
12.7	4.315	18.62	0.2317	0.05370	17.7	7.674	58.88	0.1303	0.01698
12.8	4.365	19.05	0.2291	0.05248	17.8	7.763	60.26	0.1288	0.01660
12.9	4.416	19.50	0.2265	0.05128	17.9	7.852	61.66	0.1274	0.01622
13.0	4.467	19.95	0.2239	0.05012	18.0	7.943	63.10	0.1259	0.01585
13.1	4.519	20.42	0.2213	0.04898	18.1	8.035	64.57	0.1245	0.01549
13.2	4.571	20.89	0.2188	0.04786	18.2	8.128	66.07	0.1230	0.01514
13.3	4.624	21.38	0.2163	0.04677	18.3	8.222	67.61	0.1216	0.01479
13.4	4.677	21.88	0.2138	0.04571	18.4	8.318	69.18	0.1202	0.01445
13.5	4.732	22.39	0.2114	0.04467	18.5	8.414	70.79	0.1189	0.01412
13.6	4.786	22.91	0.2089	0.04365	18.6	8.511	72.44	0.1175	0.01380
13.7	4.842	23.44	0.2065	0.04266	18.7	8.610	74.13	0.1161	0.01349
13.8	4.898	23.99	0.2042	0.04169	18.8	8.710	75.86	0.1148	0.01318
13.9	4.955	24.55	0.2018	0.04074	18.9	8.811	77.62	0.1135	0.01288
14.0	5.012	25.12	0.1995	0.03981	19.0	8.913	79.43	0.1122	0.01259
14.1	5.070	25.70	0.1972	0.03891	19.1	9.016	81.28	0.1109	0.01230
14.2	5.128	26.30	0.1950	0.03802	19.2	9.120	83.18	0.1096	0.01202
14.3	5.188	26.92	0.1928	0.03715	19.3	9.226	85.11	0.1084	0.01175
14.4	5.248	27.54	0.1906	0.03631	19.4	9.333	87.10	0.1072	0.01148
14.5	5.309	28.18	0.1884	0.03548	19.5	9.441	89.13	0.1059	0.01122
14.6	5.370	28.84	0.1862	0.03467	19.6	9.550	91.20	0.1047	0.01096
14.7	5.433	29.51	0.1841	0.03388	19.7	9.661	93.33	0.1035	0.01072
14.8	5.495	30.20	0.1820	0.03311	19.8	9.772	95.50	0.1023	0.01047
14.9	5.559	30.90	0.1799	0.03236	19.9	9.886	97.72	0.1012	0.01023
15.0	5.623	31.62	0.1778	0.03162	20.0	10.000	100.00	0.1000	0.01000

THE MAXIMUM DISRUPTIVE FIELD STRENGTH OF SOME INSULATING MATERIALS

Material	Max. disr. field strength kV [1] / mm	Material	Max. disr. field strength kV [1] / mm
Bakelite...............	32.5	Mica.................	24—41
Calan	50—64	Paper (oiled)	28—42
Calite...............	50—64	Paper (lacquered)......	42—85
Celluloid	42	Paraffin..............	17
Condensa C	21	Porcelain (hard)	14—28
Condensa F	21—28	Presspahn............	17
Condensa N	21—28	Rubber (soft)..........	27
Frequenta............	42—50	Transformer oil	17
Glass (common).......	11—12.5	Wax.................	17
Air	3.2		

[1] Direct voltage or peak value of alternating voltage of 50 c/s.

DIELECTRIC CONSTANT AND RECIPROCAL QUALITY FACTOR OF INSULATING MATERIALS AT ROOM TEMPERATURE

Material	Dielectric constant	Reciprocal quality factor $\tan \delta \times 10^4$		
		at 1,000 c/s	at 1,000kc/s (300 m)	at 10Mc/s (30 m)
Bakelite...............	3—5	50—200	160	220
Amber.................	2.9	—	300	300
Calan.................	6.6	—	2.1	2.2
Calite................	6.5	5	4—5	3—4
Celluloid	3.3—3.5	—	490	480
Condensa C	80	300—500	6.0	3.2
Condensa F	65	—	4.0	3.6
Condensa N	40	—	6.9	4.6
Ebonite...............	3	25—330	65—110	60—110
Enamel lacquer (insulation of copper wire)........	—	180	—	—
Frequenta.............	5.5—6.5	7	4.2	3.4
Glass (various kinds).....	5—16.5	20—30	4—75	—
Quartz................	4.3—4.7	1	1.1	1.1
Air	1.0	0	0	0
Mica.................	6—8	1	$<$1	—
Mycalex...............	6—9	—	15—20	18
Novotext	5.5	100	—	—
Paper, dry	1.6—2.5	15—30	150—300	—
Paper, impregnated	3.5—6	15—100	300—600	—
Paraffin...............	2—2.3	5	3	—
Pertinax	4.5—5.5	250	300	700
Porcelain..............	5—6	100—200	70—120	60—110
Presspahn.............	3.4	—	240	580
Rubber	2.5—2.8	150	—	—
Shellac...............	3—4	100—200	100	110
Trolitul	2.2—2.5	—	1.0	1.0
Tourmaline............	5.0	—	5.1	3.6
Water	80	—	—	—
Silk (insulation of copper wire)	—	400	—	—

SPECIFIC RESISTANCE, SPECIFIC CONDUCTIVITY, SPECIFIC GRAVITY AND RESISTANCE-TEMPERATURE COEFFICIENT OF MATERIALS

Material	Specif. resistance $\rho \left(\dfrac{\Omega \ mm^2}{m} \right)$	Specif. conductivity $\varkappa \left(\dfrac{\Omega \ mm^2}{m} \right)$	Resistance-temp. coeff. $\alpha \ (1/^{\circ}C)$	Spec. gravity (g/cm^3)
Silver..................	0.0147	68	0.0038	10.5
Copper.................	0.01718	58.2	0.00393	8.9
Aluminium	0.0262	38	0.0039	2.7
Duralumin	0.05	20	0.0041	2.8
Zinc..................	0.0575	17.4	0.0037	7.1
Nickel	0.0693	14.4	0.0043	8.8
Brass.................	0.07	14.3	0.002	8.2—8.7
Iron..................	0.10	10	0.005	7.6—7.9
Platinum..............	0.10—0.11	10—9	0.003	21.4
Tin...................	0.115	8.7	0.0043	7.3
New silver	0.30	3.33	0.00025	8.3—8.7
Nickelin..............	0.40	2.5	0.011	8.9
Manganin	0.43	2.33	0.00002	8.4
Constantan	0.50	2.0	0.00004	8.9
Isabellin.	0.52	1.92	—0.00002	8.0
Chrome-iron	0.6	1.66	0.0016	7.6
Invar.................	0.75	1.33	0.002	—
Mercury...............	0.95	1.05	0.0009	13.6
Chrome-nickel	1.1	0.9	0.00011	8.3
Carbon................	30	0.33	—0.0002 to —0.0008	2.25

If it is desired to express specific resistance in Ω cm $\left(\Omega \dfrac{cm^2}{cm} \right)$ the values given in the table have to be multiplied by 10^{-4}. The resistance R of a conductor is equal to $R = \dfrac{\rho \ l}{q}$, in which l is the length in m or cm and q the cross-section in mm² or cm².

Determining Temperature Rise by Means of the Temperature Coefficient

Temperature rise can be calculated with the aid of the temperature coefficient from the increase of resistance of a wire made of a known material, for instance of the copper winding of a transformer or choke. If the resistance of the wire prior to the current being passed through it is R_1 (corresponding to the ambient temperature, which is generally 20° C) and the resistance after a certain working period is R_2, then the temperature rise is:

$$\varDelta T = \frac{R_2 - R_1}{R_1 \cdot a} \ ^{\circ}C.$$

RESISTANCE OF ENAMELLED, SOFT, ELECTROLYTIC, ROUND, COPPER WIRE

Wire gauge in mm	Wire section mm²	Re-sistance Ω/m [1])	Weight g/m or kg/1,000 m	Wire gauge in mm	Wire section mm²	Re-sistance Ω/m [1])	Weight g/m or kg/1,000 m
0.020	0.00031	55.20	0.0031	0.40	0.1257	0.1383	1.14
0.025	0.00049	35.40	0.0048	0.45	0.1590	0.1092	1.45
0.030	0.00071	24.60	0.0069	0.50	0.1964	0.0885	1.78
0.035	0.00096	18.06	0.0093	0.55	0.2376	0.0731	2.16
0.040	0.00126	13.83	0.012	0.60	0.2827	0.0615	2.56
0.045	0.00159	10.92	0.015	0.65	0.3318	0.0524	2.99
0.050	0.00196	8.85	0.019	0.70	0.3848	0.0452	3.48
0.060	0.00283	6.15	0.027	0.80	0.5027	0.0346	4.54
0.070	0.00385	4.52	0.036	0.90	0.6362	0.0274	5.73
0.080	0.00503	3.46	0.047	1.0	0.7854	0.0221	7.08
0.090	0.00636	2.73	0.059	1.1	0.9503	0.01829	8.57
0.10	0.00785	2.21	0.073	1.2	1.131	0.01536	10:19
0.12	0.01131	1.537	0.105	1.3	1.327	0.01310	11.94
0.15	0.01767	0.983	0.162	1.4	1.539	0.01129	13.83
0.18	0.02545	0.682	0.232	1.5	1.767	0.00984	15.87
0.20	0.03142	0.552	0.287	1.6	2.011	0.00865	18.05
0.22	0.03801	0.457	0.347	1.8	2.545	0.00683	22.8
0.25	0.04909	0.354	0.447	2.0	3.142	0.00556	28.2
0.28	0.06158	0.282	0.559	2.2	3.801	0.00456	34.6
0.30	0.07069	0.246	0.641	2.5	4.906	0.00354	44.0
0.35	0.09621	0.1806	0.873	3.0	7.069	0.00246	63.2

[1]) For a temperature of 20° C and specific resistance 0.01736 × 10⁻⁴ Ω cm.

LIST OF PHILIPS PUBLICATIONS
Relating to Radio Valves, Radio Reception and Allied Subjects

A l e x a n d e r, J. W.: A car radio, Philips Technical Review 3 (1938), page 112.

A l m a, G. and F. P r a k k e: A new series of small radio valves, Philips Technical Review 8 (1946), page 289.

B a k k e r, C. J. and G. d e V r i e s: Amplification of small alternating tensions by an inductive action of the electrons in a radio valve with negative anode, Physica 1 (1934), pages 1045—1054.

B a k k e r, C. J.: On vacuum tube electronics, Physica 2 (1935), pages 683—697.

B a k k e r, C. J.: Some characteristics of receiving valves in shortwave reception, Philips Technical Review 1 (1936), page 171.

B a k k e r, C. J. and C. J. B o e r s: On the influence of the non-linearity of the characteristics on the frequency of dynatron and triode oscillators, Physica 3 (1936), pages 649—665.

B a k k e r, C. J.: Current distribution fluctuations in multi-electrode radio valves, Physica 5 (1938), pages 581—592.

B a k k e r, C. J. and B. v a n d e r P o l: Report on spontaneous fluctuations of current and potential, C. R. Union radiosci.int. Venice, Volume 5 (1939), pages 217—227.

B a k k e r, C. J. and G. H e l l e r: On the Brownian motion in electric resistances, Physica 6 (1939), pages 262—274.

B a k k e r, C. J.: Fluctuations and electron inertia, Physica 8 (1941), pages 23—43.

B a k k e r, C. J.: The causes of voltage and current fluctuations, Philips Technical Review 6 (1941), page 129.

B a k k e r, C. J.: Radio-Investigation of the ionosphere, Philips Technical Review 8 (1946), page 111.

B e e k, M. v a n d e: Air-cooled transmitting valves, Philips Technical Review 4 (1939), page 121.

B l o k, L.: Radio-interferences, Philips Technical Review 3 (1938), page 235.

B l o k, L.: Combating radio interference, Philips Technical Review 3 (1938), page 237.

B o e l e n s, W. W.: Valve characteristic giving linear modulation when a feedback resistor is inserted in the cathode lead, Philips Research Reports 3 (1948), pages 227—234.

B o e r, J. H. d e, and H. B r u i n i n g: Secondary electron emission VI, The influence of externally absorbed ions and atoms on the secondary emission of metals. Physica 6 (1939), pages 941—950.

B o u m e e s t e r, H. G.: Development and manufacture of modern transmitting valves, Philips Technical Review 2 (1937), page 115.

B r u i n i n g, H., J. H. d e B o e r and W. G. B u r g e r s: Secondary electron emission of soot in valves with oxide cathode, Physica 4 (1937), pages 267—275.

B r u i n i n g, H., and J. H. d e B o e r: Secondary emission of metals with a low work function, Physica 4 (1937), pages 473—477.

B r u i n i n g, H.: Secondary electron emission, Philips Technical Review 3 (1938), page 80.

B r u i n i n g, H., and J. H. d e B o e r: Secondary electron emission I, Secondary electron emission of metals, Physica 5 (1938), pages 17—30.

B r u i n i n g, H.: Secondary electron emission II, Absorption of secondary electrons III, Secondary electron emission caused by bombardment with slow primary electrons. Physica 5 (1938), pages 913—917.

B r u i n i n g, H., and J. H. d e B o e r : Secondary electron emission IV, Compounds with a high capacity for secondary electron emission V, The mechanism of secondary electron emission. Physica 6 (1939), pages 823—833 and pages 834—839.

B r u i n i n g, H.: Over de emissie van secundaire electronen door vaste stoffen. Thesis Leiden 1938, 119 pages.

B r u i n i n g, H.: Secondary emission from metals with a low work function, Physica 8 (1941), pages 1161—1164.

C a t h, P. G.: A new principle of construction for radio valves, Philips Technical Review 4 (1939), page 162.

C o e t e r i e r, F.: The multireflection tube, a new oscillator for very short waves, Philips Technical Review 8 (1946), page 257.

C o h e n H e n r i q u e z, V.: Compression and expansion in sound transmission, Philips Technical Review 3 (1938), page 204.

C o h e n H e n r i q u e z, V.: The reproduction of high and low tones in radio receiving sets, Philips Technical Review 5 (1940), page 115.

C o r n e l i u s, P.: The sensitivity of aerials to local interferences, Philips Technical Review 6 (1941), page 302.

C o r n e l i u s, P.: The aerial effect in receiving sets with loop aerial, Philips Technical Review 7 (1942), page 65.

D o r g e l o, E. G.: Several technical problems in the development of a new series of transmitter valves, Philips Technical Review 6 (1941), page 253.

D o u m a, Tj., and P. Z ij l s t r a : Recording the characteristics of transmitting valves, Philips Technical Review 4 (1939), page 56.

D ij k s t e r h u i s, P. R., and Y. B. F. J. G r o e n e v e l d : L'amplification basse fréquence par transformateurs, Q.S.T. franç. 10 (1939), pages 55—59.

D ij k s t e r h u i s, P. R., and Y. B. F. J. G r o e n e v e l d : Low-frequency amplification with transformers, Experimental Wireless 6 (1929), pages 374—379.

E l i a s, G. J., B. v a n d e r P o l and B. D. H. T e l l e g e n : Das elektrostatische Feld einer Triode, Annalen der Physik 78 (1925), pages 370—408.

E r i n g a, D.: A universal testing set for radio valves, Philips Technical Review 2 (1937), page 57.

G r o e n e v e l d, Y. B. F. J., B. v a n d e r P o l and K. P o s t h u m u s : Gittergleichrichtung, Jahrb. d. drahtl. Telegr. u. Teleph. 29 (1927), pages 139—147.

H a a n t j e s, J.: Judging an amplifier by means of the transient characteristic, Philips Technical Review 6 (1941), page 193.

H e i n s v a n d e r V e n, A. J.: Testing amplifier output valves by means of the cathode-ray tube, Philips Technical Review 5 (1940), page 61.

H e i n s v a n d e r V e n, A. J.: Output and distortion of output amplifier valves under different loads, Philips Technical Review 5 (1940), page 189.

H e i n s v a n d e r V e n, A. J.: Output stage distorsion; some measurements on different types of output valves, Wireless Engineer 16 (1939), pages 383—390 and 444—452.

H e l l e r, G.: The magnetron as a generator of ultra short waves, Philips Technical Review 4 (1939), page 189.

H e l l e r, G.: Radio sets with station dials calibrated for short waves, Philips Technical Review 4 (1939), page 284.

H e l l e r, G.: Television receivers, Philips Technical Review 4 (1939), page 342.

H e p p, G.: Measurements of potential by means of the electrolytic tank, Philips Technical Review 4 (1939), page 223.

H e y b o e r, J. P.: Five-electrode transmitting valves (pentodes), Philips Technical Review 2 (1937), page 257.

H e y b o e r, J. P.: A discharge phenomenon in large transmitter valves, Philips Technical Review 6 (1941), page 208.

J o n k e r, J. L. H., and A. J. W. M. O v e r b e e k : The application of secondary emission in amplifying valves, Wireless Engineer 15 (1938), pages 150—156.

J o n k e r, J. L. H., and M. C. T e v e s: Technical applications of secondary emission, Philips Technical Review 3 (1938), page 133.

J o n k e r, J. L. H.: Phenomena in amplifier valves caused by secondary emission, Philips Technical Review 3 (1938), page 211.

J o n k e r, J. L. H., and A. J. W. M. v a n O v e r b e e k: A new converter valve, Wireless Engineer 15 (1938), pages 423—431.

J o n k e r, J. L. H., and A. J. W. M. v a n O v e r b e e k: A new frequency-changing valve, Philips Technical Review 3 (1938), page 266.

J o n k e r, J. L. H.: Pentode and tetrode output valves, Wireless Engineer 16 (1939), pages 274—286 and 344—349.

J o n k e r, J. L. H.: Electron trajectories in multi-grid valves, Philips Technical Review 5 (1940), page 131.

J o n k e r, J. L. H.: Stroomverdeling in versterkerbuizen. Thesis Delft 1942, 184 pages.

J o n k e r, J. L. H., and B. D. H. T e l l e g e n: The current to a positive grid in electron tubes, Philips Research Reports 1 (1945), page 13.

K l e y n e n, P. H. J. A.: The motion of an electron in two-dimensional electrostatic fields, Philips Technical Review 2 (1937), page 338.

K n o l, K. S., M. J. O. S t r u t t and A. v a n d e r Z i e l: On the motion of electrons in an alternating electric field, Physica 5 (1939), pages 325—334.

K n o l, K. S., and M. J. O. S t r u t t: Ueber ein Verfahren zur Messung komplexer Leitwerte im Dezimeterwellengebiet, Physica 9 (1942), pages 577—590.

K n o l, K. S., and M. J. O. S t r u t t: A diode for the measurement of voltages on decimetre waves, Philips Technical Review 7 (1942), page 124.

L i n d e r n, C. G. A. v o n, and G. d e V r i e s: Resonance circuits for very high frequencies, Philips Technical Review 6 (1941), page 217.

L i n d e r n, C. G. A. v o n, and G. d e V r i e s: Lecher systems, Philips Technical Review 6 (1941), page 240.

L i n d e r n, C. G. A. v o n, and G. d e V r i e s: Flat cavities as electrical resonators, Philips Technical Review 8 (1946), page 149.

L o o n, C. J. v a n: Improvements in radio receivers, Philips Technical Review 1 (1936), page 264.

L o o n, C. J. v a n: A simple system of bandspread in short-wave reception, Philips Technical Review 6 (1941), page 265.

L u s s a n e t d e l a S a b l o n i è r e, C. J. d e: Ueber die Arbeitsweise von Schirm-gittersenderöhren, Hochfrequenztechn. u. Elektroakust. 39 (1932), pages 191—199.

L u s s a n e t d e l a S a b l o n i è r e, C. J. d e: Die Sekundäremission in Elektronen-röhren, namentlich Schirmgitterröhren, Hochfrequenztechn. u. Elektroakust. 41 (1933), pages 195—202.

L u s s a n e t d e l a S a b l o n i è r e, C. J. d e: Die Bestimmung des Schirmgitter-verlustes einer gesteuerten Schirmgitter-Senderöhre, Hochfrequenztechn. u.Elektroakust. 41 (1933), pages 202—203.

L u s s a n e t d e l a S a b l o n i è r e, C. J. d e: Der innere Widerstand von Schirm-gitterröhren, Hochfrequenztechn. u. Elektroakust. 41 (1933), pages 204—205.

L u s s a n e t d e l a S a b l o n i è r e, C. J. d e: Le fonctionnement des lampes d'é-mission à grille-écran, Onde électrique 12 (1933), pages 415—440.

L u s s a n e t d e l a S a b l o n i è r e, C. J. d e: The design of class B amplifiers, Wireless Engineer 12 (1935), pages 133—141.

M u l d e r, J. G. W.: Barretters, Philips Technical Review 3 (1938), page 74.

P e n n i n g, F. M.: Velocity-modulation valves, Philips Technical Review 8 (1946) page 214.

P o l, B. v a n d e r: Ueber Elektronenbahnen in Trioden, Jahrb. d. drahtl. Telegr. u. Teleph. 25 (1925), pages 121—131.

P o l, B. v a n d e r: The non-linear theory of electric oscillations, Proc. Inst. Radio Engineers 22 (1934), pages 1051—1086.

P o l, B. v a n d e r, and T h. J. W e y e r s: Fine structure of triode characteristics, Physica 1 (1934), pages 481—496.

P o l, B. v a n d e r, and H. B r e m m e r: The propagation of wireless waves round the earth, Philips Technical Review 4 (1939), page 245.

P o s t h u m u s, K.: Kurzwellenröhren, Ref. u. Mitt. int. Kongr. Kurzwellen, Vienna 1937, pages 78—88.

P o s t h u m u s, K.: The hum due to the magnetic field of the filaments in transmitting valves, Philips Technical Review 5 (1940), page 100.

P r a k k e, F., J. L. H. J o n k e r and M. J. O. S t r u t t: A new "all glass" valve construction, Wireless Engineer 16 (1939), pages 224—230.

S c h o u t e n, J. F.: The perception of pitch, Philips Technical Review 5 (1940), page 290.

S l o o t e n, J. v a n: The communal aerial, Philips Technical Review 1 (1936), page 246.

S l o o t e n, J. v a n: The stability of a triode oscillator with grid condenser and leak, Wireless Engineer 16 (1939), pages 16—19.

S l o o t e n, J. v a n: Input capacitance of a triode oscillator, Wireless Engineer 17 (1940), pages 13—15.

S l o o t e n, J. v a n: Receiving aerials, Philips Technical Review 4 (1939), page 320.

S l o o t e n, J. v a n: The functioning of triode oscillators with grid condenser and grid resistance, Philips Technical Review 7 (1945), page 40.

S l o o t e n, J. v a n: Stability and instability in triode oscillators, Philips Technical Review 7 (1945), page 171.

S m e l t, J.: Glass for modern electric lamps and radio valves, Philips Technical Review 2 (1937), page 87.

S t e t t l e r, O.: The octode, a new mixing valve for superheterodyne receivers, Bull. Assoc. Suisse des Electr. 25 (1934), pages 441—443.

S t r u t t, M. J. O.: Gleichrichtung, Hochfrequenztechn. u. Elektroakust. 42 (1933), pages 206—208.

S t r u t t, M. J. O.: Radioempfangsröhren mit grossem inneren Widerstand, Hochfrequenztechn. u. Elektroakust. 43 (1934), pages 18—22.

S t r u t t, M. J. O.: On conversion d tectors, Proc. Inst. Radio Engr. 22 (1934), pages 981—1008.

S t r u t t, M. J. O.: Anode bend detection, Proc. Inst. Radio Engr. 23 (1935), pages 945—958.

S t r u t t, M. J. O.: Mixing valves, Wireless Engineer 12 (1935), pages 59—64.

S t r u t t, M. J. O.: Whistling notes in superheterodyne receivers, Wireless Engineer, 12 (1935), pages 194—197.

S t r u t t, M. J. O., and A. v a n d e r Z i e l: Messungen der charakteristischen Eigenschaften von Hochfrequenz-Empfangsröhren zwischen 1,5 und 60 Megahertz Elektr. Nachr. Techn. 12 (1935), pages 347—354.

S t r u t t, M. J. O.: Diode frequency changers, Wireless Engineer 13 (1936), pages 73—80.

S t r u t t, M. J. O.: Performance of some types of frequency changers in all-wave receivers, Wireless Engineer 14 (1937), pages 184—192.

S t r u t t, M. J. O., and A. v a n d e r Z i e l: Einfache Schaltmassnahmen zur Verbesserung der Eigenschaften von Hochfrequenzverstärkerröhren in Kurzwellengebiet, Elektr. Nachr. Techn. 13 (1936), pages 260—268.

S t r u t t, M. J. O., and A. v a n d e r Z i e l: Erweiterung der bisherigen Messungen der Admittanzen von Hochfrequenzverstärkerröhren bis 300 Megahertz, Elektr. Nachr. Techn. 14 (1937), pages 75—80.

S t r u t t, M. J. O.: Moderne Mehrgitterelektronenröhren, Schweiz. Archiv. angew. Wiss. und Techn. 2 (1936), pages 183—199, 230.

S t r u t t, M. J. O.: Les performances de certains types de lampes changeuses de fréquence dans les récepteurs toutes ondes, Onde Electrique 16 (1937), pages 29—44.

S t r u t t, M. J. O.: Verzerrungseffekte bei Mischröhren, Hochfrequenztechn. u. Elektroakust. 49 (1937), pages 20—23.

S t r u t t, M. J. O., and A. v a n d e r Z i e l: Die Ursachen für die Zunahme der Admittanzen moderner Hochfrequenz-Verstärkerröhren im Kurzwellengebiet, Electr. Nachr. Techn. 14 (1937), pages 281—293.

S t r u t t, M. J. O.: Characteristic constants of h.f. pentodes, Measurements at frequencies between 1,5—300 Mc/s, Wireless Engineer 14 (1937), pages 478—488.

S t r u t t, M. J. O.: Mesures des constantes caractéristiques de quelques penthodes haute fréquence pour des fréquences de 1,5—300 mégacycles par seconde, Onde Electrique 16 (1937), pages 553—577.

S t r u t t, M. J. O.: Die charakteristischen Admittanzen von Mischröhren für Frequenzen bis 70 MHz, Elektr. Nachr. Techn. 15 (1938), pages 10—17.

S t r u t t, M. J. O., and A. v a n d e r Z i e l: Messungen der komplexen Steilheit moderner Mehrgitterröhren im Kurzwellengebiet, Elektr. Nachr. Techn. 15 (1938), pages 103—111.

S t r u t t, M. J. O.: and A. v a n d e r Z i e l: Einige dynamische Messungen der Elektronenbewegung in Mehrgitterröhren, Elektr. Nachr. Techn. 15 (1938), pages 277—283.

S t r u t t, M. J. O.: Electron transit time effects in multi-grid valves, Wireless Engineer 15 (1938), pages 315—321.

S t r u t t, M. J. O., and A. v a n d e r Z i e l: On electronic space charge with homogeneous initial electron velocity between plane electrodes, Physica 5 (1938), pages 705—717.

S t r u t t, M. J. O., and A. v a n d e r Z i e l: The causes for the increase of the admittances of modern high-frequency amplifier tubes on short waves, Proc. Inst. Radio Engrs. 26 (1938), pages 1011—1032.

S t r u t t, M. J. O., and A. v a n d e r Z i e l: Some dynamic measurements of electronic motion in multigrid valves, Proc. Inst. Radio Engrs. 27 (1939), pages 218—225.

S t r u t t, M. J. O., and A. v a n d e r Z i e l: The behaviour of amplifier valves at very high frequencies, Philips Technical Review 3 (1938), page 103.

S t r u t t, M. J. O.: Etages à haute fréquence, étages changeur de fréquence et détecteur des récepteurs de télévision, Onde Electrique 18 (1939), pages 14—26 and pages 83—91.

S t r u t t, M. J. O., and K. S. K n o l: Messungen von Strömen, Spannungen und Impedanzen bis herab zu 20 cm Wellenlänge, Hochfrequenztechn. u. Elektroakustik 53 (1939), pages 187—195.

S t r u t t, M. J. O.: Hochfrequenz-, Misch- und Gleichrichterstufen von Fernsehenpfängern, Schweiz. Arch. angew. Wiss. und Techn., Sonderheft, Vortr. und Disk. Ber. Fernseh-Tagung Zürich 1938, pages 26—36.

S t r u t t, M. J. O., and A. v a n d e r Z i e l: Ueber die Elektronenraumladung zwischen ebenen Elektroden unter Berücksichtigung der Anfangsgeschwindigkeit und Geschwindigkeitsverteilung der Elektronen, Physica 6 (1939), pages 977—996.

S t r u t t, M. J. O., and A. v a n d e r Z i e l: Kurzwellen-Breitband-Verstärkung, Elektr. Nachr. Techn. 16 (1939), pages 229—240.

S t r u t t, M. J. O., and A. v a n d e r Z i e l: A new push-pull amplifier valve for decimetre waves, Philips Technical Review 5 (1940), page 172.

S t r u t t, M. J. O., and A. v a n d e r Z i e l: A variable amplifier valve with double cathode connection suitable for metre waves, Philips Technical Review 5 (1940), page 357.

S t r u t t, M. J. O., and A. v a n d e r Z i e l: The noise in receiving sets at very high frequencies, Philips Technical Review 6 (1941), page 178.

S t r u t t, M. J. O., and A. v a n d e r Z i e l: The diode as a frequency-changing valve especially with decimetre waves, Philips Technical Review 6 (1941), page 285.

S t r u t t, M. J. O., and A. v a n d e r Z i e l: Die Folgen einiger Elektronenträgheitseffekte in Elektronenröhren I, Theoretische Erläuterungen, Physica 8 (1941), pages 81—108.

S t r u t t, M. J. O., and A. v a n d e r Z i e l: Die Folgen einiger Elektronenträgheitseffekte in Elektronenröhren II, Anwendungen und numerische Ergebnisse, Physica 9 (1942), pages 65—83.

S t r u t t, M. J. O., and A. v a n d e r Z i e l: Verringerung der Wirkung spontaner Schwankungen in Verstärkern für Meter- und Dezimeterwellen, Physica 9 (1942), pages 1003—1012.

S t r u t t, M. J. O., and K. S. K n o l: A diode for the measurement of voltages on decimetre waves; Philips Technical Review 7 (1942), page 124.

S u c h t e l e n, H. v a n: The electrometer triode and its applications, Philips Technical Review 5 (1940), page 54.

T e l l e g e n, B. D. H.: Endverstärkerprobleme, Jahrb. d. drahtl. Telegr. u. Teleph. 31 (1928), pages 183—190.

T e l l e g e n, B. D. H.: Die Endröhre, Funkmagasin 2 (1929), pages 689—692.

T e l l e g e n, B. D. H.: Inverse feed-back, Philips Technical Review 2 (1937), page 289.

T e l l e g e n, B. D. H., and V. C o h e n H e n r i q u e z: Inverse feed-back, its application to receivers and amplifiers, Wireless Engineer 14 (1937), pages 409—413.

T e l l e g e n, B. D. H., and J. H a a n t j e s: Gegenkopplung, Electr. Nachr. Techn. 15 (1938), pages 353—358.

T e v e s, M. C.: The photo-electric effect and its application in photo-electric cells, Philips Technical Review 2 (1937), page 13.

T e v e s, M. C.: A photocell with amplification by means of secondary emission, Philips Technical Review 5 (1940), page 253.

T r o m p, T h. P.: Technical problems in the construction of radio valves, Philips Technical Review 6 (1941), page 317.

V e e g e n s, J. D., and M. K. d e V r i e s: A simple high-frequency oscillator for the testing of radio receiving sets, Philips Technical Review 6 (1941), page 154.

V e r m e u l e n, R.: Octaves and decibels, Philips Technical Review 2 (1937), page 47.

V e r m e u l e n, R.: The relationship between fortissimo and pianissimo, Philips Technical Review 2 (1937), page 266.

V r i e s, G. d e: Electromagnetic cavities, Philips Technical Review 9 (1947), page 73.

W e e l, A. v a n: An experimental transmitter for ultra-short-wave radio-telephony with frequency modulation, Philips Technical Review 8 (1946), page 121.

W e e l, A. v a n: An experimental receiver for ultra-short-wave radio-telephony with frequency modulation, Philips Technical Review 8 (1946), page 193.

W e e l, A. v a n: Developments in radio-receiver circuits for the ultra-short-wave range, Philips Research Reports 3 (1948), pages 191—212.

W e e l, A. v a n: A new principle for transceivers, Philips Research Reports 3 (1948), pages 361—370.

W e y e r s, Th. J.: Frequency modulation, Philips Technical Review 8 (1946), page 42,

W e y e r s, Th. J.: Comparison of frequency modulation and amplitude modulation, Philips Technical Review 8 (1946), page 89.

Z a a l b e r g v a n Z e l s t, J. J.: Constant amplification independent of variable circuit elements, Philips Technical Review 9 (1947), page 309.

Z i e g l e r, M.: Space charge depression of shot effect, Physica 2 (1935), pages 413—414.

Z i e g l e r, M.: Shot effect of secondary emission, Physica 2 (1935), pages 415—416.

Z i e g l e r, M.: Shot effect of secondary emission I, Physica 3 (1936), pages 1—11.

Z i e g l e r, M.: Shot effect of secondary emission II, Physica 3 (1936), pages 307—316.

Z i e g l e r, M.: The causes of noise in amplifiers, Philips Technical Review 2 (1937), page 136.

Z i e g l e r, M.: Noise in amplifiers contributed by the valves, Philips Technical Review 2 (1937), page 329.

Z i e g l e r, M.: Noise in receiving sets, Philips Technical Review 3 (1938), page 189.

Z i e l, A. v a n d e r: Fluctuations in electrometer triode circuits, Physica 9 (1942), pages 177—192.

Z i e l, A. v a n d e r, and A. V e r s n e l: Induced grid noise and total-emission noise, Philips Research Reports 3 (1948), pages 13—23.

Z i e l, A. v a n d e r, and A. V e r s n e l: Measurements of noise factors of pentodes at 7.25 m wavelength, Philips Research Reports 3 (1948), pages 121—129.

Z i e l, A. v a n d e r, and A. V e r s n e l: The noise factor of grounded-grid valves, Philips Research Reports 3 (1948), pages 255—270.

Some Books on Electronic Valves and their Applications

A r g u i m b e a u, L a w r e n c e B a k e r: Vacuum Tube Circuits, 668 pages, John Wiley & Sons, Inc., New York, 1948.

B a r k h a u s e n, H.: Lehrbuch der Elektronenröhren.
Vol. 1: Allgemeine Grundlagen, 171 pages, 118 fig., 1936.
Vol. 2: Verstärker, 289 pages, 127 fig., 1933.
Vol. 3: Rückkoppelung, 174 pages, 85 fig., 1935.
Vol. 4: Gleichrichter und Empfänger, 294 pages, 147 fig., 1937.
Published by S. Hirzel, Leipzig.

B a t c h e r, R. R., and W. M o u l i c: The Electronic Engineering Handbook, 456 pages, Electronic Development Associates, New York, 1944.

B r a i n e r d, J. G., G. K o e h l e r, H. J. R e i c h and L. F. W o o d r u f f: Ultra-High-Frequency Techniques, 570 pages, Chapman & Hall Ltd., London, 1943.

B r u e c h e, E., and A. R e c k n a g e l: Elektronengeräte, 447 pages, edited by Springer, Berlin 1941.

C a m p b e l l, N. R., and D. R i t c h i e: Photoelectric cells, Isaac Pitman & Sons, New York 1934.

C h a f f e e, E. L.: Theory of Thermionic Vacuum Tubes, McGraw-Hill Book Co., New York 1933.

D o w, W i l l i a m G.: Fundamentals of Engineering Electronics, 604 pages, John Wiley & Sons, New York and Chapman & Hall Ltd., London, 1937.

F i n k, D. G.: Principles of Television Engineering, 541 pages, 311 fig., McGraw-Hill Book Co., New York and London, 1940.

F i n k, D. G.: Radar Engineering, McGraw-Hill Book Co., New York and London, 1948.

G r e e n w o o d, I v a n A. Jr., V a n c e H o l d a m Jr. and D u n c a n M a c r a e Jr.: Electronic Instruments, 721 pages, McGraw-Hill Book Co., New York and London, 1948.

H a r v e y, A. F.: High-frequency Thermionic Tubes, 235 pages, 99 fig., Chapman & Hall Ltd., London, 1944.

H e n n e y, K e i t h: Radio Engineering Handbook, McGraw-Hill Book Co., New York, 1935.

H e n n e y, K e i t h: Electron Tubes in Industry, 2nd edition, McGraw-Hill Book Co., New York, 1937.

H o r n u n g, J. L.: Radar Primer, 218 pages, McGraw-Hill Book Co., New York and London, 1948.

H u n d, A.: Frequency Modulation, 375 pages, 113 fig., McGraw-Hill Book Co., New York and London, 1942.

K a m m e r l o h e r, J.: Elektronenröhren und Verstärker, 326 pages, 290 fig., (vol. II of the series "Hochfrequenztechnik"), C. F. Winter'sche Verlagshandlung, Leipzig, 1939.

K a m m e r l o h e r, J.: Gleichrichter, 386 pages, 284 fig., (vol. III of the series "Hochfrequenztechnik"), C. F. Winter'sche Verlagshandlung, Leipzig, 1942.

K l o e f f l e r, R o y c e G.: Industrial Electronics and Controls, 478 pages, John Wiley & Sons, Inc., New York, 1949.

K o l l e r, L. R.: The Physics of Electron Tubes, 2nd edition, McGraw-Hill Book Co., New York, 1937.

L a d n e r, A. W., and C. R. S t o n e r: Short-wave Wireless Communication, 452 pages, 248 fig., Chapman & Hall Ltd., London, 1936.

M c A r t h u r, E. D.: Electronics and Electron Tubes, John Wiley & Sons, New York, 1936.

M a r k u s, J o h n and V i n Z e l u f f: Handbook of Industrial Electronic Circuits, 272 pages, McGraw-Hill Book Co., New York and London, 1948.

Members of the staff of the Department of Electrical Engineering, Massachusetts Institute of Technology: Applied Electronics, 772 pages, John Wiley & Sons, New York, and Chapman & Hall Ltd., London, 1943.

M i l l m a n, J., and S. S e e l e y: Electronics, McGraw-Hill Book Co., New York and London, 1941.
M o r e c r o f t, J o h n H.: Electron Tubes and their Application, 458 pages, 537 fig. Chapman & Hall Ltd., London.
R e i c h, H. J.: Theory and Application of Electron Tubes, 716 pages, McGraw-Hill Book Co., New York and London, 1944.
R e i m a n n, A. L.: Thermionic Emission, John Wiley & Sons, New York, 1934.
R o t h e, H., and W. K l e e n: Grundlagen und Kennlinien der Elektronenröhren, 325 pages, 196 fig., first book of the series "Bücherei der Hochfrequenztechnik", edited by Prof. Dr. J. Zenneck, Akademische Verlagsgesellschaft Becker & Erler, Leipzig, 1940.
R o t h e, H., and W. K l e e n: Elektronenröhren als Anfangsstufenverstärker, 303 pages, 197 fig., second book of the series "Bücherei der Hochfrequenztechnik", edited by Prof. Dr J. Zenneck, Akademische Verlagsgesellschaft Becker & Erler, Leipzig, 1940.
R o t h e, H., and W. K l e e n: Elektronenröhren als End- und Sendeverstärker 141 pages, 118 fig., third book of the series: "Bücherei der Hochfrequenztechnik", edited by Prof. Dr. J. Zenneck, Akademische Verlagsgesellschaft Becker & Erler, Leipzig, 1940.
R o t h e, H., and W. K l e e n, Elektronenröhren als Schwingungserzeuger und Gleichrichter, fourth book of the series: "Bücherei der Hochfrequenztechnik", edited by Prof. Dr. J. Zenneck, Akademische Verlagsgesellschaft Becker & Erler, Leipzig, 1941.
S a n d e m a n, E. K.: Radio Engineering, 775 pages, John Wiley & Sons, New York, 1948.
S a n d r e t t o, P. C.: Principles of Aeronautical Radio Engineering, 414 pages, 223 fig., McGraw-Hill Book Co., New York and London, 1942.
S a r b a c h e r, R. J., and W. A. E d s o n: Hyper and Ultra High Frequency Engineering, 632 pages, John Wiley & Sons, New York, and Chapman & Hall Ltd., London, 1943.
S c h i n t l m e i s t e r, J.: Die Elektronenröhre als physikalisches Messgerät, 179 pages, 119 fig., published by Springer, Vienna, 1942.
S p a n g e n b e r g, K a r l R.: Vacuum Tubes, 860 pages, McGraw-Hill Book Co., New York and London, 1948.
S t r u t t, M. J. O.: Moderne Mehrgitterelektronenröhren, Bau - Arbeitsweise - Eigenschaften - Elektrophysikalische Grundlagen, 2nd edition, 283 pages, published by Julius Springer, Berlin, 1940.
S t r u t t, M. J. O.: Moderne Kurzwellenempfangstechnik, 245 pages, published by Julius Springer, Berlin, 1939.
S t r u t t, M. J. O.: Verstärker und Empfänger, 384 pages, volume 4 of the series: Lehrbuch der drahtlosen Nachrichtentechnik, edited by N. von Korshenewsky and W. T. Runge, published by Springer, Berlin 1943.
S t r u t t, M. J. O.: Ultra- and Extreme-short-wave Reception, Principles, Operation and Designs, 387 pages, published by D. van Nostrand Cy. Inc., Toronto, New York, and London 1947.
T e r m a n, F. E.: Radio Engineering, McGraw-Hill Book Co., New York and London, 1938.
T e r m a n, F. E.: Fundamentals of Radio, 458 pages, 278 fig., McGraw-Hill Book Co., New York and London, 1938.
T e r m a n, F. E.: Measurements in Radio Engineering, 400 pages, 210 fig., McGraw-Hill Book Co., New York and London, 1935.
V a l l e y, G e o r g e E. a n d H e n r y W a l l m a n: Vacuum Tube Amplifiers, 743 pages, McGraw-Hill Book Co., New York and London, 1948.
V i l b i g, F.: Lehrbuch der Hochfrequenztechnik, 3rd edition, vol. I, 650 pages, 766 fig., vol. II., 616 pages, 891 fig. and 2 tables, Akademische Verlagsgesellschaft Becker & Erler, Leipzig, 1942.
W a l k e r, R. C.; Photoelectric Cells in Industry, 517 pages, Sir Isaac Pitman & Sons, Ltd., London, 1948.
Westinghouse Electric Corporation, Electronic Engineers of the: Industrial Electronics Reference Book, 680 pages, John Wiley & Sons, New York, and Chapman & Hall Ltd. London, 1948.
Z i n k e, O.: Hochfrequenz-Messtechnik, 223 pages, 221 fig., published by S. Hirzel, Leipzig, 1938.

Symbols

1. Symbols Relating to Valve Electrodes

Anode .... a
Anode of diode .... d
Anodes of double and multiple diodes .... d_1, d_2 etc.
    The number denotes the position of the diode anode reckoned from the cathode input side. The diode anode d_1 is the one nearest to the pinch. If there is only one diode plate the figure 1 is omitted.
Heater element (filament)... f
Grid .... g
    In multi-grid valves: g_1, g_2, etc., the number indicates the position counting from the cathode. Where there is only one grid the figure is omitted.
Indirectly-heated cathode .... k
Metallizing .... m
Internal screening in the valve..................................... s
Fluorescent screen of fluorescent-screen indicator or cathode-ray tube ... l
Deflecting plate of a cathode-ray tube .... D
Equivalent electrodes are distinguished by an accentuation mark....... a, a′, a″
    In secondary-emission valves the primary cathode is indicated by k_1 and the secondary by k_2.

2. Symbols Relating to Valve Systems

    In multiple valves the electrodes of the individual valve-systems are indicated as follows:
for a diode... by D
,, ,, triode .... ,, T
,, ,, tetrode .... ,, Q
,, ,, pentode.. ,, P
,, ,, hexode and heptode .... ,, H
,, ,, octode .... ,, O
,, ,, rectifying valve .... ,, R

3. Symbols for Voltages, Currents, Capacities, etc.

Voltage (V)

Anode voltage .... V_a
Anode voltage in cold state, or for $I_a = 0$........................ V_{ao}
Diode voltage .... V_d
    Where there is more than one diode: V_{d1}, V_{d2}, etc.
Heater voltage (filament voltage)................................... V_f
Voltage between filament and cathode .... V_{fk}
Grid voltage.. V_g
    Where there is more than one grid: V_{g1}, V_{g2} etc.
Effective (r.m.s.) value of grid alternating voltage .... $V_{g\ eff}$
Grid voltage in cold state, or for $I_a = 0$ .... V_{go}
Input alternating voltage .... V_i or $V_{i\ eff}$
Output alternating voltage .... V_o or $V_{o\ eff}$
Voltage of supply source, or battery voltage........................ V_b
Noise voltage .... V_N

Current (I)

Anode current .... I_a
Anode quiescent current (in push-pull stages or in oscillator valves) .... I_{ao}

522

Anode current at max. grid control .... $I_{a\,max}$
Diode current .... I_d
    Where there is more than one diode: I_{d1}, I_{d2}, etc.
Heater current.. I_f
Grid current... I_g
    Where there is more than one grid: I_{g1}, I_{g2}, etc.
Cathode current ($= I_a + I_{g1} + I_{g2}$, etc.)............................ I_k
Noise current .... I_N

Power (P)

Anode dissipation .... P_a
Grid dissipation.. P_g
    Where there is more than one grid: P_{g1}, P_{g2}, etc.
Power output with max. control, where distortion is n% or where there
    is grid current .... P_o (n%)

Capacity (C)

Capacity of anode with respect to all other electrodes .... C_a
Capacity of grid with respect to all other electrodes .... C_g
    Where there is more than one grid: C_{g1}, C_{g2}, etc.
Capacity between anode and grid 1 .... C_{ag1}
Capacity between grid 1 and grid 3 .... C_{g1g3}
Capacity between grid 1 and grid 4 .... C_{g1g4}
Capacity between grid 2 and grid 4 .... C_{g2g4}
Capacity between diode anodes d_1 and d_2 .... C_{d1d2}
Capacity between cathode and diode anode d_1 .... C_{kd1}
Capacity between grid and cathode.............................. C_{gk}
Capacity between anode and cathode .... C_{ak}
Capacity between anode and grid 4 .... C_{ag4}

Resistance (R)

External (load) resistance (in the anode circuit) .... R_L
Resistance in the cathode lead................................. R_k
External resistance between filament and cathode .... R_{fk}
External resistance in the grid circuit .... R_{gk}
    Where there is more than one grid: R_{g1k}, R_{g2k}, etc.
Internal resistance (of the anode) .... R_a
Equivalent noise resistance.................................... R_{eq}

Amplification factor

Amplification factor (control grid v. anode) .... μ
Amplification factor of control grid with respect to screen grid .... μ_{g1g2}
    The voltage amplification of a valve in a certain circuit is indicated
    by the quotient of the output voltage divided by the input voltage
    (V_o/V_i).

Transconductance (mutual conductance)

Transconductance .... g_m
Transconductance at commencement of oscillation .... g_{mo}
Conversion conductance.. g_c
Dynamic transconductance .... g_{md}

Efficiency

Efficiency .... η

INDEX

VALVE-TYPE INDEX